Biology and Ecology of Earthworms

Biology and Ecology of Earthworms

Third edition

C.A. Edwards
Ohio State University, Columbus, Ohio, USA

and

P.J. Bohlen
Institute of Ecosystem Studies, Millbrook, New York, USA

CHAPMAN & HALL
London · Glasgow · Weinheim · New York · Tokyo · Melbourne · Madras

Published by Chapman & Hall, 2–6 Boundary Row, London SE1 8HN, UK

Chapman & Hall, 2–6 Boundary Row, London SE1 8HN, UK

Blackie Academic & Professional, Wester Cleddens Road, Bishopbriggs, Glasgow G64 2NZ, UK

Chapman & Hall GmbH, Pappelallee 3, 69469 Weinheim, Germany

Chapman & Hall USA, 115 Fifth Avenue, New York NY 10003, USA

Chapman & Hall Japan, ITP-Japan, Kyowa Building, 3F, 2-2-1 Hirakawacho, Chiyoda-ku, Tokyo 102, Japan

Chapman & Hall Australia, 102 Dodds Street, South Melbourne, Victoria 3205, Australia

Chapman & Hall India, R. Seshadri, 32 Second Main Road, CIT East, Madras 600 035, India

First edition 1972

Second edition 1977

© 1996 C.A. Edwards and P.J. Bohlen

Typeset in 10/12pt Palatino by Saxon Graphics Ltd, Derby

Printed in Great Britain by St Edmundsbury Press, Bury St Edmunds, Suffolk

ISBN 0 412 56160 3

A catalogue record for this book is available from the British Library

Library of Congress Catalog Card Number: 95-71089

∞ Printed on permanent acid-free text paper, manufactured in accordance with ANSI/NISO Z39.48-1992 and ANSI/NISO Z39.48-1984 (Permanence of Paper).

Contents

Preface

In the 17 years since the second edition of the *Biology of Earthworms* was published, there has been an enormous expansion of interest and research in earthworms world-wide. This has arisen, at least partially, from the progressive appreciation by scientists and the general public of the key role of earthworms in the promotion of soil fertility and environmental improvement. This realization has resulted not only from the very considerable increase in research into the ecology and biology of earthworms, but also from the application of the results of this research into the management of earthworm populations in the field and the use of earthworms in waste management and environmental monitoring.

This progress in research and publications on earthworms exceeded even the optimistic forecasts that Dr J. E. Satchell made in the Foreword to the first edition (1972). In the first edition, 565 references were cited, and these were a fairly exhaustive coverage of the existing scientific literature on earthworms at that time. Based on an analysis of the frequency of appearance of new publications in time, up to 1972, Dr Satchell forecast that by 1992 a further 640 papers would have to be reviewed to update the first edition. By the time the second edition appeared, in 1977, the number of references quoted was already 674. These updated references were chosen more selectively than in the first edition and probably represented less than half of the available literature at the time and concentrated only on the more important publications.

Another important book on earthworm ecology, by Dr K.E. Lee published in 1985, *Earthworms: Their Ecology and Relationships with Soils and Land Use*, cited 814 references relevant to the more limited scope of this book, which did not address research into earthworm morphology, taxonomy, diversity, biology or physiology. It is reasonable to assume that more than twice the number of references quoted in that book had been published by that time. In support of this conclusion, Satchell and Martin (1985) published an exhaustive review of the available literature on earthworms up to 1985, with about 1360 references quoted.

In the present third edition, nearly 1500 references are cited and it is the authors' impression that this may represent only about one-third of the total available references in the scientific literature. Those quoted specifically in this edition concentrate on the more significant and important publications on earthworms. The rate of expansion of research on earthworms is so rapid that it is becoming increasingly difficult to cover the entire subject in a single volume; indeed, it seems likely that any further revisions of this book will involve more than one volume unless the single volume is very large and expensive. As conclusive evidence of the rapidly expanding literature on earthworms, at the Fifth International Symposium on Earthworm Ecology (ISEE 5), held in Columbus, Ohio, USA in 1994, and organized by the authors of this book, more than 160 manuscripts were received for publication, covering most of the themes addressed in this book. These papers will appear as a special issue of the journal *Soil Biology and Biochemistry*, probably in early 1996. Despite the wealth of understanding generated by this rapid expansion of research, many critical questions about earthworm biology and ecology still remain unanswered. We hope that this third edition will point the way to future important research and serve as an indispensable reference for workers in the field.

In the years since the second edition was published there have been a number of major International Symposia on Earthworm Ecology, including the first (ISEE 1) held in Grange-over-Sands, England (Satchell, 1983), the second (ISEE 2) held in Bologna, Italy (Bonvicini-Pagliai and Omodeo, 1987), the third in Hamburg, Germany in 1987 (ISEE 3), the fourth in Avignon, France (ISEE 4) in 1990 (Kretzschmar, 1992). A workshop on the role of earthworms in the stabilization of organic residues was held in Kalamazoo, Michigan in 1979 (Appelhof, 1981). There was also a symposium on *Earthworms in Waste and Environmental Management* held in Cambridge, England (Edwards and Neuhauser, 1988) and one on *Earthworm Ecology in Forest, Rangeland and Crop Ecosystems in North America* held in Georgia, USA in 1993 (Hendrix, 1995). A conference on the ecotoxicology of earthworms was held at Sheffield University, UK in 1991 (Greig-Smith *et al.*, 1992).

We would like to express our gratitude to those of our colleagues who have provided valuable advice, assistance and other contributions which have facilitated the production of this third edition of our book. In particular, we should like to thank Mr J. R. Lofty, coauthor of the first two editions, who had a major role in preparing both these editions, but was unable to make any contribution to this new third edition. We should also like to thank Miss Amy Martin and Mr Matt Hill for their help in preparing and editing the manuscript and Myrtis Smith for reviewing the references. We still depend heavily on diagrams prepared by Mr Arthur Whiting, Mr John Bater and Ms Barbara Burrows (née

Jones). We thank Dr Sam James for reading Chapter 2 critically and providing extremely valuable suggestions and comments, Dr John Reynolds for providing us with copies of the four parts of *Oligochaeta Nomenclaturum*, and his journal *Megadrilogica*, Dr Patrick Lavelle for sending us a comprehensive collection of his publications, Drs Rob Parmelee and John Blair for helping with literature surveys for Chapters 8, 9 and 10, and Dr Ken Lee for his general interest and support in our project. We also thank Dr Paul Hendrix for providing us with a current review of earthworms and soil structure and Mr George Brown for helping with the literature review for Chapter 9.

Acknowledgements for diagrams redrawn from various sources are given below the figures. In particular, we should like to thank Dr B.G.M. Jamieson for Fig. 2.1 which effectively summarizes a great deal of information.

We apologize for having to curtail the treatment of taxonomy and diversity and to remove the simplified key to common genera of terrestrial earthworms that appeared in the second edition. There remains controversy on various aspects of the taxonomy and evolutionary history of the various groups of earthworms and, with the gradual expansion of research into diversity and geographical distribution of earthworms, it is becoming increasingly difficult to provide relatively simplistic keys with broad applications. Increasingly, regional keys to the various groups are becoming necessary and, in some cases, available. There is a pressing need for more earthworm taxonomists if we are to continue to make progress in our understanding of the various groups of earthworms. We have also had to remove the section on 'Simple experiments and field studies with earthworms' in the interests of space, so we have to refer the reader interested in such experiments to the second edition.

The third edition has greatly extended treatments of earthworm community ecology, interactions between earthworms and micro-organisms, and the importance of earthworms in environmental management and their use in organic waste management. We have also added an extensive appendix which summarizes the toxicity of a wide range of chemicals to earthworms. It is our impression from reaction to several printings of the first two editions that were fully sold out, that we are reaching a wide audience, ranging from high school and college students, teachers, university faculty, zoologists, biologists, ecologists, soil scientists, agriculturalists and farmers. Hence, we have attempted to combine a straightforward and integrative approach to reviewing the earthworm literature with scientific accuracy and appropriate acknowledgements. We hope that the book continues to arouse world-wide interest in these ubiquitous and important animals and contributes to initiating critical research in the field of earthworm ecology and biology.

C.A.E
P.J.B
Ohio State University
November 1994

Earthworm morphology

1

The principal systematic features of earthworms are that they are bilaterally symmetrical, externally segmented, with a corresponding internal segmentation. They have no skeleton and a thinly pigmented cuticle, bearing setae on all segments except the first two; with an outer layer of circular muscles and an inner layer of longitudinal muscles. They are hermaphrodite and have relatively few gonads, which are situated in definite segmental positions. When mature, a swollen area of the epidermis called a clitellum, located in particular segments, forms a cocoon in which the eggs or ova are deposited, and this is then passed over the anterior segment. The eggs are usually fertilized and the young develop within the eggs without a free larval stage, the newly hatched worms resembling adults.

Structurally, earthworms have large coelomic cavities containing coelomocytes, a closed vascular system with at least a dorsal and a ventral trunk and a ventral nerve cord. The alimentary canal is basically an anterior–posterior tube with excretion through the anus or specialized organs called nephridia; respiration is mainly cuticular. These systems are discussed further by Wallwork (1983).

1.1 SEGMENTATION: EXTERNAL

Earthworms are divided externally into bands or segments along the length of the body by furrows or intersegmental grooves, which coincide with the positions of the septa dividing the body internally. The segments vary in width, usually being widest in the anterior and clitellar regions. Segments are numbered arbitrarily from the front to the rear, and the grooves are designated by the numbers of the segments on either side, e.g. 3/4, 10/11, etc. Often the external segments are subdivided by one or two secondary grooves, particularly in the anterior part, but these are superficial divisions which are not reflected in the internal anatomy. The mouth opens on the first segment, or peristomium, which bears on its dorsal surface the prostomium, a lobe overhanging the mouth (Plate 1). The prostomium varies in size, and in some worms it may be so small that it cannot be distinguished. The way in which the peristomium and prostomium are joined differs between species and is a useful systemic character. The connection is termed zygolobous, prolobous, epilobous or tanylobous, depending on the demarcation of the prostomium (Fig. 1.1). Some of the aquatic worms (Naididae and Lumbriculidae) have the prostomium extended forward into a proboscis.

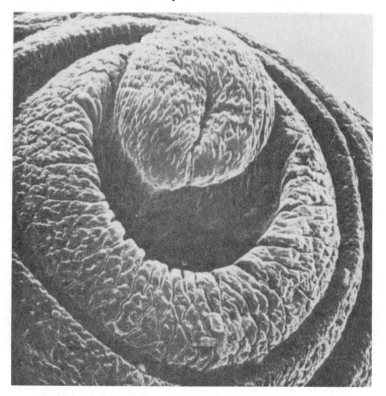

Plate 1   Prostomium of *Aporrectodea caliginosa*

Sometimes there is unusual segmentation in the region just before the anus. *Pontoscolex* (Glossoscolecidae) has a swollen posterior end, with narrow segments; it has been suggested this is a tactile sensory region adapted for gripping the sides of the burrow to avoid being pulled from the ground.

1.2 CHAETOTAXY

The setae, which are bristle-like structures borne in follicles on the exterior of the body wall, can be extended or retracted by means of protractor

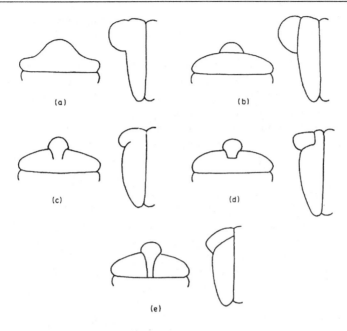

Figure 1.1   Various forms of prostomium (cephalization). (a) Zygolobous; (b) pro-lobous; (c) and (d) epilobous; (e) tanylobous.

and retractor muscles which are attached to the base of the follicles and pass through the longitudinal muscle layer into the circular muscle layer below. As the setae are used to grip the substrate, their principal function is locomotory. Different species of oligochaetes have setae of varying shapes – either rod-, needle- or hair-like. Rarely, the distal end of a seta is forked. The shape of setae varies with their position, the most common form being those of *Lumbricus*, which are sigmoid and about 1 mm long. Often setae are enlarged both at the anterior and posterior ends, as in *L. terrestris*; the setae in the region of the genital pores (particularly the male pores) are sometimes modified in size and shape, and situated on raised papillae. The genital setae, which can be up to 7 mm long in some species of Lumbricidae, are usually grooved along their length, and may have a hook-like process at the distal end. These setae assist the physiological processes which take place at copulation by providing physical stimuli to the partner, while other setae have been shown to assist in holding copulating earthworms together by gripping, clasping or penetrating the skin (Feldkamp, 1924).

The setae are arranged in a single ring around the periphery of each segment, their number and distribution being typically termed either lumbricine or perichaetine. The lumbricine arrangement, typical of the Lumbricidae, consists of eight setae per segment in ventral and latero-ventral pairs. If the distance between the setae in each pair is very small, they are termed 'closely paired', if wider apart, they are termed 'widely paired', or if they are very far apart so that the pairing is not obvious, they are termed 'distant'. The distance between each pair and between neighboring pairs is constant for each species. The setae are designated by letters, a, b, c and d, beginning with the most ventral setae on each side (Fig. 1.2). The distance between the setae is important as a systemic character and is usually expressed as the ratio of distances between setae. For example, '$aa:ab:bc:cd:dd = 16:4:14:3:64$' is the ratio in the post-clitellar region of *L. terrestris*. Alternatively, the distances may be expressed as an equation, $aa = 4ab$; or as $ab<bc>cd$. The distance between the two most dorsal setae on either side is usually compared with the circumference of the body (μ) at that point, thus $dd = 1/2\mu$. For many species of the Megascolecidae, particularly *Megascolex, Pheretima* and *Perionyx*, and certain other species, the setal arrangement is termed perichaetine. Here the setae are arranged in a ring right around the segment, although usually with a large or smaller break in the mid-dorsal and mid-ventral regions. (Both types of setal arrangement are shown in Fig. 1.2.) These worms have more than eight setae per segment, often 50–100, sometimes more. Setal distributions intermediate between the lumbricine and perichaetine arrangements may be 12, 16, 20 or 24 setae per segment, fairly distinctly arranged in 6, 8, 10 or 12 pairs, or the arrangement may be lumbricine in the anterior and appear perichaetine in the posterior part of the body. The setae are desig-

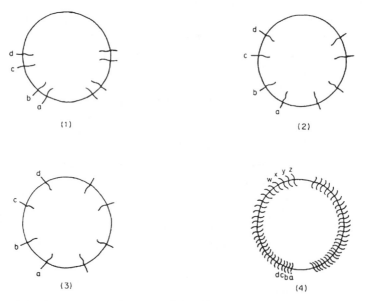

Figure 1.2   Arrangement of setae in Oligochaeta. 1,2,3, Lumbricine arrangement; 4, perichaetine arrangement. 1, Closely paired; 2, widely paired; 3, distant paired.

nated by the letters *a, b, c, d* – beginning with the most ventral one on each side; and *z, y, x, w* – beginning with the most dorsal one on each side, irrespective of how many there are in between.

1.3 GENITAL AND OTHER APERTURES

Earthworms are hermaphrodite and have both male and female genital openings to the exterior, consisting of paired pores on the ventral or ventro-lateral side of the body. In lumbricid worms the male pores are situated ventro-laterally on the 15th, or occasionally on the 13th segment. Each pore lies in a slit-like depression, which in some species is bordered by raised lips or glandular papillae, often extending on to the neighboring segments. In other families, the male pores may be on quite different segments, and in some, particularly the Megascolecidae, the male pores may be associated with one or two pairs of prostatic pores; these are openings of the ducts of accessory reproductive bodies known as prostates, which are usually absent from lumbricid species. Both the male and prostatic pores may be on raised papillae or ridges, or may open directly on to the surface. The male and prostatic pores are sometimes combined as one opening, but when separate they are usually joined by longitudinal seminal grooves, on either side of the ventral surface of the body.

Usually earthworms have two or more pairs of spermathecal pores, with a maximum of seven pairs in some species, but in others they are absent (*Bimastos tenuis, B. eiseni*). Earthworms belonging to the Enchytraeidae, and two or three other families, have only a single pair of pores. Spermathecae and their pores are not always paired. For instance, some megascolecid worms, particularly the genus *Pheretima*, have a single series of pores, situated in the mid-ventral line. The same genus has species with many spermathecae per segment, for example, *P. stelleri* has as many as 30. Spermathecal pores are usually intersegmental and are most often situated in the ventral or latero-ventral position, but sometimes they are close to the mid-dorsal line.

The female pores are most commonly a single pair, situated either in an intersegmental groove or on a segment, their position often being diagnostic of a particular family. Thus, in the Enchytraeidae they are in groove 12/13, and in the Lumbricidae, Megascolecidae and Glossoscolecidae they are on segment 14. Sometimes the female pores are united into a single median pore.

The dorsal pores, which are small openings situated in the intersegmental grooves on the mid-dorsal line occur in most terrestrial oligochaetes, but not in aquatic and semi-aquatic species. These pores communicate with the body cavity and the coelomic fluid. There are usually no dorsal pores in the first few intersegmental grooves. The position of the first dorsal pore is used as a systematic character at the species level, although these pores may be hard to distinguish in some worms. Some species have pores which are readily visible, and some, e.g. *Aporrectodea rosea*, have pores surrounded by a dark pigmented zone.

The other openings in the body wall are the nephridiopores, which are very small and often difficult to see. They are situated just posterior to the intersegmental grooves on the lateral aspect of the body, and usually extend in a single series along the body on either side. The nephridiopores are the external openings of the nephridia (the excretory organs of the earthworm) and, like the dorsal pores, are normally kept closed by sphincter muscles.

1.4 THE CLITELLUM AND ASSOCIATED STRUCTURES

The clitellum is the glandular portion of the epidermis that is associated with cocoon production. It is either a saddle-shaped or annular structure, the former being most usual among lumbricids. It usually appears swollen, although sometimes it can be differentiated externally only by its color, being paler or darker than the rest of the body, or even a different color. In some megascolecids it appears only as a well-defined constriction (Stephenson, 1930). When swollen, the intersegmental grooves are often indistinct or entirely obscured, particularly on the dorsal surface.

The position of the clitellum and the number of segments over which it extends differ considerably among oligochaetes. Lumbricidae have the clitellum on the anterior part of the body, behind the genital pores, beginning between segments 22 and 38, and extending over about 4–10 segments posteriorly (Fig. 1.3). Megascolecidae have the clitellum further forward, beginning at or in front of segment 14, thus including the female pore, and posteriorly it may also include the male pore. Some aquatic or semi-aquatic worms, and also members of the Enchytraeidae, have a clitellum which is only a very temporary development during the period of cocoon formation, while even in the Lumbricidae it is often only conspicuous during breeding phases.

Most earthworms possess various markings at sexual maturity, in the form of tubercles, ridges and papillae on the anterior ventral surface, and these differ greatly in number and form in different species of oligochaetes. The tubercula pubertatis are glandular thickenings on the ventral surface, either on or near the clitellum. Lumbricidae have these in the form of a pair of more or less oval longitudinal ridges, which are sometimes partly divided by intersegmental grooves, or they are separate papillae on either side of the ventral surface of the clitellum. The tubercula pubertatis usually extend over fewer segments than those occupied by the clitellum, except in a few species of lumbricids where they extend beyond the clitellum. Those species without spermathecae, e.g. *B. eiseni*, usually have no tubercula pubertatis. It has been suggested that these organs assist in keeping worms together during copulation (Benham, 1896). Worms of

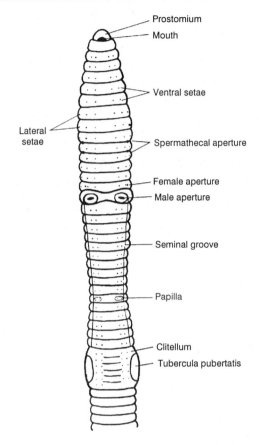

Prostomium
Mouth
Ventral setae
Lateral setae
Spermathecal aperture
Female aperture
Male aperture
Seminal groove
Papilla
Clitellum
Tubercula pubertatis

Figure 1.3 Ventral view of the anterior region of L. terrestris (after Stephenson, 1930).

other families do not always have the tubercula pubertatis situated on the clitellar segments, for example, in the Megascolecidae they are situated on either side or on both sides of the clitellum, or on the clitellar segments. They may be transverse ridges or small separate pads, symmetrically (sometimes asymmetrically) arranged about the mid-ventral line.

The position of the clitellum is used as a diagnostic character, particularly for lumbricids, because the position and number of segments occupied by the clitellum is, with small variations, constant for each species. Its position is defined by the number of the first and last segments occupied, with the range of variations in position, as, for example: 26, 27–31, 32.

1.5 PIGMENTATION

The color of earthworms depends mostly on the presence or absence of pigment, which is either in the form of granules or in pigmented cells in the subcuticular muscle layer. Pigmented worms, such as *L. terrestris*, are usually red, brown, a combination of these colors, or even greyish or greenish. Occasionally, the color is at least partly due to the presence of yellow coelomic fluid or green chloragogenous cells near the surface. The ventral surface of these worms is usually much lighter in color than the dorsal surface, although some deeply pigmented megascolecids are pigmented equally on both dorsal and ventral surfaces. The pigments, which are mainly porphyrins, probably originate as breakdown products of the chloragogenous cells (Stephenson, 1930). Unusually, the pigment is not evenly distributed, but appears as dark segmental bands separated by lighter intersegmental zones, as in some strains of *Eisenia fetida*. Lightly pigmented or unpigmented worms often appear reddish or pink, due to the hemoglobin of the blood in the surface capillary vessels showing through the transparent body wall. If the body wall is opaque, unpigmented worms are whitish. In a number of species the cuticle is strongly iridescent and causes the worms to appear bluish or greenish. This is particularly noticeable in species of *Lumbricus* and *Dendrobaena*. The color of pigmented worms, when preserved in formalin, is fairly stable but the reds and pinks of unpigmented worms usually fade rapidly.

1.6 THE BODY WALL

The body wall consists of an outer cuticle, the epidermis, a layer of nervous tissue, circular and longitudinal muscle layers, and finally the peritoneum, which separates the body wall from the coelom. The cuticle, which is a very thin (7 μm in *L. terrestris*), noncellular layer, is colorless and transparent, consisting of two or more layers, each composed of interlacing collagenous fibers, with several homogeneous nonfibrous layers beneath. It is perforated by many small pores which are larger and more irregularly distributed about the middle of each segment than in its anterior part, and are absent from the posterior part of each segment. The cuticle is thinnest where it overlays the epithelial sense organs, and here it is perforated by many very small pores through which project fine hairs from the sensory cells.

The epidermis consists of a single layer of several different kinds of cells (Fig. 1.4). The supporting cells, which are columnar in shape, are the main structural cells of the epidermis, and have processes which extend into the muscle at their bases. These cells secrete material to form the cuticle. Short and round, or club-shaped, basal cells lie on the inner wall of the epidermis, and are also known as 'replacing cells', although it is doubtful whether they can replace the epithelial cells.

Two forms of glandular cells present are the mucous or goblet cells and albumen cells. Mucous cells secrete mucus over the surface of the cuticle

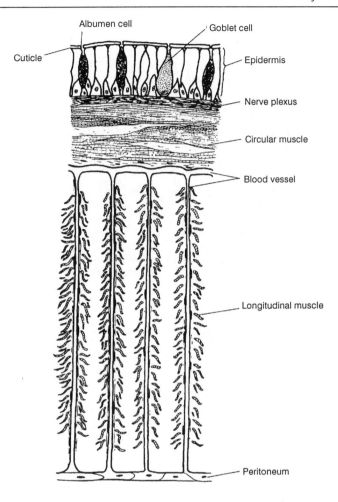

Cuticle — Albumen cell — Goblet cell — Epidermis — Nerve plexus — Circular muscle — Blood vessel — Longitudinal muscle — Peritoneum

Figure 1.4   Transverse section of a portion of the body wall (after Grove and Newell, 1962).

to prevent desiccation and to facilitate movement through soil. The function of the albumen cells in not known.

Large numbers of sensory cells, grouped together to form sense organs which respond to tactile stimuli, are scattered throughout the epidermis (Fig. 1.5). These are more numerous on the ventral than on the dorsal surface. Photoreceptor cells, capable of distinguishing differences in light intensity, occur in the basal part of the epidermis, and are most numerous on the prostomium and first segment and on the last segment. There are few or none in the epidermis of the middle segments. The epithelium of the buccal cavity bears groups of sensory cells which can be stimulated by chemical substances associated with taste. The prostomium contains

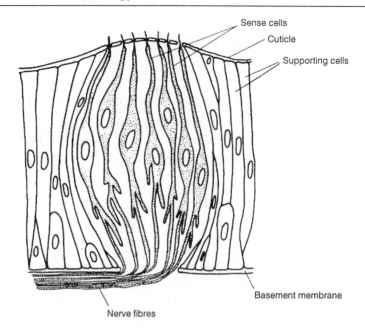

Figure 1.5　Vertical section of an epidermal sense organ (after Grove and Newell, 1962).

receptors which can detect sucrose, glucose and quinine, as well as many other chemicals (Laverack, 1960a).

The epidermis is bounded at its inner surface by a basal membrane within which lie two muscle layers. The circular muscle layer consists of muscle fibers extending around the circumference of the body, except at the intersegmental positions. The muscle fibers are arranged in an irregular manner, aggregated into groups, each surrounded by a sheath of connective tissue (Fig. 1.4).

The longitudinal muscles, which are in a much thicker layer than the circular muscles, are continuous throughout the length of the body, and arranged in groups or blocks around the body. *Lumbricus* has nine such blocks of muscles, two dorsal, one ventral and three ventro-lateral on either side. They have ribbon-like fibers which are arranged into U-shaped bundles in each block; these are surrounded by sheaves of connective tissue, so that the mouth of the 'U' is towards the coelom. Usually the ends of the 'U' are closed, forming a box-like structure, but some species, e.g. *Eisenia fetida*, have both ends open, so that the bundles are disposed in radial columns. Between each bundle lies a double layer of connective tissue containing blood vessels. The inner surface of the longitudinal muscle layer is separated from the coelom by a layer of coelomic epithelial cells, the peritoneum.

1.7 THE COELOM

The coelom is a large cavity that extends through the length of the body, and is filled with coelomic fluid. It is surrounded on the outer side by the peritoneum of the body wall and on the inner side by the peritoneum covering the alimentary canal. Transverse septa divide it into segmental portions. The peritoneum covering these septa is similar in structure to that covering the inner surface of the muscle layers; in a few species the peritoneum on the septa is so very much thickened that it almost fills the coelom in this region. The septa usually correspond to the external seg- mental grooves but often do not occur in the first few segments of the body, and in some species they are missing in other parts, such as the oesophageal region of some species of *Pheretima*. Some species have septa that do not correspond with the intersegmental grooves in the anterior end but are displaced backwards, sometimes by as much as a segment. Two adjacent septa are sometimes fused at their junction with the body wall. The septa differ in thickness, depending on their position in the body, those in the anterior of the body being markedly thickened and more muscular. The degree of thickening, and the position and number of these septa are used as systemic features for many species of earth- worm. Septa are constructed from muscle fibers, mostly derived from the longitudinal muscle layer, together with some circular muscles on the posterior face, with connecting tissue and blood vessels. The septa are

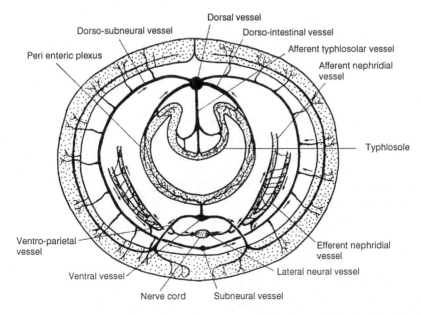

Figure 1.6 Diagram to show the main segmental blood vessels in the intestinal region (after Grove and Newell, 1962).

usually perforated by pores which allow the coelomic fluid to pass freely between segments, although some septa effectively isolate each body segment and make it virtually self-sufficient with its own supply of blood, central and peripheral nervous system, excretory system and even gonads. Bahl (1919) showed that these pores are surrounded by sphincter muscles in species of *Pheretima*, although Stephenson (1930) stated that sphincters are not present in other genera of the Megascolecidae. Other authors have suggested that most earthworms have septa with at least one sphinctered opening.

There are also bands and sheets of mesenteric membranes between the body wall and the gut, forming pouches and dividing off some organs into separate chambers. Some peritoneal cells are modified in the region of the intestine as chloragogen cells, forming chloragogenous tissue. Chloragogenous cells are characterized by the presence of yellow or greenish-yellow globules, the chloragosomes.

The coelomic fluid is a milky white liquid which is sometimes colored yellow by eleocytes, cells containing oil droplets, as in *Dendrobaena subrubicunda*. The coelomic fluid of *E. fetida* smells of garlic, hence the name of this species. The consistency of the coelomic fluid differs between different species of earthworms, and also depends upon the humidity of the air in which the worms live; thus, it is thicker and more gelatinous in worms in dry situations than in those from wetter habitats.

The coelomic fluid contains many different kinds of particles in suspension. The inorganic inclusions are mainly crystals of calcium carbonate, but the corpuscular bodies in the coelomic fluid of lumbricid worms include phagocytic amoebocytes, which feed on waste materials; vacuolar lymphocytes (small disk-shaped bodies which do not occur in worms with large numbers of eleocytes); and mucocytes (lenticular bodies which give the coelomic fluid a mucilaginous component). Other inclusions in the coelomic fluid include breakdown products of the corpuscular bodies, protozoan and nematode parasites, and bacteria. 'Brown bodies' are aggregated dark-colored masses or nodules usually found in the coelom at the posterior end of the body. They consist of disintegrated solid debris, such as setae and the remains of amoebocytes, and also cysts of nematodes and *Monocystis.*

Many earthworms eject coelomic fluid through the dorsal pores, in response to mechanical or chemical irritation, or when subjected to extremes of heat or cold. Some species, such as *Megascolides australis*, can eject fluid to a height of 10 cm, and *Didymogaster sylvaticus* (known as the squirter earthworm) to a height of 30 cm. Coelomic fluid is also expelled through the dorsal pores at times of stress and may have several functions such as preventing desiccation, promoting cutaneous respiration or providing protection from predators (Vail, 1972).

1.8 THE ALIMENTARY CANAL

The alimentary canal or gut of earthworms is basically a tube extending from the mouth to the anus, although it is differentiated into a buccal cavity, pharynx, esophagus, crop, gizzard and intestine (Fig. 1.7). The short buccal cavity flanked by a tube-like prostomium begins at the mouth and occupies only the first one or two segments, with one or two diverticula or evaginations. The epithelium lining the buccal cavity is not ciliated, except occasionally in a small dorsal diverticulum, but very young worms can have patches of ciliated epithelium on the ventral surface. Overlying the epithelium, except in the ciliated portions, is a thick cuticle.

The pharynx, which is not always differentiated clearly from the buccal cavity, extends backwards to about the sixth body segment. The dorsal surface of the pharynx is thick, muscular and glandular, and contains the pharyngeal glands, which secrete mucus and appear as a whitish lobed mass. Some species, e.g. *Bimastos parvus*, have pharyngeal glands which extend back beyond the pharynx and are attached to the septa. Earthworms use the pharynx as a suction pump, so that muscular contractions of its walls draw particles through from the mouth, and some species evert the pharynx in this process.

All terrestrial oligochaetes have an esophagus, which opens from the pharynx as a narrow tube, modified posteriorly as a crop and gizzard. Calciferous glands situated in evaginations or folds in the esophageal wall open into the esophagus. Lumbricidae have these folds developed strongly with their tips fused together, to form esophageal pouches, which separate the glands from the esophageal cavity. Esophageal pouches are usually paired and lateral, but may be unpaired and ventral in some megascolecid species. In *Pheretima* spp. they are reduced to low ridges of calciferous tissues. In some species of lumbricids, such as *Lumbricus castaneus*, *Lumbricus rubellus* and *Dendrobaena* spp., the glands connect with the lumen of the esophagus via a duct.

At the posterior end of the esophagus is the crop, a thin-walled storage chamber situated in front of the gizzard, which is muscular and lined with a thicker cuticle than the crop. Muscular contractions of the gizzard grind up the food into small particles with the aid of mineral particles taken in with the food. The crop and gizzard are situated much further forward in some of the other species of earthworms than in the Lumbricidae; they are immediately behind the pharynx in megascolecid species which commonly have 2–10 gizzards, each occupying one segment.

One species of megascolecid, *Pheretima californica*, has no gizzard and a crop situated in the middle of the esophagus. It has no calciferous glands in the strict sense of the term, although there are folds in the wall of the posterior half of the esophagus. The intestine has two conical caecae which originate in segment 26 and extend forwards into three or four segments (El-Duweini, 1965). Megascolecids are extremely variable in terms

(a)

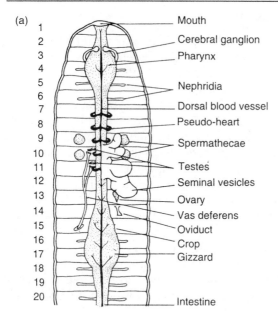

1
2
3
4
5
6
7
8
9
10
11
12
13
14
15
16
17
18
19
20

Mouth
Cerebral ganglion
Pharynx

Nephridia
Dorsal blood vessel
Pseudo-heart
Spermathecae
Testes
Seminal vesicles
Ovary
Vas deferens
Oviduct
Crop
Gizzard

Intestine

(b)

Mouth
Cerebral ganglion
Pharynx
Proventriculus
Dorsal blood vessel
Gizzard
Septal nephridia
Spermathecae
Testes
Seminal vesicles
Ovary
Pseudo-heart
Vas deferens
Oviduct
Esophageal sacs
Prostate
Intestine

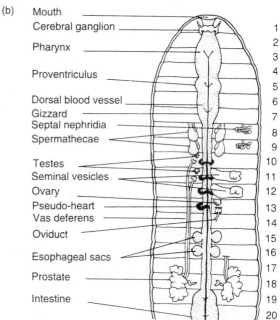

1
2
3
4
5
6
7
8
9
10
11
12
13
14
15
16
17
18
19
20

Figure 1.7   Digestive, circulatory and reproductive systems of earthworms (♂/♀ reproductive structures not bilaterally complete; all nephridia not shown). (a) Diagrammatic lumbricid (based on *L. terrestris*). (b) Diagrammatic megascolecid (based on *Pheretima* spp.).

of gizzard development, some species having as many as 10 gizzards.

The rest of the alimentary canal is the intestine, which is a straight tube for most of its length, slightly constricted at each septum. Most of the digestion and absorption of food materials takes place in the intestine. The internal surface of the intestine has many small longitudinal folds and its surface area is increased by a large fold in the medial dorsal line, the typhlosole, which projects from the dorsal wall (Fig. 1.6); this differs in relative size in different species, being largest in the lumbricid earthworms. Such increases in surface area are common in animals that consume food mostly of plant origin; but the typhlosole may be small in some other species.

The epithelial lining of the intestine is composed mainly of glandular cells, and nonglandular ciliated cells. The ciliated cells of *Lumbricus* species are also contractile, acting upon the gland cells which they surround, causing them to open or close. The intestine has two muscular layers, an inner circular and an outer longitudinal layer, the reverse order to that in the body wall.

1.9 THE VASCULAR SYSTEM

Earthworms have a closed vascular system, although it is not possible to distinguish 'arteries' and 'veins'. Instead, there is a system of vessels that distribute blood in various directions. There are three principal blood vessels, one dorsal and two ventral, that extend almost the entire length of the body, joined in each segment by blood vessels which ring the peripheral region of the coelom, and the body wall (Figs 1.6, 1.7 and 1.8). The largest of the longitudinal vessels, the contractile dorsal vessel, is associated closely with the gut for most of its length, except in the most anterior

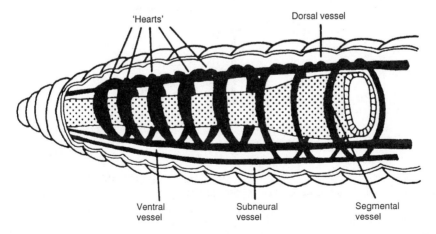

Figure 1.8a   The circulatory system of a lumbricid (lateral view; from Wallwork, 1983).

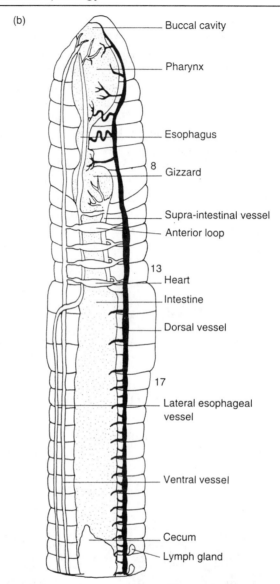

(b)

Buccal cavity

Pharynx

Esophagus

8 Gizzard

Supra-intestinal vessel
Anterior loop

13 Heart
Intestine

Dorsal vessel

17

Lateral esophageal
vessel

Ventral vessel

Cecum
Lymph gland

Figure 1.8b   The circulatory system of *Pheretima hupeiensis* (from Grant, 1956).

portion, where it is separated from the gut by a mesentery. Some species of Megascolecidae and Glossoscolecidae have paired dorsal vessels for part or all of the length of the body. The ventral vessel, which is narrower than the dorsal vessel, lies immediately below the gut, and is suspended from it by a mesentery. It conveys blood from the anterior part of the body to the posterior. The subneural vessel is even smaller than the ventral vessel. It lies beneath the ventral nerve cord, with which it is closely associated for its entire length. Some earthworm species do not have a subneural vessel, but most species (except some Megascolecidae) have two much narrower vessels which lie either side of the ventral nerve cord throughout its length (the latero-neurals). Megascolecid and some glossoscolecid earthworms, but not lumbricids, have a supra-intestinal vessel (some genera, e.g. *Pheretima*, have two such vessels); this lies along the dorsal wall of the gut in the anterior segments and is part of the complex of blood vessels serving the alimentary canal. Other longitudinal blood vessels are the extra-esophageal (or lateral-esophageal) vessels which lie along either side of the gut, from the pharynx to the first of the dorso-subneural vessels.

The paired commissural vessels pass round the body in each segment from the dorsal vessel (and supra-intestinal vessel in those species which have it) to the ventral or subneural vessels. Some of the anterior commissures are enlarged, contractile, and with valves termed 'hearts' or 'pseudo-hearts'; *Lumbricus* has five pairs of such vessels (situated in segments 7 to 11), but some species have more and others less; megascolecid worms have between two and five pairs of 'hearts'. For instance, *Pheretima californica* has four 'hearts'. In addition to these dorso-ventral anterior 'hearts', some species have 'intestinal hearts' which connect the supra-intestinal vessels with the ventral vessel, and also other 'hearts' originating in branches from the dorsal and supra-intestinal vessels; *Pheretima* has 'hearts' of both these types. The 'hearts' direct the flow of blood through the body by means of valves similar to those in the dorsal vessel. The lateral-esophageal-subneural commissures link the lateral-esophageal and subneural vessels in the anterior segments of the body. Behind segment 12, a pair of dorso-subneural commissures, lying on the posterior face of each septum, link the dorsal vessel with the subneural vessel in each segment. A pair of ventro-parietal vessels on the anterior face of each segment branch off from the ventral vessel to the body wall and end in capillaries. Some species of Megascolecidae have supra-intestino-ventral commissures. The commissural or septal vessels are always associated with the faces of septa, and sometimes the 'hearts' are attached by mesenteric-like folds of the peritoneum to the faces of the septa. Between the alimentary epithelium and the outer muscular and peritoneal layers is a vascular network or plexus, the peri-enteric or alimentary plexus, which communicates with the dorsal and ventral vessels by two pairs of dorso-intestinal and three pairs of ventro-intestinal vessels per segment. The dorsal vessel supplies blood to the typhlosole via three small typhlosolar vessels per segment.

The body wall has capillaries which follow the muscle bundles and eventually connect with the subneural vessel via the lateral-esophageal-subneural commissures in the front of the body, or via the dorso-sub-neural commissures in the rear. These capillaries also connect with the supra-intestinal vessel and the supra-intestino-ventral commissure behind the 12th segment of *Lumbricus*. The paired dorso-subneural com-missures run between the dorsal and subneural vessels in every segment behind the 12th.

The efferent nephridial vessels join the dorsal-subneural, and the smaller and more numerous afferent nephridial vessels come from the ventro-parietals. The reproductive organs receive most of their blood supply from the ventro-parietal vessels.

1.10 THE RESPIRATORY SYSTEM

Terrestrial oligochaetes have few specialized respiratory organs. Most res-piration is through the body surface which is kept moist by the mucous glands of the epidermis, the dorsal pores of which exude coelomic fluid and the nephridial excretions through the nephridiopores.

This method of gaseous exchange depends upon a network of small blood vessels buried in the body wall of terrestrial earthworms, so that oxygen dissolved in the surface moisture film can permeate through the cuticle and the epidermis to the thin walls of these vessels, where it is taken up by the hemoglobin in the blood and passed around the body. The hemoglobin in earthworms has a very high affinity for oxygen. As in other animals, carbon monoxide can block the functioning of hemoglobin in earthworms (Gardiner, 1972). Some species of *Lumbricus* also have looped capillaries extending from the ventro-parietal blood vessels into the epidermal layer. Combault (1909) stated that *Aporrectodea caliginosa* f. *trapezoides* has special respiratory regions of the body wall, where the cuti-cle is thinner, the epithelial cells shorter, there are fewer glandular cells in the epidermis, and there is a special subcutaneous blood supply. In order for the body surface to act as a respiratory organ, it has to be kept moist; this is achieved by secretions mainly from mucous glands. The mecha-nism for removal of carbon dioxide from the earthworm body is poorly understood. Carbonic acid formed in the calciferous glands can combine with calcium and pass out of the body with the feces. The high affinity of earthworm hemoglobin for oxygen allows earthworms to live even in poorly ventilated soils. However, when soil is flooded after rain, earth-worms may experience low oxygen tensions. This may cause them to leave their burrows in search of atmospheric oxygen.

1.11 THE EXCRETORY SYSTEM

The following description of the excretory system refers to *L. terrestris*, and detailed differences in other species are discussed later. The nephridia, which are the main organs of nitrogenous excretion in oligochaetes, are paired in each segment except the first three and the last. The internal opening from the coelom into each nephridium is just in front of a septum, and is a funnel-shaped nephrostome (Fig. 1.9a, b)

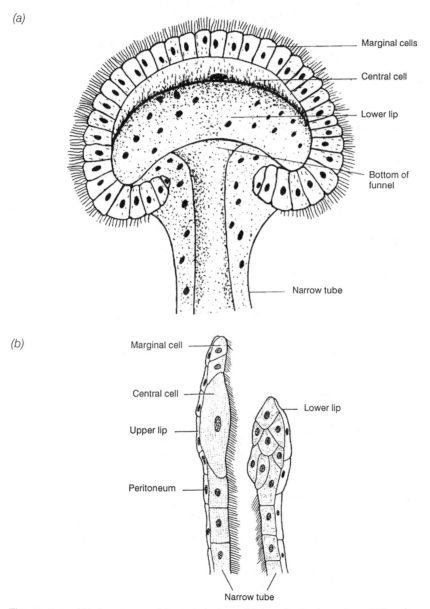

Figure 1.9 Nephrostome: (a) ventral view; (b) longitudinal section. (After Grove and Newell, 1962 (based on Goodrich).)

which leads to a short pre-septal canal that penetrates the septal wall into the segment behind, where the main part of the nephridium lies. The nephridium continues as a long post-septal canal with three loops, which can be distinguished as four sections: a very long 'narrow tube'; a shorter ciliated 'middle tube'; a 'wide tube'; and finally, the 'muscular tube' or reservoir, which opens to the exterior at the nephridiopore (Fig. 1.10).

The mouth of the nephrostomal funnel is flattened into two lips, an upper posterior lip and a lower anterior lip. The upper lip is much larger than the lower lip, and forms a semicircular expansion overhanging and projecting beyond the lower lip (Fig. 1.9b). The upper lip is composed of a radiating circle of marginal cells that are ciliated on their edges and inner faces, the cilia beating towards the canal. The lip is peripheral to a large crescent-shaped central cell with nucleate marginal cells and cilia on its lower face. The dorsal wall of the pre-septal canal extends to the lower boundary of the central cell, forming a part of the upper lip which has no nuclei or cilia; neither has the lower lip, which is much thinner than the upper lip. The short pre-septal canal, which leads from the nephrostome to the wall of the associated septum, is intercellular, like most of the nephridial tubes, i.e. the lumen perforates a single column of cells which are called 'drainpipe cells', and which have cilia in two longitudinal rows, one on each lateral wall.

Most of the post-septal part of the nephridial tube is secretory, except for the last portion (the muscular tube), and the lumen of the whole of the secretory section is intracellular, also with 'drainpipe cells'. The first part of the post-septal canal is the 'narrow tube', which has thin transparent walls containing granular protoplasm, nucleated at intervals (Fig. 1.10). The only parts of this tube that are ciliated are at the beginning, about

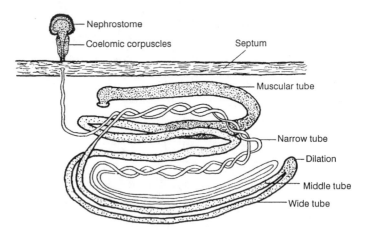

Figure 1.10   Diagram of a nephridium (after Grove and Newell, 1962 (altered after Benham))

half-way along its length and at the end. After the narrow tube is the 'middle tube' which is shorter and wider, with thicker brown-colored walls, and two rows of cilia along its whole length. This leads into the unciliated 'wide tube', which begins at a point where one of the loops bends sharply back on itself, via the ampulla, a dilated chamber with very large 'drainpipe' cells. The inner surface of the ampulla is covered with a large number of rod-like bacteria (the function of these has been investi- gated by Knop, 1926) and the last part of the wide tube has a ringed appearance. The last part of the nephridial duct is the 'muscular tube', which is much wider than the preceding tubes, is unciliated, and is the only part to have an intercellular lumen. This tube has a lining of flat epithelial cells surrounded by a muscular coat in its wall. It opens to the exterior by the nephridiopore through a sphincter muscle.

Nephridia of this pattern, i.e. opening externally on to the body surface (**exonephridia**) are found in most lumbricids, but in *Allolobophora antipae*, all the nephridia from segment 35 backwards open into two longitudinal canals that discharge into the posterior part of the gut (**enteronephridia**) (Fig.1.11a). The system in *Megascolex* spp. is similar. The nephridia of *Lampito mauritii* empty into a median excretory duct which leads to the posterior portion of the intestine (Fig. 1.11b) .

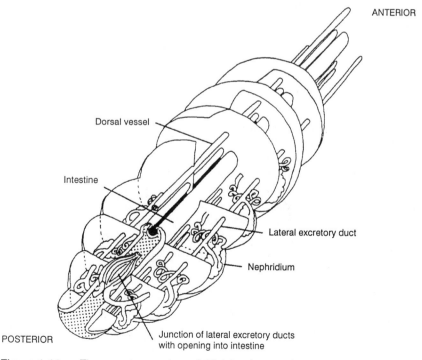

ANTERIOR

Dorsal vessel

Intestine

Lateral excretory duct

Nephridium

POSTERIOR

Junction of lateral excretory ducts with opening into intestine

Figure 1.11a   The excretory system of *Allolobophora antipae*.

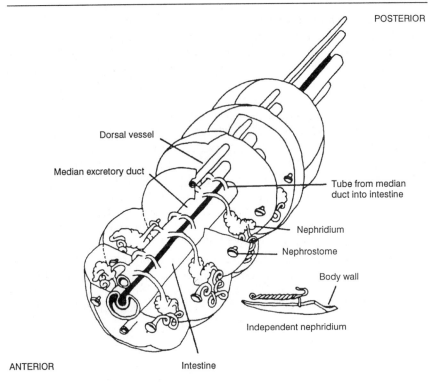

POSTERIOR

Dorsal vessel

Median excretory duct

Tube from median duct into intestine

Nephridium

Nephrostome

Body wall

Independent nephridium

ANTERIOR

Intestine

Figure 1.11b   The excretory system of *Lampito mauritii* (Wallwork, 1983).

Large, paired nephridia such as described above, are known as **meganephridia**, but some earthworms also have **micronephridia**, which are much smaller and more numerous. For instance, species of *Pheretima* have 40–50 micronephridia attached to each septum, except those septa in the anterior part of the body. Micronephridia are much simpler in form, and hang freely in the coelom from the ends of their funnels. Leading from a short, narrow tube is the main part of the micronephridial tube, a long, spirally twisted loop which has a ciliated lumen only in certain parts. The terminal ducts of these tubes pass into two septal excretory canals which curve around the posterior faces of the septa, and then pass to supra-intestinal excretory ducts that lie on either side of the dorsal surface of the gut. Some other megascolecid worms also have exonephric micronephridia.

Other kinds of nephridia end in a closed tube instead of a nephrostome; they are very small, exonephric, and attached to the body wall, with as many as 200–250 per segment. Species of *Pheretima* have this type of nephridia as well as septal micronephridia. They also have tufted nephridia, which are bush-like organs consisting of bunches of nephridial loops on the walls of the alimentary canal in segments 4, 5 and 6, one pair per segment, the ducts from the loops joining up and emptying into the alimentary canal. Some megascolecid species have tufted nephridia throughout the body; other species have them only in parts of the body. All earthworms have nephridia of one or more of the types described above, but the majority of species have open-funnelled exonephric meganephridia similar to those of *Lumbricus*.

The terms meganephridia and micronephridia are purely descriptive, but are disliked by some taxonomists who prefer the terms **holonephridia** and **meronephridia**, although these terms are not synonymous. Holonephridia are always single and relatively few in number in each segment. Meronephridia are multiple, although they are derived from a single rudiment in each segment. They are usually very numerous in each segment and much smaller than holonephridia. The urine of earthworms may contain ammonia, urea and, to a limited extent, uric acid in different proportions according to species, the season and the diet of the earthworm.

1.12 THE NERVOUS SYSTEM

Earthworms are very sensitive to chemicals, to touch and to light, and react to avoid adverse conditions. The ventral nerve cord runs beneath the gut close to the ventral wall of the coelom, from the last segment of the body to the 4th segment from the front. Anteriorly, it passes into a subesophageal ganglion, then bifurcates into the circumpharyngeal (circumesophageal) connectives, which pass up around either side of the esophagus, and meet as the bilobed cerebral ganglion on the dorsal surface of the pharynx in segment 3 (Fig. 1.12). Structurally, the ventral nerve cord is really a backward extension of the two circumpharyngeal connectives, fused together. The ventral nerve cord is swollen into segmental ganglia in each segment and, from the 5th segment backwards, each of these ganglia has two pairs of segmental nerves branching from it in the posterior part of each segment, with a third pair just in front of each ganglion. These segmental nerves extend around the body wall, at first in the longitudinal muscle layer, and then in the circular muscle layer, ending near the mid-dorsal line, thus forming almost a complete ring (in some megascolecid species, e.g. *Pheretima*, the ring is completely closed.) Each nerve ring has branches to the muscles and to the epidermis. Every segmental nerve has a small subsidiary branch, close to the nerve cord, which passes down into the circular muscle layer, and then between this and the longitudinal muscle layer, ending in branches to the epidermis just before it reaches the mid-ventral line. Each septum is supplied by a pair of nerves (septal nerves) arising near the junction of the posterior segmental nerves and the ventral nerve cord (Fig. 1.13).

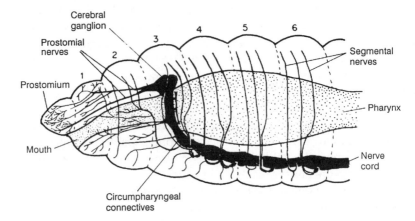

Figure 1.12   Nervous system of *L. terrestris* from the side (after Hess, 1925a).

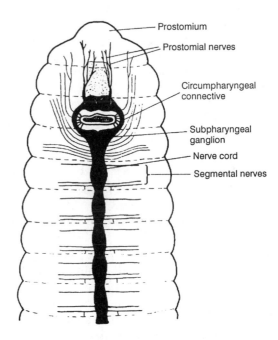

Figure 1.13   Nervous system of *L. terrestris* from above (after Grove and Newell, 1962).

The distribution of the nerves in the anterior four segments of the body differs from that in the other segments. Segment 3 has the typical distribution of three pairs of segmental nerves, but they originate from segment 4, which also possesses three pairs, so that six pairs of segmental nerves come from the ventral nerve cord in segment 4. Segment 2 is supplied with two pairs of nerves which originate from the junction of the circumpharyngeal connectives with the ventral nerve cord in segment 3, the larger posterior pair dividing, to give the normal three nerve rings in segment 2. The first segment is supplied by a pair of nerves originating in the lateral portions of the subpharyngeal connectives, which branch into two, shortly after they leave these nerves, and ramify through the first segment without forming nerve rings. A small nerve comes from the most ventral of these two branches to supply the ventral surface of the buccal cavity. The prostomium is innervated by two prostomial nerves, which originate from the front of the cerebral ganglion; these are the only nerves coming from this ganglion. A branch is given off from each prostomial nerve to supply the roof of the buccal cavity. The prostomial nerves then ramify through the prostomium, ending in the prostomial epithelium as nodule-like enlargements which then become part of the subepidermal nerve plexus. The last (caudal) segment of *Lumbricus* normally has six pairs of septal nerves instead of three (Hess, 1925a), arranged as if they were in two successive normal segments. The most posterior of these six pairs of nerves are lateral terminations of the nerve cord itself.

The system of nerves supplying the gut is sometimes called 'the sympathetic system', and consists of a plexus of nerves which lie between the epithelium and the circular muscle layer of the alimentary canal throughout its length. Six small nerves from each circumpharyngeal connective unite in the external wall of the pharynx as a pharyngeal ganglion, from which nerve fibers pass forward and backwards into the plexus. The muscles of the intestinal wall are supplied from a nerve plexus in the septum in each segment.

The wall of each blood vessel is covered with a network of nerve fibers and cells. Smallwood (1923, 1926) mapped the nerve supply to the major blood vessels, such as the dorsal and ventral vessels, but found no nerves to the hearts.

Before bifurcating, branches of the lateral nerves which link with the septal plexuses supply the longitudinal muscles of the body wall. The circular muscles are supplied by branches from nerve rings embedded in the body wall, which contribute to an intermuscular nerve plexus. Between the basal membrane of the epidermis and the circular muscle layer is the subepidermal nerve plexus, similar to the subepithelial plexus in the wall of the alimentary canal. Nerve fibers from this plexus, which is linked with branches from the nerve ring, end between the epidermal cells, or enter the sensory cells. The peritoneum has an extensive nerve net, derived from the nerves of the body wall.

1.12.1 THE FINE STRUCTURE OF THE NERVOUS SYSTEM

The ventral nerve cord is covered externally by a peritoneum, and immediately beneath this is a layer of longitudinal muscle fibers within which lies a thin layer of tissue described as a neural lamella (Schneider, 1908). Beneath these layers, surrounding the nervous tissue, is the fibrous epineurium, which extends inwards into the nervous tissue and also outwards into the lateral nerves. This layer, which is ectodermal in origin, is made up of cells that resemble connective tissue. Three dorsally situated giant fibers, one median and two lateral, are separated from each other and from the rest of the nervous tissue by fibrous septa or lamellae (Stough, 1926). Some *Lumbricus* species have two other smaller giant fibers in the ventral portion of the cord. There is also a double vertical septum dividing the cord into two lateral halves, which communicate in each segment by three large fenestrae through which pass fibrillae. The central portions of the nerve cord are partitioned further by extensions of the main horizontal and vertical septa. The nervous tissue in the cord is ramified by supporting tissue, called the neuroglia, which has both protoplasmic and fibrous cells, and a close network of fibers derived from processes of the cells. The rest of the central part of the nerve cord is occupied by the neuropile, which is a felt-like mat of fine anastamosing fibers derived from processes of the nerve cells. The nerve cells proper occupy the ventral and lateral portions of the nerve cord. They are termed bipolar or tripolar depending on the number of filamentous extensions, each cell always having more than one process, although they may appear monopolar if the two processes of a bipolar are so close together that they cannot be distinguished even after staining. The processes are distinguished as the axon, which is long, and one or more dendrites, which branch within the nervous system. Some of the cells of this region are large, pear-shaped and unipolar. These giant cells, or ganglion cells, link up with the giant fibers by long processes. The central portion of the nerve cord consists of a mass of fibrous material, forming felt-like masses with numerous perforations.

The cerebral ganglion, which contains no motor cells, is characterized by an external layer of small spindle-shaped ganglion cells, with a few large cells interspersed between them. The neuropile, which is very dense, is connected to the ganglion cells and contains the endings of the neural paths from the nerve cord. The histology of the main nerve rings has been described in detail (Dawson, 1920). Scattered along their length are nerve cells, which are most numerous on the posterior ring of each segment; bipolar nerve cells occur in the nerve plexus of the buccal cavity, unipolar 'ganglion cells' in the epithelium of the prostomium, and in the pharyngeal and intestinal plexuses.

The sense organs are of two main types, photoreceptor organs and epithelial sense organs. The photoreceptor organs of lumbricids are 22–63 μm long, elongated and conical. Those occurring at the ends of the nerves to the skin are smaller than those in the epidermis (Hess, 1925b) and are aggregated. One, or sometimes two, nerves pass into each cell, and large neural fibers also extend from the nerves into the cells. Small nerves continuous with the neural fibers extend throughout the protoplasmic contents of the cells as a neurofibrillar network. In each cell there is an ellipsoidal or elongated rod-shaped body, the optic organelle, with an outer surface consisting of interconnecting fibrillae, the retinella, and inside, a transparent hyaline substance.

The epithelial sense organs are groups of 35–45 elongated cells, broader at their bases, with distal ends terminating in sensory hairs projecting through thin regions of the cuticle. The bases of the cells end as processes, one of which extends as a nerve fiber along the basal membrane and joins the nearest epidermal nerve. Proprioceptor cells situated in the muscle layers act as tension or stretch receptors.

1.13 THE REPRODUCTIVE SYSTEM

Oligochaetes are hermaphrodite, and have more complicated genital systems than most unisexual animals. The reproductive organs, which are confined to comparatively few segments in the anterior portion of the body (Figs 1.7, 1.14 and 1.15), include the male and female organs and associated organs, the spermathecae, the clitellum and other glandular structures.

The paired ovaries, which produce oocytes, are roughly pear-shaped in *Lumbricus* (or fan-shaped in *Pheretima*), and are attached by their wider ends to the ventral part of the posterior face of septum 12/13, hanging freely in segment 13, in most terrestrial species. The ovisacs are backward-facing evaginations of the anterior face of the septum immediately behind the ovaries, and open into the dorsal wall of the ovarian funnels. They narrow

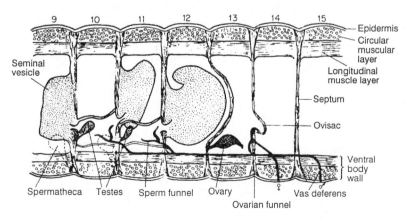

Figure 1.14   Diagram of the reproductive system from the side (after Stephenson, 1930).

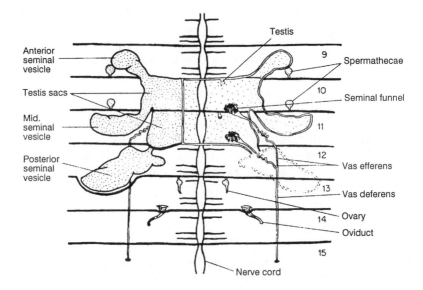

Figure 1.15 Dissection of the reproductive system from above (after Grove and Newell, 1962).

posteriorly to form the oviducts, which in turn open on to the ventral surface of the body, their position differing between earthworm families.

The basic male organs are the testes. Most species of Lumbricidae, Megascolecidae and Glossoscolecidae have two pairs of testes (holoandric), but some species of Lumbricidae and also the Ocnerodrilidae have only a single pair (meroandric). The testes are lobed organs attached to the posterior faces of septa 9/10 and 10/11 of *Lumbricus* and projecting from the septal walls into two median testis sacs, one in segment 10 and one in segment 11. The testis sacs are separate compartments within segments 10 and 11, lying below the ventral vessel and enclosing the ventral nerve cord. Some species, e.g. *Pheretima* spp., have much more extensive sacs which contain the hearts, dorsal vessel, esophagus and seminal vesicles. These sacs are filled with nutrient fluid in which lie the developing male cells. The testis sacs communicate with the seminal vesicles (vesiculae seminalis) which are storage sacs for the developing male cells.

The seminal vesicles are the largest and most conspicuous organs of the reproductive system, and are immediately obvious when earthworms are dissected as white masses on either side of the alimentary canal. They are sacs, for the most part divided by connective tissue into intercommunicating compartments. *Lumbricus* has three pairs of seminal vesicles in segments 9, 10 and 11, two pairs of which communicate anteriorly and posteriorly with the anterior testis sacs, and the third pair with the posterior testis sac in the segment immediately in front. The Megascolecidae have from one to four pairs of seminal vesicles, *Pheretima* spp. having two pairs. Seminal vesicles are never in the same segment as the testes or the testis sacs. The seminal vesicles differ in size and are sometimes very large, extending backwards from the testis sacs through a number of segments, for example, they extend back as far as segment 20 in *Lumbricus castaneus.*

The sperm or funnel sacs (Fig. 1.14) are set into the posterior walls of the segments that contain them by their narrow portions. These sperm funnels are much folded, ciliated on their inner surfaces and open into coiled or straight male ducts, the vasa efferentia. The anterior and posterior vasa efferentia on each side pass backwards and join to become a vas deferens, leading to the ventral exterior surface. Some species, e.g. *Pheretima posthuma,* have the ducts associated very closely, although not actually united, and each vas deferens opens to the exterior as a male pore.

The prostates or prostatic glands are large glands associated with the posterior ends of the vasa deferentia. Their function is to produce a fluid in which sperm cells can be transferred between worms during copulation. In the Megascolecidae they are of two types, tubular, or racemose (as in *Pheretima*) where the glandular tissue is compact with branching canals opening into the associated ducts. Other families have prostates in the form of muscular finger-like processes, or convoluted tubes. Prostate glands are rare in the family Lumbricidae.

Most oligochaetes have spermathecae, which are almost always paired organs: *L. terrestris* has two pairs, situated in segments 9 and 10, but other species may have more, up to a maximum of seven pairs, as in *Bimastos,* and some fewer than this or even none. The spermathecae are attached to the body wall by short stalk-like ducts. Many species of megascolecids have one or more diverticula from the spermathecal ducts.

The clitellum consists of thickened glandular epidermis, particularly on its dorsal and lateral portions. A section through the clitellum shows gland cells in three layers; those nearest the surface are mucous cells, which are similar to the mucus-secreting goblet cells in the ordinary epidermis. Also reaching the surface, but extending deeper into the clitellar tissue, are long, slender and often convoluted cells, containing large granules. The bulk of the clitellar tissue in the deeper layer down to the basal membrane is made up of albumen-secreting gland cells, terminating distally in long slender ducts which open on to the body surface. Epidermal columnar cells also occur. The posterior segments of the clitellum are less glandular than the anterior segments, with no cocoon-secreting and albumen-secreting cells, and fewer mucous cells.

Earthworm diversity and geographical distribution   2

2.1 SYSTEMATIC AFFINITIES AND EVOLUTIONARY DESCENT

Earthworms belong to the class Oligochaeta and are related to the Polychaeta (bristle worms) and the Hirudinea (leeches). Polychaeta and Hirudinea are exclusively marine or freshwater invertebrates, whereas species of oligochaetes (except for a few species secondarily adapted) inhabit soil or fresh water. Polychaetes may be considered as the older group because their larval development is more primitive than the oligochaetes, which have an embryo in a cocoon supplied with yolk or albumen. However, the greatest differences between the classes are in the structure of their genital organs. Thus, the genitalia of the Polychaeta, which have separate sexes, are simple, with extensive production of sexual cells from the coelomic epithelium, and sexual products that are expelled into the sea by rupturing the body wall. The oligochaetes, by contrast, are hermaphrodite, with their sexual organs confined to two or three segments and a very specialized and complex mechanism of fertilization and dispersal of eggs.

Possibly the Polychaeta are ancestral to the Oligochaeta, or they may both be derived from a common aquatic ancestor. Stephenson (1930) suggested that the Oligochaeta evolved from the Polychaeta, but branched off before the modern families of the latter appeared. The Hirudinea possess characters that occur to some degree in many or all of the Oligochaeta, and are hence closely related. Stephenson (1930) and Michaelsen (1926) believed that the leeches were derived secondarily from the Oligochaeta, probably from the primitive aquatic family, Lumbriculidae, but Brinkhurst (1992) considered that they were not related, although he later stated (1994) that the evolutionary position of the Hirudinea was ambiguous due to convergence. Michaelsen (1919) considered that the Hirudinea and the Oligochaeta were sufficiently closely related to be grouped together as orders in the Class Clitellata

(Jamieson, 1988). Although there is merit in giving the Polychaeta, Hirudinea and Oligochaeta equal ranking, Gates (1972a) considered that there were advantages in grouping the worms and leeches together.

Fossil records of oligochaetes are sparse, so there is little paleontological information concerning the history and development of the order. The generic name *Protoscolex* (Ulrich) was used to describe four species of fossil segmented worms without setae or appendages that occurred during the Upper Ordovician period in Kentucky, USA.

Bather (1920) also described a species, *Protoscolex latus*, from the Upper Silurian period in Herefordshire, England, which apparently bore papillae in one or two rows on each segment. He also placed *Pronaidites*, another fossil worm, which he considered to belong to the Oligochaeta, into this genus, suggesting that the papillae have some connection with setae. However, many authors are by no means convinced that these fossil worms are related to the oligochaetes. Benham (1922) found sporangia and vascular bundles of fern in the guts of some species of earthworms, but he did not think that the appearance of dicotyledonous plants was a necessary precursor for the evolution of oligochaetes, although he believed that they may have increased considerably in numbers and species once dicotyledonous plants appeared.

The early oligochaetes probably lived in mud rather than water (as presumably did their polychaete-type ancestor), becoming transiently terrestrial when the mud dried up periodically. They then became separated gradually into two groups, one purely terrestrial, the other aquatic (in fresh water), so that some aquatic families, such as the Aeolosomatidae, Naididae and Tubificidae, have probably never passed through a terrestrial phase in their developmental history.

As discussed earlier, Stephenson (1930) considered that the common ancestor of the terrestrial Oligochaeta belonged to the aquatic Lumbriculidae, which is one of the most primitive of the oligochaete families. Of the modern families that have been described, he considered the Moniligastridae to be the most primitive, and indeed, the structure of some members of this family, such as possession of a single-layered clitellum for example, is more similar to that of the aquatic worms than to that of any of the other terrestrial families. The Megascolecidae and Eudrilidae have more advanced characteristics, but they still retain one primitive feature in the position of the spermathecae and male or prostatic pores which come into contact during mating, thus allowing for the direct transfer of sperm. The Glossoscolecidae, Lumbricidae, Hormogastridae and Microchaetidae have fewest primitive features and may be considered to have evolved later than the other families, the Lumbricidae probably being most recent. Lumbricids never have a direct alignment of genital openings during copulation, so that a more elaborate mechanism is required to ensure the effective transference of sperm from one individual to another. Jamieson (1988) prepared a comprehensive cladogram summarizing the key characters of the families (Fig. 2.1).

2.2 FAMILIES, GENERA AND SPECIES

Many authors have produced classifications of the Oligochaeta, but it was not until 1900 that Michaelsen produced the system that is the basis of the modern taxonomy of this group, and divided them into 11 families, containing about 152 genera and 1200 species. Since then, Michaelsen (1921) reorganized his own classification into 21 families in two suborders, and Stephenson (1930) simplified this arrangement into 14 families, which differed little from Michaelsen's original grouping. A division of families into the Microdrili, consisting of small, mainly aquatic worms (including the terrestrial Enchytraeidae) and the Megadrili (larger, mostly terrestrial worms) is now very outdated, but may be useful in defining the scope of this book, which is concerned mainly with the terrestrial worms, and therefore does not include the Microdrili (although the Enchytraeidae are terrestrial they are not dealt with in this book at any length).

Of Stephenson's 14 families, seven – the Aeolosomatidae, Naididae, Tubificidae, Pheodrilidae, Enchytraeidae, Lumbriculidae and Branchiobdellidae – were placed in the Microdrili group and the remaining seven families in the Megadrili. Stephenson's classification of the megadrile families (1930) was as follows:

1. Family Alluroididae.
2. Family Haplotaxidae.
3. Family Moniligastridae (Syngenodrilinae, Moniligastrinae).
4. Family Megascolecidae (Acanthodrilinae, Megascolecinae, Octochaetinae, Ocnerodrilinae).
5. Family Eudrilidae (Parendrilinae, Eudrilinae).
6. Family Glossoscolecidae (Glossoscolecinae, Sparganophilinae, Microchaetinae, Hormogastrinae, Criodrilinae).
7. Family Lumbricidae.

Two of the megadrile families, the Alluroididae and the Haplotaxidae, are composed of species that are either aquatic or subaquatic, and are therefore considered, together with the Microdili, to be outside the scope of this book, which concentrates on terrestrial species.

Since Stephenson (1930), a number of authors have attempted to revise the classification of various megadrile families, particularly the Glossoscolecidae, Megascolecidae and the Moniligastridae (Jamieson, 1971a,b,c). Of Stephenson's two moniligastrid subfamilies, the Syngenodrilinae and Monogastrilinae, the former was placed in the Alluroididae (Gates, 1959), thus raising the remaining subfamily to family status.

Stephenson divided the Glossoscolecidae into five subfamilies, Glossoscolecinae, Sparganophilinae, Microchaetinae, Hormogastrinae and Criodrilinae. Gates raised all the glossoscolecid subfamilies to family status, on the grounds that anatomical differences not investigated previously, together with phylogenetic relationships with other families, do not indicate a sufficiently close relationship for them to be all included in a single family. However, Jamieson (1988) concluded that there was a total lack of affinity between Criodrilidae, Sparganophilidae and Glossoscolecidae.

The classification of the megascolecid earthworms has always been much more controversial than that of other oligochaete families, and more recently four new systems of classification have been proposed, those of Omodeo (1958), Gates (1959), Lee (1959) and Jamieson (1971a,b,c, 1985), all of which replace that of Stephenson (1930) (Fig. 2.1). Omodeo recognized the taxonomic groups by the position and numbers of the calciferous glands, and on this basis raised one group to family status. Lee used the number and position of the male pores and position of the nephridiopores as a key characteristic. Gates considered the structure of the prostatic glands and excretory system, and position of the calciferous glands to be important and raised all the main groups to family status. In an attempt to assess the relative merits of the three earlier systems, Sims (1966) investigated the classification of a selection of megascolecid genera using computer techniques, by arranging them into groups with mutual characteristics using a dendrogram and vector diagram. Sims found that the pattern of the arrangement of the genera coincided to a large extent with the classification proposed by Gates, and he disagreed with those proposed by Lee and Omodeo.

In his classification of the megascolecids, Sims (1967) agreed with the definitions of the groupings proposed by Gates, but recognized only two families, the Megascolecidae (*sensu stricto*) and the Acanthodrilidae, dividing this last family between the Acanthodrilinae, Octochaetinae and Ocnerodrilinae.

Jamieson (1971a,b) proposed an alternative classification to that of Omodeo, Gates, Lee and Sims, by combining certain of the basic material used by the above workers, together with evidence from previously neglected morphology of the excretory system. This was modified by Brinkhurst and Jamieson (1972) and again by Jamieson (1978) into a classification that has been widely accepted. It is not within the scope of this book, which deals with the biology and ecology of earthworms, to judge the individual merits of the various systems of classification that have been proposed.

The major divisions of the oligochaetes according to Gates (1959) were as follows:

1.   Family Moniligastridae.

2. Family Megascolecidae.
3. Family Ocnerodrilidae.
4. Family Acanthodrilidae.
5. Family Octochaetidae.
6. Family Eudrilidae.
   Subfamily Parendrilinae
   Subfamily Eudrilinae
7. Family Glossoscolecidae.
8. Family Sparganophilidae.
9. Family Microchaetidae.
10. Family Hormogastridae.
11. Family Criodrilidae.
12. Family Lumbricidae.

Jamieson (1988) reviewed the overall phylogeny and higher classification of the Oligochaeta based on a cladistic analysis. He placed all the megadrile families that were predominantly or wholly terrestrial into a new cohort (Fig. 2.1).

Cohort: Terrimegadrili

Anus terminal, rarely dorsal. Supra-esophageal vessel on esophagus only. Male pores in 18; posterior in clitellum. Agrosome long, >2 μm.

1. Superfamily: Ocnerodriloidea

Family: Ocnerodrilidae

Tubular prostates.

2. Superfamily: Eudriloidea

Family: Eudrilidae

Subneural present. Euprostates. In more advanced forms: spermathecae with oviducal link; spermathecae post-testicular. Eudrilid spermathecae.

3a. Superfamily: Lumbricoidea

Longitudinal tubercula pubertatis. Ovary single-stringed. Supra-esophageal vessel lost. First esophageal gizzard segment 6. Male pores migrate posteriorly, to 21 or 22.

Family: Kynotidae

Prostate-like setal glands. Nephridial bladders and caeca. Muscular male bursa. Male pores migrate forward to 16. Segmental testis sacs. Male pores anterior on clitellum.

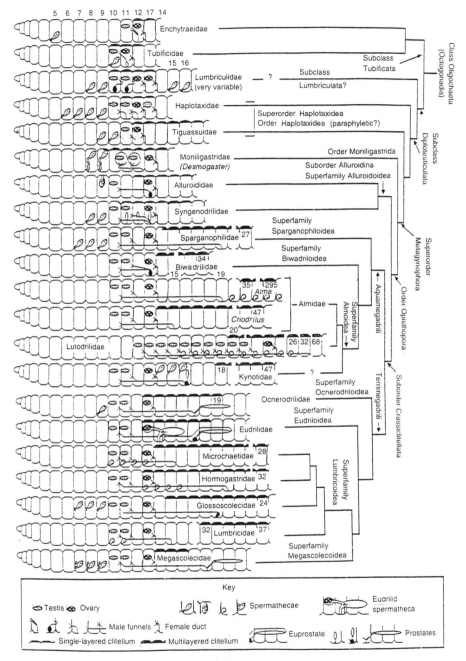

Figure 2.1 Cladogram representing the key characters of the families, showing the taxonomy adopted here and including some of the anatomy of the constituent families. (After Jamieson, 1988.)

Family: Komarekionidae

Prostate-like setal glands. First intestinal segment 13.

Family: Ailoscolecidae

Longitudinally grooved penial setae. Clitellum increased to 23 segments.

Family: Microchaetidae

Penial setae. Prostate-like setal glands. Gizzard in segment 7.

Family: Hormogastridea

Esophageal gizzards in three segments. Male pores migrate forward to 15.

Family: Glossoscolecidae

Anterior testes lost (some spp.). Supra-esophageal on esophagus only. Segmental testis sacs.

Family: Lumbricidae

Esophageal gizzards lost. cd=ab. Male pores anterior on clitellum.

3b. Superfamily: Megascolecoidea

Family: Megascolecidae

Dorsal pores. Penial setae. Hearts migrate posteriorly, to 13. cd>ab. Spermathecae with seminal diverticulum. Clitellum reduced in length, to 3.5–5 segments.

Subfamily: Ocnerodrilinae

Extramural calciferous glands on 9, or 9 and 10, not arranged as in the Acanthodrilinae.

Subfamily: Acanthodrilinae

Last hearts not in 11. Male pores usually acanthodrilin (male pores one pair with two pairs of prostatic pores, in the next anterior and posterior segments). Sometimes with a single pair of three pairs of prostatic pores, in which case one pair lies in or near the male pore segment and may be united with the male pores though never united on 18 if there is only a single pair of prostates. Nephridia: holonephridia, mixed holonephridia and meronephridia, or meronephridia only. Prostates tubular, or (*Exxus*) two pairs of racemose prostates in the acanthodriline arrangement.

Subfamily: Megascolecinae

Last hearts rarely in 11. Male terminalia acanthodrilin or with only one pair of prostates, the pores of which are near to or coincident with the male pores on 18 or its homeotic equivalent. Prostate tubular to racemose. Purely holonephric or meronephric, or with meronephridia in varying number of segments anterior to holonephridia; acanthodrilin forms meronephric throughout and posteriorly with stomate nephridia median to astomate micromeronephridia.

The two most important families ecologically in Europe, North America, Australia and Asia are the Megascolecidae and the Lumbricidae. The megascolecids and their close relatives comprise more than half the known species, and this group includes worms that are very widely distributed outside the Palearctic zone, with two genera, *Pheretima* and *Dichogaster*, that together probably contain more species than any other oligochaete genera. To break this down even further, Sims and Easton (1972) reviewed the genus *Pheretima* and divided it into eight genera: *Archipheretima, Pitbemera, Ephemitra, Metapheretima, Planapheretima, Amynthas, Metaphira,* and *Pheretima.* However, the most important family in terms of human welfare is undoubtedly the Lumbricidae, generally considered to be the most recently evolved family. This family of worms is of particular importance because it is the dominant endemic family in the Palearctic zone, including Europe, where until recent times many advances in agricultural practice have originated. Because of their ability to colonize new soils and become dominant, to the near exclusion of local endemic species, the Lumbricidae have followed the spread of human colonization from the more developed countries around the world. It has been said (Michaelsen, 1903) that only the genus *Pheretima* and its relatives can oppose an invasion by lumbricids successfully. Thus, the earthworm populations in crop-growing areas in temperate regions are far more likely to consist mainly of species of the Lumbricidae than of members of any other family, and because of this ubiquitous distribution, a large proportion of the work on the biology and ecology of earthworms has been based on species of this family. Many of the discussions in this book concern the Lumbricidae, although members of other families, particularly megascolecids, are discussed and referred to whenever relevant.

It is not within the scope of this book to give more than a brief outline of earthworm diversity. The following discussion includes most of the genera of earthworms, with an indication of their distribution, and the families to which they are currently assigned, together with a description of the family taxonomic features.

The characters used to define genera of the Lumbricidae are mostly internal structures, such as the number and distribution of the organs of the reproductive system, including the testis sacs, spermathecae and seminal vesicles, and also the position of the gizzard. Externally, the positions of the male pores, spermathecal and first dorsal pores and the spacing of the setae are considered to be generic and specific characters. Other features that are used when describing a genus fully include cephalization (mode of attachment of prostomium to the peristomium), position of the female pores, the presence or absence of pigment and the segmental position of the clitellum.

There are about 220 species of lumbricids (Cernosvitov and Evans, 1947), of which 19 are common in Europe, and have spread throughout the world, mostly through the agency of man. External characters such as cephalization, the position and extent of the clitellum and tubercula pubertatis, the position of the first dorsal pore, the prominence of papillae, the position and spacing of genital setae, the spacing of setae, color, length of body and number of segments of the adults are all used in describing species. Some species have one or more variants, forms or subspecies which differ in only one or two very minor characters from the typical form, and sometimes the variant replaces the typical form in a particular area. The number of species in the family Lumbricidae is small compared with that of other families, and diagnostic differences between genera and species are not great. This family is still being studied extensively taxonomically and further divisions, particularly at species level, may be recognized. There have been many changes over the past 20 years and not all of them are accepted by all earthworm taxonomists. In 1969, Gates redefined two lumbricid earthworm genera, *Bimastos* and *Eisenoides*, the latter erected for two North American species which had for years been attributed to *Allolobophora*, *Eisenia* and *Helodrilus* (Gates, 1969b).

Bouché (1972) retained a part of Omodeo's taxonomic system (genera *Dendrobaena*, *Octolasion*, *Octodrilus*, *Eiseniella*, *Lumbricus*) and retained the subgenera *Dendrobaena* and *Dendrodrilus*. In addition, he described the genus *Kritodrilus*. He included in his *Eisenia* genus two species which have subsequently been placed in different genera (*Allolobophoridella eiseni* and *Bimastos parvus*).

Bouché made dramatic changes when it came to *Allolobophora* and *Eophila*. Based on species found in France, he classified them as *Allolobophora*, *Helodrilus* and the new genus *Nicodrilus* (junior synonym of *Aporrectodea*) with subgenera *Nicodrilus* and *Rhodonicus*. Some species of *Allolobophora* were classified as *Aporrectodea*. The genus *Eophila* was put into three new genera: *Orodrilus*, *Prosellodrilus* and *Scherotheca*.

Gates (1972b) demonstrated, using morphological characters, that the *A. trapezoides* complex, with three to four species, included in what is termed *Allolobophora caliginosa* were in fact distinct species i.e. *trapezoides*, *tuberculata* (Eisen, 1874), *turgida* (Eisen, 1873), *longa* (Ude), *limicola* (Michaelsen, 1890), *nocturna* (Evans, 1946) and *icterica* (Savigny). Many

European authors prefer to use *caliginosa* for *turgida* and several of the other species as forms or varieties of *caliginosa*, e.g. *trapezoides* and *tuberculata*. Gates placed *Allolobophora longa*, *Allolobophora caliginosa*, *Allolobophora trapezoides* and *Allolobophora tuberculata* species in the genus *Aporrectodea*, and was able to show in biogeographical surveys throughout North America that not only were these separate species, but that they had distinct distributional patterns.

The centenary (1981) of the publication of Darwin's book *The Formation of Vegetable Mould through the Action of Worms* was the occasion for a symposium held in the UK to discuss various aspects of earthworm ecology and taxonomy (Satchell, 1983). Sims (1983) discussed the *Allolobophora* problem and the confusion over *A. caliginosa*. Sims also devoted considerable space to the *Octolasion* problems (orthography, nomenclature and taxonomy). In an attempt to obtain universality and remove confusion in earthworm taxonomy, Easton (1983) produced a checklist of valid lumbricid species through December 1981.

In 1976, Reynolds and Cook produced the original volume of *Nomenclatura Oligochaetologica – A Catalogue of Names, Descriptions and Type Specimens of the Oligochaeta*. This was the first time that the names of all oligochaetes, along with the citation of their descriptions and the location(s) of their type specimens, were assembled in one place. Three additional supplements have been produced, which update the original book (Reynolds and Cook, 1981, 1989, 1993). The four publications list 7254 species of oligochaetes and 739 genera, but, of course, include many aquatic species.

The latest classification of terrestrial species by Reynolds and Cook (1993) had the following divisions:

1. Order Moniligastridae
   Family Moniligastridae
2. Order Haplotaxidae
   Suborder Enchytraeina
     Superfamily Enchytraeoidea
       Family Enchytraeidae
       Family Propappidae
   Suborder Lumbricina
     Superfamily Alluroidoidea
       Family Alluroididae
     Superfamily Criodriloidea
       Family Criodrilidae
     Superfamily Lumbricoidea
       Family Lumbricidae
       Family Komarekionidae
       Family Diporochaetidae

Family Ailoscolecidae
Family Hormogastridae
Family Lobatocerebridae
Superfamily Sparganophiloidea
   Family Sparganophilidae
Superfamily Biwadriloidea
   Family Biwadrilidae
Superfamily Glossoscolecoidea
   Family Glossoscolecidae
   Family Kynotidae
   Family Microchaetidae
Superfamily Almoidea
   Family Almidae
   Family Lutodrilidae
Superfamily Megascolecidea
   Family Megascolecidae
   Family Acanthodrilidae
   Family Octochaetidae
   Family Eudrilidae
Superfamily Ocnerodriloidea
   Family Ocnerodrilidae

It is outside the scope of this book to provide detailed taxonomic keys to genera and species and the reader is referred to Fender (1992), Schwert (1992) and Ernst (1995) for keys to US species, to Gates (1972a) for tropical species, to Sims and Gerard (1985) for keys to British species, to Bouché (1972) for keys to French species, to Lee (1959) for keys to New Zealand species, to Baker and Barrett (1994) for common Australian species and to Jamieson (1971a,b,c) for keys to megascolecids.

2.3 GEOGRAPHICAL DISTRIBUTION

Earthworms occur all over the world, but only rarely in deserts, areas under constant snow and ice, mountain ranges and areas almost entirely lacking in soil and vegetation. Such features are natural barriers against the spread or migration of earthworm species, and so are the seas, because most species of earthworms cannot tolerate salt water even for short periods. Nevertheless, some species of earthworm are widely distributed, and Michaelsen has used the term 'peregrine' to describe such species, whereas the other species that do not seem able to spread successfully to other areas to any great extent have been termed 'endemic' species. The peregrine species are discussed in more detail later in this chapter.

Both the Megascolecidae and Lumbricidae are far-ranging in distribution. A few species belonging to the Lumbricidae are believed to be endemic to North America, south-east of the Mississippi, Europe, Asia

Minor, the Caucasus, Siberia and Japan. Many authors consider that this area, larger than that occupied by any other family, precludes the concept presented by Michaelsen (1910) and Stephenson (1930) that the Lumbricidae is the youngest of the oligochaete families. The area occupied by the endemic lumbricids forms a band around the northern hemisphere, and its southern boundary is formed by seas, deserts or other dry areas and is mainly semi-tropical (Julin, 1949). However, the northern limit, apart from the large area of mountainous plateaus and arid regions in central and north-eastern Asia, does not appear to be a physical one. Michaelsen (1903) pointed out that this northern limit is close to the southernmost limits of the furthest southern extension of the Quaternary ice caps, not only in Europe, but also in North America. North of this limit, earthworm populations were exterminated by the physical action of the ice sheets. As the ice receded, earthworms slowly colonized northwards in the wake of the retreating ice cap. This advance was very slow and by the time European colonists began settling in Canada, neither lumbricids nor megascolecids had reached the nineteenth parallel of latitude. All earthworms in Canada today are peregrine, and anthropochorous (i.e. spread by man).

In the United States, about 33 lumbricid species have been recorded; only the genera *Eophila* and *Helodrilus* not being represented. Two genera, *Bimastos* and *Eisenoides*, are endemic American, according to Gates (1969b). Gates created the genus *Eisenoides* to include two species (*carolinensis* and *lönnbergi*) from the European genus *Eisenia* and he (1972b) considered that the genus *Bimastos* originated in North America, and that many of its species were eliminated during the glacial epoch.

In Europe, the effect of the glaciation has been that those countries, partly or wholly north of the ice limit, have an oligochaete fauna with fewer species (with a higher proportion of peregrine species) than those south of this limit. For example, Czechoslovakia has 37 recorded species (17 peregrine), Switzerland 38 species (18 peregrine), Italy 57 species (17 peregrine), all south of the limit of the ice cap. Of those countries north or partly north of this limit, England has 25–30 recorded species (17 peregrine), Germany 24 species (19 peregrine), and Norway 30 species (25 peregrine).

Peregrine species of the Lumbricidae are also found distributed widely throughout the rest of the world, particularly in the temperate zone of the southern hemisphere. They are found in Mexico, Central America, South America, South Africa, India, Australia and Hawaii, as well as many of the scattered islands of the Atlantic and Indian Oceans. The agency of this spread has been mainly passive, usually by unintentional transportation by man.

It has been often stated that lumbricids have an ability to adapt to new environments more than any other oligochaetes, and that once introduced into an area 'frequently cause the disappearance of the endemic earthworm fauna' (Stephenson, 1930). Gates (1972a) postulated that no lumbricid species have been known to have colonized tropical lowlands anywhere, and when deliberately introduced in large numbers, have never survived. The evidence is that anthropochorous lumbricids are climatically restricted to areas that are within the climatic range of the northern temperate zone. They also have the ability to live in those environments that are constantly disturbed by the cultural and agricultural activities of man. The term hemerophilic has been coined for those worms that can tolerate human presence and activities, hemerophobic for those that cannot.

In the United States, there are comparatively few representatives of the other megadrile families amongst the endemic fauna other than Lutodrilidae and Komarekionidae. The acanthodrilid genus *Diplocardia* is truly American and widespread. The family Sparganophilidae is represented by five or more genera, and in the region between California and British Columbia are a number of species which Omodeo (1963) wrongly placed in the acanthodrilid genus *Plutellus* and which may even belong in other families.

In Europe, species of the Sparganophilidae are to be found in many countries, including England, and species of the Hormogastridae are endemic in the southern part of the continent. Included in this family is *Hormogaster redii* f. *gigantes* which can attain a length of 75 cm. Members of this family are found in Sardinia, Corsica, Italy, Sicily, southern France and North Africa.

The most localized megadrile families are undoubtedly the Eudrilidae, whose members, with one notable exception, are confined to Central Africa; the Lutodrilidae, found only in Louisiana, east of the Mississippi River; the Biwadrilidae, found only in Japan; and the Kynotidae, found only in Madagascar. *Eudrilus eugeniae* is the only peregrine eudrilid species, and is so common in the southern United States that it is familiarly known as the African night-crawler. In the southern part of the African continent are two groups of endemic worms, the Microchaetidae, and members of the Acanthodrilidae. Both these groups are being exterminated gradually in this particular area by human activities; the microchaetids are influenced by the lowering of the water table and the formation of desert conditions in the savannas, and the acanthodrilids by the removal of indigenous forests. There are introduced species of at least seven genera of the Lumbricidae, the megascolecid genus *Pheretima* (*sensu lato*) and at least one species each of the families Octochaetidae and Glossoscolecidae. There is no evidence to indicate any form of competition between endemic and introduced species (Ljungström, 1972a).

Members of the family Glossoscolecidae, with the exception of the peregrine anthropochorous genera *Pontoscolex*, *Diacheta* and *Onychochaera*, are confined to Central America and a few West Indian islands, and South America excluding that part south of the River Plate. *Pontoscolex corethrurus* is now distributed widely in tropical areas throughout the world. Also included in the megadrile fauna of tropical and subtropical South America are species of the family Ocnerodrilidae, and the microchaetid genus *Drilocrius*. Probably, South America is one of the two endemic habitats of the Ocnerodrilidae, the other being western Central Africa.

The remaining megadrile families that have not been discussed so far have a wide-ranging distribution and can be endemic as well as peregrine; they are based mainly in eastern countries. The family Moniligastridae has a very large range, encompassing South-East and eastern Asia, South India, Manchuria, Japan, the Philippines, Borneo and Sumatra. The majority of this area is colonized by only one genus, *Drawida*. It inhabits a larger area than the megascolecid genus *Pheretima*, and Gates (1972a) suggests that careful surveys throughout Asia and Malaysia could· be expected to reveal numbers of *Drawida* species equalling, if not exceeding, the number of *Pheretima* species.

Other than the Moniligastridae, South-East Asia and Australasia are dominated by earthworms belonging to the classical Megascolecidae, i.e. the families Megascolecidae, Acanthodrilidae, Ocnerodrilidae and Octochaetidae. Megascolecid earthworms of the genus *Pheretima* s.l. (indigenous in China, Philippines, Malaysia, Indonesia and Papua New Guinea) have also been transported to many tropical, subtropical and temperate regions. Peregrine species are found in Australasia, South and Central America, North America and the West Indies. The endemic earthworms of Australasia are megascolecids, including some species of *Pheretima*. Species of the family Acanthodrilidae, which, other than *Plutellus*, are not found in South-East Asia, occur in Australia, New Caledonia and New Zealand (the genus *Deinodrilus* is confined to this last country). Members of this family predominate in South America, south of the River Plate, and are also to be found in North and Central America. A few earthworms (species unknown) have been reported from the South Shetland Islands in Antarctica.

There seems little doubt that the world-wide distribution of the different families of earthworms gives important evidence of geological history and land movements.

Michaelsen (1903) first used the word 'peregrine' to describe the distribution of some earthworm species that are dispersed over a wide range of geographically remote localities. He included species (e.g. *Pontodrilus* spp. and some *Microscolex* spp., widely distributed on isolated islands and land masses) that are known, or are presumed, to be euryhaline and to have been dispersed by drifting with debris in ocean currents. Lee (1985) excluded the euryhaline species and restricted the use of the term peregrine to species that have been transported by man or whose dispersal has been facilitated by man.

The most widespread peregrine species are confined, or nearly so, to man-modified habitats. Gates (1970) distinguished these as anthro- pochorous species, contrasting them with allochthonous species that are not so constrained, and attributing their success to their rather unusual ability not only to withstand but also to take advantage of human distur- bance of the environment. It is these species, more than any others, that are important in maintaining soil fertility in agricultural soils.

Lee (1987) reviewed the distribution of peregrine species. Nearly 100 species, or 3% of all earthworms, including representatives of all families, are peregrine. The most widespread peregrine species comprise:

1. 20–30 species of Lumbricidae, particularly of the genera *Lumbricus*, *Aporrectodea*, *Allolobophora*, *Eisenia*, *Eiseniella*, *Dendrobaena*, *Dendrodrilus*, *Bimastos* and *Octolasion*. These species have been spread from northern and western Europe by man, within the past few hun- dred years, so that they are now the dominant earthworms of agricul- tural lands and gardens throughout temperate and some upland tropical regions of the world (Lee, 1985). They cannot have been widespread in the European regions from which they spread for more than 10–15 000 years, i.e. since the retreat of the Pleistocene ice sheets. Their occurrence there must also be regarded as being due to their great aptitude for occupying new habitats, and perhaps to their introduction by man. The origin of their actual distribution must lie to the south of the Pleistocene ice sheets, perhaps in southern Europe and Asia.

2. Representatives of several megascolecid families, in particular:
    (a) *c.* 15–20 species of the genera that constitute a *Pheretima* group that includes *Amynthas* spp. apparently originated from eastern and south-eastern Asia, and are now established through most of the tropical regions of the world, with some species also well established in temperate regions;
    (b) *Microscolex dubius* and *M. phosphoreus*, possibly of South American origin, but which are now established throughout the southern temperate zone and also in North America and in Europe, mainly in agricultural and pastoral areas;
    (c) *Dichogaster bolaui, D. saliens* and *Eudrilus eugeniae*, probably origi- nally from western Africa, but (especially *D. bolaui* and *E. eugeniae*) now widely distributed throughout tropical and warmer temperate regions;
    (d) *Ocnerodrilus occidentalis*, probably originally from central America, but now pantropical and sometimes found in temperate regions.

3. The glossoscolecid *Pontoscolex corethrurus*, originally from South America, now widely established in tropical and in some warmer temperate regions, e.g. Florida.
4. The moniligastrid *Drawida bahamensis*, probably originally from eastern Asia and now pantropical.

Special characteristics of peregrine species include: potential for hermaphroditism, adoption of polyploid inheritance, tolerance of environmental variability, habitat specificity, opportunism in choice of food, ability to withstand chemical stress, association with cultivated soils and ecological plasticity (Lee, 1987).

Fragoso *et al.* (1992) provided data on 176 earthworm species and 60 earthworm communities from tropical America (six countries) and Africa (two countries). Peregrine species with pantropical distributions were dominant in disturbed environments because they had greater tolerances than native species towards environmental factors. Nevertheless, a few native species with rather broad regional distributions were able to live in the disturbed environments. In many of the disturbed environments, earthworm communities had a lower diversity but larger biomass due to the colonization of a few peregrine species, e.g. *Pontoscolex corethrurus* or *Polypheretima elongata*.

In recent years, our knowledge of earthworm diversity and geographical distribution has been greatly enhanced by modern taxonomic techniques, which have been used for earthworms to confirm morphological identifications. For instance, starch gel electrophoretic determination using three enzymes (LAP, GPI, EST) has proved to be suitable for all species except those belonging to the genus *Lumbricus* (Bogh, 1992). Further refinement of these techniques may help to resolve existing controversies on the origins of particular groups and taxonomic relationships among different groups of earthworms. There are relatively few earthworm taxonomists world-wide and there is an urgent need for more information on the diversity and distribution of terrestrial earthworms.

Earthworm
biology

3

3.1 LIFE CYCLES

It must be emphasized that our present knowledge of the life cycles of even quite common species of earthworms is still very inadequate, and there are many species about which we know very little.

Earthworms are semi-continuous or continuous breeders, producing ova at most times in the year (Olive and Clark, 1978). The ova of earthworms are contained in cocoons (oothecae), which differ in shape with species, and which most lumbricids deposit near the soil surface. The cocoons of common lumbricids are illustrated in Fig. 3.1. If the soil is very moist, earthworms will deposit their cocoons near the surface, placing them much deeper in the soil when it is dry. Cocoons can be produced at any time in the year, but most species of earthworms produce cocoons when the temperature, soil moisture, food supplies and other environmental factors are suitable. However, there is good evidence that peak cocoon production occurs in the spring or early summer in the northern hemisphere. It has been demonstrated clearly in cultures (Evans and Guild, 1948a), and in the field (Gerard, 1967) that seasonal fluctuations of the soil climate can cause the number of cocoons produced by different species of earthworm to vary from year to year.

Since cocoons are produced from the clitellum, it is probable that the size of a cocoon is correlated with the size of the worm. Lavelle (1981) plotted cocoon size against fresh weight of adults for 11 European lumbricid species and tropical species of several families and reported a strong positive correlation.

Evans and Guild (1948a) reported that, in cultures of common species of lumbricids, kept for a year at what they considered to be optimum soil moisture, the number of cocoons produced by each species paralleled closely the seasonal changes in soil temperature (Fig. 3.2). Fewest cocoons were produced in the winter months and there was a temperature thresh-

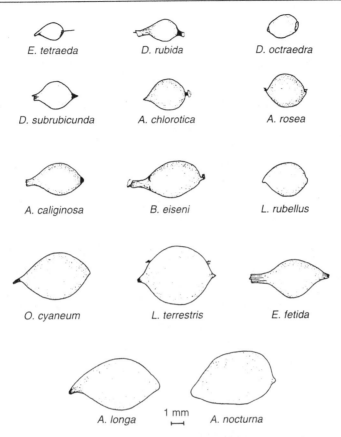

Figure 3.1   Cocoons of common lumbricids (from various sources).

old of about 3 °C, below which no cocoons were produced. Many cocoons were produced from the end of February to July when temperatures were rising, and the greatest numbers were produced between May and July. Thereafter, the numbers of cocoons produced decreased quite rapidly as temperatures fell, particularly for some species. Gerard (1967) confirmed these conclusions by sampling in the field for the cocoons of *Aporrectodea caliginosa* and *Allolobophora chlorotica*. In this study, most cocoons occurred in soil samples taken in May and June. More cocoons were probably produced in July and later, but by this time the cocoons would be hatching faster than they were being produced, so the numbers found in soil samples would decrease. Very few cocoons occurred in samples taken from August to November. Although some cocoons were found in the December, January and February samples, these had probably accumulated over time because cocoons produced during the colder months would not hatch until the soil warmed up in the spring.

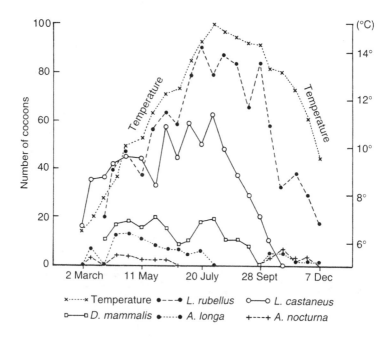

Figure 3.2 Seasonal production of cocoons (after Evans and Guild, 1948a).

Some lumbricids, such as *Aporrectodea longa* and *Allolobophora nocturna*, have an obligatory diapause during the summer months, and so produce cocoons only from mid-March to June or early July, and October to November. Other lumbricids belonging to species of the genera *Eisenia* and *Octolasion* and other species of *Allolobophora* may have a facultative diapause during dry periods, and this also interrupts the production of cocoons.

The number of cocoons produced in a season differ greatly with both species and climate (Evans and Guild, 1948a). In one study, between 3 and 79 cocoons were produced by individuals of eight different earthworm species in culture (Wilcke, 1952). In another study, *A. caliginosa*, *A. longa* and *O. cyaneum* produced between 3 and 13 cocoons/year. *Allolobophora chlorotica* produced 25–27 and *L. rubellus*, *L. castaneus* and *Dendrodrilus rubidus* 42–106 cocoons/year (Satchell, 1967). Edwards (1988) reported that *Dendrobaena veneta* could produce 84 cocoons/year; *Eudrilus eugeniae*, 188; *E. fetida*, 198; and *Perionyx excavatus*, 1014. Phillipson and Bolton (1977) reported that *A. rosea* produced only 3.13 cocoons/year but Nowak (1975) found that Polish earthworms produced between 26 and 42 cocoons/year.

Satchell (1967) pointed out that there was a direct correlation between the number of cocoons produced by any species and how much this species is exposed to adverse environmental factors such as desiccation, extremes of temperature and predation. In other words, those species of earthworms that may be exposed regularly to environmental hazards normally tend to produce more cocoons to enable them to survive adverse conditions. Thus, those species that live in, or can move into, the deeper soil layers and are protected from adverse conditions, e.g. *L. terrestris*, *A. longa* and *O. cyaneum*, produce fewest cocoons, whereas those species that live near the surface, such as *L. castaneus*, *L. rubellus* and *Dendrobaena subrubicunda*, which are exposed more directly to these factors, produce very many more. *Lumbricus castaneus* survives the summer drought in western Europe solely as eggs protected in cocoons, i.e. it avoids climatic stress by surviving as an inactive stage (Wallwork, 1983). Because individuals of any species of earthworm are influenced by the prevailing environmental conditions, they produce fewer cocoons not only when the soil is too dry, but also when it is too wet (Evans and Guild, 1948a). Since cocoons can resist desiccation and cold, they may enable earthworm populations to survive adverse conditions. Another factor that can influence the numbers of cocoons produced is the state of nutrition of the adults that produce them; for instance, earthworms fed on materials such as sewage sludge, straw or farmyard manure produced less than one-tenth of the cocoons that were produced by worms fed on bullock or horse droppings, or peat (Evans and Guild, 1948b).

In the semi-arid steppe grassland of Russia, *Eisenia nordenskioldi* restricts cocoon production to the period of the summer rains (Pokarzhevskii and Titisheva, 1982). In the southern hemisphere, cocoon production is also seasonal, as Madge (1969) showed for African eudrilids. In India, cocoon production is related to the monsoons, with maximum cocoon production in late October and early November (Dash and Senapati, 1982). *Millsonia anomala* in the Ivory Coast in Africa can produce cocoons at any time in the year, but there are two peaks of cocoon production which coincide with soil drying at the end of the wet season, resulting in two distinct generations per year (Lavelle, 1978). In South Australia, the highest populations of the introduced species *A. trapezoides* and *A. caliginosa* occurred from winter to early spring in five pastures (Baker *et al.*, 1992a). Adults were restricted to winter and spring, when breeding occurred. Subadults and juveniles, but few adults, survived the summer.

The time cocoons take to hatch varies considerably among species, ranging from 3 weeks for *Bimastos zeteki* (Murchie, 1960) to 5 months for *Eisenia rosea* (Wilcke, 1952). The overall time earthworms take to reach sexual maturity from hatching differs greatly between species and has been investigated in detail for only a few species, mostly in culture. Temperature also affects the time for development before the cocoons hatch. For instance, Gerard (1967) found that cocoons of *A. chlorotica* hatched after 36 days at 20 °C, 50 days at 15 °C and 112 days at 10 °C. Table 3.1 provides some data on the periods of growth of some species

Table 3.1 Time to maturity in weeks of several earthworm species (from Evans and Guild, 1948)

Month when growth began	A. caliginosa	A. chlorotica	A. longa	D. subrubicunda	A. rosea	E. fetida	L. castaneus	L. rubellus
November 1944	–	42	–	–	–	74	–	–
December	–	39	38	–	–	–	–	–
January 1945	–	35	–	33	–	70	–	–
February	–	30	–	29	–	74	–	–
March	–	29	–	27	–	–	–	–
April	58	–	71	27	–	–	–	–
May	53	–	70	22	62	–	–	–
June	56	–	64	–	61	–	–	–
July	54	–	58	–	53	47	18	37
August	–	–	54	–	50	–	–	–
September	–	–	51	–	–	–	–	–
February 1946	–	–	40	42	–	–	–	–
March	–	–	–	38	–	–	25	–
April	–	–	–	33	–	–	25	–
May	–	–	–	32	–	–	25	–

kept in culture (Evans and Guild, 1948a). Edwards (1988) reported that the time that cocoons of *E. fetida* took to hatch was 32–73 days; *E. eugeniae*, 13–27 days; *P. excavatus*, 16–21 days; and *D. veneta*, 40–126 days. Michon (1954) showed that the type of food available also influenced the length of the maturation period. The incubation period of cocoons of *Pheretima hilgendorfi* can vary from 244 to 264 days, which is considerably longer than that of most lumbricid species. Clearly, the production and time for maturation of cocoons varies with species, population density, age structure, available food, moisture and temperature.

The life span of mature lumbricids in the field is probably often no more than a few months (Satchell, 1967), because they are exposed to many hazards, although their potential longevity is 4–8 years (Lakhani and Satchell, 1970). In a field site, Satchell (1967) reported that *L. terrestris* matured in 1 year but environmental factors greatly affected the time taken for growth to maturity. *Allolobophora chlorotica* took from 29 to 42 weeks to mature at field temperatures, those worms that matured during the winter months taking longest, but Gerard (1967) calculated that *A. chlorotica*, hatching in a field site in late July, took only 21 weeks to mature. Graff (1953a) recorded that *A. chlorotica* took 17–19 weeks to mature when kept at 15 °C, and Michon (1954) reported that the same species matured in 13 weeks at 18 °C. In protected culture conditions, individuals of *A. longa* have been kept for 10¼ years, *E. fetida* for 4½ years and *L. terrestris* for 6 years (Korschelt, 1914), although in the field they are unlikely to attain such ages. Longer times to reach maturity must be expected if conditions are poor. Michon (1954) studied the life span of other species and reported that they ranged from 15 months for *A. chlorotica*, to 31 months for *A. longa*, in a group of 10 species kept at 18 °C, with an average life span for all species of 2 years. Data for the times of development of some lumbricid worms were given by Wilcke (1952) (Table 3.2). Earthworms stop breeding some time before they die. For *D. subrubicunda* the active breeding period, which is when the clitellum is prominent, is only half the adult life span: 100–350 days out of a total of 550–600 days (Michon, 1954).

The life cycles of species from few other families of earthworms have been studied in any detail, but they probably differ from those of lumbricids mainly in time of development. Bahl (1922) reported that although *Pheretima* spp. produced cocoons throughout the year in culture, in the subtropical climate of India, most cocoons were produced from March to June, and relatively few during the rainy months of July and August. In Japan, cocoons of *Pheretima hupeiensis* hatched mainly in May; immature worms grew rapidly but hibernated in both the immature and mature stages (Watanabe, 1975). The length of earthworms can range from less than a centimeter to 2 m for some tropical species. The dimensions of British lumbricids range from 2 to 30 cm (Arthur, 1965) (Fig. 3.3).

Table 3.2   Time of development for some lumbricid worms (from Wilcke, 1952)

Species	No. of cocoons per worm per year	Incubation time of cocoon (weeks)	Period of growth of worm (weeks)	Total time for development (weeks)
E. fetida	11	11	55	66
D. subrubicunda	42	8 ½	30	38 ½
L. rubellus	106	16	37	53
L. castaneus	65	14	24	38
E. rosea (A. rosea)	8	17 ½	55	72 ½
A. caliginosa	27	19	55	74
A. chlorotica	27	12 ½	36	48 ½
A. terrestris f. longa (A. longa)	8	10	50	60

The life cycles of earthworms not in the Lumbricidae have also been studied by Lavelle (1974, 1978). The numbers and time of cocoon production, growth and longevity did not differ significantly from those of lumbricid earthworms, other than that relatively fewer cocoons were produced (Table 3.3). Cocoon production and growth were correlated with both temperature and moisture.

The life cycle of *P. hupeiensis* was studied in North America (Grant, 1956). Mature worms occurred from June to October. Young worms first appeared in July and reached largest numbers in September, probably becoming immatures in October and overwintering in this form. All the available evidence was that *P. hupeiensis* has an annual life cycle, with most of the mature individuals dying soon after reproduction occurs. Lavelle

Table 3.3   Life cycle of nonlumbricid species of earthworms from a tropical habitat (adapted from Lavelle, 1979)

Species	No. of cocoons produced annually	Time for growth (months)	Maximum longevity (years)
Dichogaster agilis	7.5–15.8	15	2–2.5
Chuniodrilus zielae	5.9	18	1.5–2.0
Millsonia anomala	5.6–14.2	20	2.0–2.5
Millsonia lamtoiana	2.4–2.8	24	2.0–3.0
Agastrodrilus opisthogynus	0.4–1.9	24	3.0–4.0
Dichogaster terrae-nigrae	1.3–2.4	36	4.0
Millsonia ghanensis	0.8–1.4	42	4.0–5.0

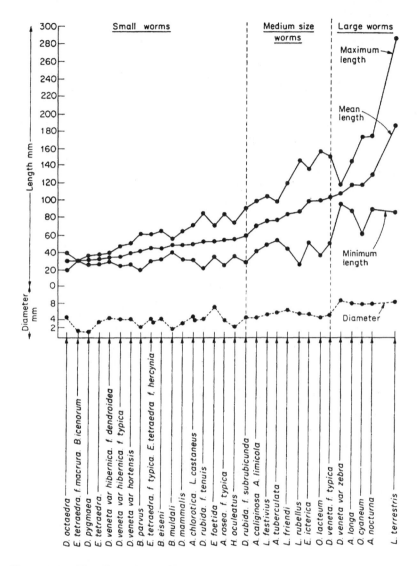

Figure 3.3   The dimensions of British lumbricids (modified from Arthur, 1965).

(1971) studied the life cycle of the acanthodrilid, *Millsonia anomala* in West Africa in some detail. Cocoons hatched during the wet seasons after 18–30 days of incubation. The young worms took 10–14 months to mature and they may not have lived much longer than this, because survival over dry seasons was mainly as immature stages. A typical structure of a population pyramid in July was 1.7% over 24 months old, 67% between 16 and 19 months old, 18.3% between 11 and 12 months old, 48.30% between 6 and 8 months old and 25% between 1 and 2 months old.

Madge (1969) studied the life history of a eudrilid species in Nigeria. He concluded that the first generation of *Hyperiodrilus africanus* was produced during the dry season and that these developed into mature individuals during the following dry season, and produced cocoons at the beginning of the next wet season, to produce a second generation. Fully clitellate individuals of the acanthodrilid worm *Diplocardia egglestoni* occurred only during May and June in an area of Michigan, USA during a dry summer (Murchie, 1958b). Similarly, clitellate adults of *Bimastos zeteki* were found from May until mid-July. The cocoons of this species were abundant in late May and June, and some were found in late summer. There seemed to be evidence of considerable numbers of this species overwintering as immatures (Murchie, 1960). The breeding period of the marsh-dwelling worm *Criodrilus lacuum* (Glossoscolecidae) was restricted to about 2 months, from the end of April to the end of June, and Stephenson (1929) suggested that such a short breeding period may be common among marsh-dwelling oligochaetes. *Alma* spp., a microchaetid worm from Egypt, lives in the mud of ditches, and produces cocoons before the mud dries up during the hot season; this enables it to survive adverse conditions. Little is known of other aspects of the life cycles of these less-studied families.

Hamoui (1991) reported that *Pontoscolex corethrurus* (Mueller) from Brazil took 14–19 months after hatching before cocoons were produced and thereafter cocoons were produced for a further 12–16 months. Murchie (1960) calculated that individuals of the species *B. zeteki* do not become sexually active for 12–24 months after hatching or, if they emerged at the beginning of a dry period, they might take as long as 22–26 months. Obviously, the growth periods of different species of earthworms differ greatly, and can vary considerably for a given species depending on environmental conditions.

3.2 REPRODUCTION

3.2.1 SPERMATOGENESIS

Earthworms are hermaphrodites with separate testes and ovaries that function simultaneously. Groups of spermatogonia are formed from follicles (masses of reproductive cells) in the testes, and pass into the seminal vesicles; these contain male cells at all stages of development. These cells eventually form spermatozoa which, at first, float freely in the seminal vesicles, but pass quickly into the testis sacs and attach themselves to the surface of the sperm funnels, where they remain until copulation occurs. During mating they are swept by the cilia of the funnel into the vas efferens and then to the exterior through the male pore via the vas deferens.

3.2.2 OOGENESIS

The first stage of the development of the ova normally occurs in the basal part of the ovaries, with the formation of oogonia, which divide and form oocytes. These do not divide again but increase in size and accumulate yolk. The oocytes are shed from the ovaries into the ovisacs, where reduction division takes place, by the rupture of the peritoneum. When ripe, the ova are discharged from the oviducts through the female pores, and pass into the future cocoon which is secreted by the clitellum. However, the ova of eudrilid earthworms are transferred to the ovisacs at a much earlier stage of development, so that their ovaries are much smaller, except during the very early stages of sexual development.

3.2.3 COPULATION AND FERTILIZATION

Most earthworm species reproduce by cross-fertilization, although some species can also produce cocoons parthenogenetically. *Lumbricus terrestris* mates on the soil surface but many other species mate below ground. Most species mate periodically throughout the year, except when environmental conditions are unsuitable or when they are estivating or in diapause.

Methods of copulation are not identical for all species. When individuals of *L. terrestris* mate, two worms, which are attracted to each other by glandular secretions, lie with the ventral parts of their bodies together, and their heads pointing in opposite directions (Fig. 3.4). They come into close contact in the region of the spermathecal openings where the clitellar region of one worm touches the surface of the other. While copulating, the worms do not respond readily to external stimuli such as touch and light. Large quantities of mucus are secreted so that each worm becomes covered with a slime tube between segment 9 and the posterior border of the clitellum, the two slime tubes adhering to, but remaining independent of, each other.

A seminal groove (normally seen as a pigmented line) extends from the male pore to the clitellum, but this groove may not be very obvious in some species. Each seminal groove is a depression of the outer body wall formed as a series of pits by the contraction of muscles, the arciform muscles, which lie in the longitudinal muscle layer. These muscles do not contract simultaneously but successively, beginning at segment 15, each contraction forming a pit. These pits carry seminal fluid as droplets from the male pore to the clitellar region, where it collects, and eventually enters the spermathecae of the opposing worm. The exact way in which this transfer occurs is not completely understood. Individuals of *Eisenia fetida* have been seen to clasp and release each other several times, and such movements may assist the entry of the sperm into the spermathecae. Tembe and Dubash (1961) described the copulation of a species of

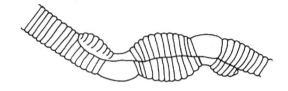

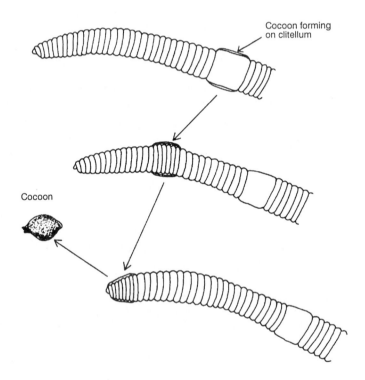

Figure 3.4   Copulation and stages in cocoon production in Lumbricidae (modified from Tembe and Dubash, 1961).

Pheretima that has three or four pairs of spermathecae. The male pores first come into contact with the hindmost pair of spermathecal apertures and discharge seminal fluid and prostatic fluid into them. Each worm then moves backwards, and the seminal fluid is discharged into the next pair of spermathecae, until all have been 'charged'. Different methods of sperm transference have been observed in other species; for instance, the octochaetid worm *Eutyphoeus waltoni* has spermathecal openings and male pores on raised papillae. The latter, termed penes, are inserted into the spermathecal openings (Bahl, 1927). The eudrilid species, *Schubotziella dunguensis* has a single median male pore formed from two fused copulatory pouches, and a single median spermatheca; during copulation, the two pouches (which communicate with the male ducts) evert, and are inserted into the opening of the spermathecae. All these species that transfer the spermatozoa by direct methods do not form mucous tubes as do lumbricids.

A number of species transfer the sperm cells between individuals in spermatophores, ensuring that the ova are fertilized as they are produced. The production of spermatophores may be facultative or obligate (Bouché, 1975b). This was first observed in *Bimastos antiguus michalisi*, from Albania, but has been reported for more than 20 species of earthworms from the genera *Aporrectodea, Bimastos, Dendrobaena, Dendrodrilus, Eisenia, Eisenoides, Kritodrilus, Lumbricus, Murchieona, Octolasion* and *Satchellius*.

Some eudrilid worms are unusual in that they seem to have a mechanism for internal fertilization. Sims (1964b) described the fertilization of one *Hyperiodrilus* sp. where fertilization is internal in a special chamber (the bursa propulsoria) that is connected to both the spermathecal system and the ovisac.

It is important that copulating worms keep close together, and the contours of the ventral surfaces at the points of closest contact help to achieve this. Some setae in segments 6–10 are shorter, thicker, and less curved than normal. They flex inwards and grip the partner, and the long, pointed and grooved setae on the ventral surface of the clitellar region pierce the body wall of the opposing worm. After copulation, which may take as long as an hour, the worms separate, and each clitellum produces a secretion which eventually hardens over its outer surface. When this is hard, the worm moves backwards, so drawing the tube over its head, and when the worm is completely free, the ends of the tube close, to form the cocoon which is roughly lemon-shaped, but with a shape typical of each species (Fig. 3.1). The cocoon contains a nutritive albuminous fluid, produced by the clitellar gland cells, the ova and the spermatozoa which were discharged into it as the tube passed the spermathecal openings. Cocoons continue to be formed until all the stored seminal fluid has been used up. Fertilization, which occurs in the cocoon, is external.

The cocoon wall consists of interwoven fibrils, that are soft when first formed, but later become harder and very resistant to drying and damage. The ends of cocoons are extended into processes or tufts which may be stem-like, conical, or umbrella-like, according to species. Cocoons vary in color from whitish (when formed) to yellow, greenish or brownish, and differ greatly in size. The very large Australian earthworm, *Megascolides australis*, probably produces the largest cocoons, which may measure up to 75 × 20 mm. However, the size of the earthworm does not always correlate with the size of the cocoon; for example, *L. terrestris* produces cocoons measuring about 6 × 4.5–5.0 mm, whereas species of *Pheretima*, which are about the same size as *L. terrestris*, produce much smaller cocoons, about 1.8–2.4 × 1.5–2.0 mm.

The number of fertilized ova in each cocoon ranges from 1 to 20 for lumbricid worms (Stephenson, 1930) but usually only one or two survive and hatch. Out of 14 species of lumbricid earthworms tested in culture, it was reported that *E. fetida* was the only species that commonly produced more than one worm from a cocoon, averaging 3.3 hatchlings per cocoon (Evans and Guild, 1948a; Edwards, 1988). In a detailed study of the hatching of cocoons of *E. fetida* and *Bimastos tumidus* (Vail, 1974) some of these variabilities were elucidated. Of 250 *E. fetida* cocoons, 21.5% did not hatch; 14.7% produced one young; 25.0% produced two young; 21.5%, three young; 14.1%, four young; 2.0%, five young; and 1.2%, six young. Of those that hatched, the average number of young was between 2 and 6.

In other work, the hatching of 100 *B. tumidus* cocoons was studied; of these 19 failed to hatch; 6 produced one young; 17, two young; 28, three young; 24, four young; 3, five young; and 3, six young; with an average number of young per cocoon of 3.11 (Vail, 1974). Edwards (1988) reported that cocoons of *Eudrilus eugeniae* produced an average of 2.3 hatchlings per cocoon.

Some species can reproduce parthenogenetically. *Aporrectodea, Allolobophora* and *Lumbricus* species can breed sexually only, but *Dendrobaena* species can sometimes be parthenogenetic and *Octolasion* spp., *Eiseniella* spp., *Aporrectodea rosea* and *Dendrobaena rubida* f. *tenuis* are usually parthenogenetic (Satchell, 1967). The male organs of *A. rosea* are sterile, so that the production of all cocoons must be parthenogenetic. Asexual reproduction by fragmentation and regeneration occurs in some species of aquatic Microdrili but does not seem to occur in terrestrial earthworms. Parthenogenesis is sometimes considered to be linked with the absence or regression of secondary sexual organs (spermathecae and prostates).

3.2.4 PARTHENOGENESIS

Most megadrile earthworms reproduce by normal cross-fertilization, and parthenogenesis has tended to be somewhat discredited or associated with some form of parasitism. However, beginning with the work of Gavrilov (1939, 1960) and Muldal (1952b), parthenogenesis has begun to be accepted as a relatively widespread phenomenon. Muldal (1952b) stated that there was no reason to believe that cross-breeding is obligatory and parthenogenesis rare on the basis of chromosomal studies of lumbricids. Other workers have reported various forms of parthenogenesis in a wide range of species (Kobayashi, 1937; Gavrilov, 1939; Reynolds, 1974;

Gates, 1974), although not all of these have been confirmed and may be overemphasized.

Reynolds (1974) discussed the phenomenon at length, and he defined 35 North American species as having sexual reproduction (amphimictic), 11 as probably having sexual reproduction, 4 as usually having sexual reproduction but with facultative parthenogenesis, 1 as having possibly parthogenetic reproduction and 30 as having mainly parthenogenetic reproduction. Some writers consider these to be overestimates. Other species that have been described as having parthenogenetic reproduction include: *Microscolex dubius, Plutellus papillifer, P. umbellulariae, Pontodrilus bermudensis, Udeina avesicula* (Ljungström, 1969), *Pontoscolex corethrurus* (Ljungström, 1970), *Pheretima alexandri, P. anomala, P. campanulata, P. defecta, P. elongata, P. exigus, P. glabra, P. hilgendorfi, P. meridiana, P. rimosa, P. virgo, P. infantilis, P. varians, P. voeltzkowi, P. zoysiae* (Gates, 1960, 1961), *Eukerria hortensis, E. subandina* (Gavrilov, 1960, 1967), *Malabaria levis, M. sulcata* (Gates, 1966). However, some workers have questioned whether it has been fully established through experimental evidence that many of these species do in fact have mainly parthenogenetic reproduction.

Probably one of the more important indicators of parthenogenesis is male sterility, and evidence of this includes: retention of testes and seminal vesicles in a juvenile state in adults, and absence of iridescence of the male funnels, ducts and spermathecae, which demonstrates the aggregation of mature sperm. A lack of externally adhesive spermatophores is not a good indication, because they have been found even in obligatory parthenogenetic species of lumbricids.

Reproduction cannot occur without a clitellum, ovaries, oviducts and possibly the ovisacs, but the male organs are not essential, so the absence in various degrees of testes, seminal vesicles, seminal receptacles, vas deferens, copulatory chambers, penes, prostates, spermathecae, tubercula pubertatis and genital and penial setae may be indications of occasional or obligatory parthenogenesis (Gates, 1972a).

Evans and Guild (1948a) also investigated whether some lumbricid earthworms could produce viable cocoons without mating and fertilization, by rearing isolated individuals of 19 species in culture. In general, *Allolobophora* spp. produced no cocoons even though they were observed for periods of up to a year. Individuals of *L. castaneus* and *L. rubellus* began producing cocoons 5–6 months after they became mature, but those of *L. rubellus* were not fertile, and, out of a total of 1704 cocoons produced by *L. castaneus*, only 3 produced young worms. The other two *Lumbricus* species produced some cocoons, but none were viable. Some individuals of the other species tested produced cocoons at different intervals after they became sexually mature. Most individuals of *D. rubida* and *D. subrubicunda* produced viable cocoons soon after they became mature, but a few individuals of the latter species did not produce cocoons for 9 months.

Most unmated worms of the species *E. fetida* produced cocoons within 9 months of maturity, although in the same period mated worms produced more cocoons. Only one individual of *Octolasion cyaneum* and one of *O. lacteum* produced any viable cocoons. The same authors cross-mated individuals of *L. rubellus* and *L. festivus*, which are closely related species, and similar in size. An average of 5–6 cocoons were produced by each worm and, 3 months after mating, hybrid worms emerged from some of the cocoons.

3.2.5 POLYPLOIDY

In Lumbricidae, about half of the karyotypes that have been investigated are polyploid. In the other families of the Class, the data are few and discontinuous, but it is clear that hermaphroditism is not a precondition for the occurrence of polyploidy; in fact another condition seems to be necessary to ensure evolutionary success of the polyploid emerging races, i.e. parthenogenesis. When polyploidy is closely associated with parthenogenesis, it results in an unexpected amount of heterozygosity which is advantageous for the individuals, conferring on them high resistance to environmental stresses.

Many authors have demonstrated that potentially parthenogenetic polyploid earthworm populations tend to have a much wider geographical distribution, but it is difficult to say if this is a result of parthenogenesis or polyploidy. Asexual reproduction confers a biological advantage, since a single thelytokous individual can easily colonize any area to which it is introduced. An important question is the elucidation of the ecological relations between the different karyological strains within the same species. Can they coexist in the same area or does the polyploid replace the diploid race? There are reports supportive of both hypotheses. Undoubtedly, better knowledge of the autoecology of the different karyological races is needed to clarify this question.

3.3 QUIESCENCE, DIAPAUSE AND ESTIVATION

At some times in the year, the soil close to the surface may become too dry, too cold, or too warm for earthworms to survive in it. They have several ways of surviving such adverse periods. For instance, cocoons can resist desiccation and extreme temperatures much better than worms, and the cocoons hatch when conditions become more favourable. Alternatively, worms may migrate to deeper soil where the moisture and temperature conditions are better. For instance, Madge (1969) observed that two species of eudrilid worms in tropical Africa moved down into deeper soil layers during the dry season. *Lumbricus terrestris* can move readily to deeper soil through its more or less permanent vertical burrow system. Many species of earthworms that normally live near the surface move from the top layers of the soil and become comparatively inactive in deeper soil during adverse periods. Three states of such inactivity can be distinguished:

1. *Quiescence*: in which the worm responds directly to adverse conditions, particularly drought and high temperatures, and becomes active again as soon as conditions become favorable. Bouché (1972) distinguished two types of quiescence: (a) anhydrobiosis – a response to dehydrative aggregates and loss of water; and (b) hibernation – a response to low soil temperatures. Earthworms are coiled in small cells below the frozen layers.
2. *Facultative diapause*: which is also caused by adverse environmental conditions, but does not terminate until a certain critical time after conditions become favorable.
3. *Obligatory diapause*: which occurs at a certain time or times each year, independent of current environmental conditions but usually in response to a certain sequence of environmental changes or to some internal mechanism. These stimuli are usually such that adverse conditions tend to occur during the period of diapause.

The term 'estivation' has been used to cover any or all of these states, but it applies properly only to the passing of summer or dry seasons in a dormant state. In the temperate zones, earthworms usually become quiescent during warm spells. Many immature worms of the species *A. chlorotica*, *A. caliginosa* and *A. rosea* become quiescent when the soil is either too dry or too cold (Evans and Guild, 1947c), but more earthworms become quiescent during the summer than during the winter, probably because earthworms can tolerate wet and cold conditions better than hot and dry ones (Gerard, 1967). The difference between quiescence and facultative diapause is probably only one of degree. Gerard (1967) thought that quiescent behavior enabled worms to exist for long periods on their reserves, when there was little food available in their environment. Lee (1951) stated that in many pasture soils in New Zealand that are subject to severe drought in summer, earthworms move to deeper soil layers and become quiescent. Murchie (1956) reported that *D. egglestoni* can enter a period of quiescence or facultative diapause in response to drought or low soil temperatures. It has been claimed that red-pigmented worms, which include many species of the lumbricid genera *Lumbricus* and *Dendrobaena*, do not go into diapause (Michon, 1957), but there is some evidence that *L. terrestris* does go into diapause.

At the beginning of diapause, worms stop feeding, empty their guts, and construct a small round or oval cell lined with mucus. Then they roll into a tight ball, the body often forming one or two knots in the process, with the two ends of the worm tucked into the center of the ball. The coiled shape and the mucus-lined cell reduce water loss to a minimum.

Aporrectodea longa (Morgan and Winters, 1991) and *A. nocturna* have been reported to go into obligatory diapause (Satchell, 1967), as have *L. festivus*, *L. rubellus* and some species of *Eophila*. These species, which begin to go into estivation as early as May, lose all secondary sexual characters such as the clitellum during diapause, and usually come out of diapause spontaneously in September or October. By contrast, *A. caliginosa*, *A. chlorotica* and numbers of the genera *Eisenia* and *Octolasion* probably undergo facultative diapause. Morgan and Winters (1991) reported that they considered that *A. longa* had an obligatory diapause in the summer from May to August/September. Prior to diapause, this species acquired energy stores in the form of lipids and they discussed the ecophysiological significance of these changes. Byzova (1977) considered that for *A. caliginosa*, glycogen reserves are stored in chlorogogenous tissues before diapause but that this can be used up in 2 months, thereby limiting the potential time of diapause. Michon (1954) stated that these species do not diapause if they are kept at 9 °C and are given adequate amounts of food, but other work by Doeksen and van Wingerden (1964) has not confirmed this. Doeksen and van Wingerden (1964) considered that diapause in lumbricids was induced by substances excreted by the worms themselves. On the basis of laboratory experiments, they concluded that during the active period the concentration of these substances increases until the worms eventually enter diapause, after which excretion of these substances decreases. From the available evidence, it seems likely that the differences recorded between obligatory diapause, facultative diapause and quiescence in European lumbricids represent a continuous gradation in response among species, dependent upon the severity of the stress (Lee, 1985).

3.4 GROWTH

According to Hyman (1940), earthworms continue to grow throughout their lives by continually adding segments proliferated from a 'growing zone' just in front of the anus. However, Sun and Pratt (1931) reported that earthworms emerged from the cocoon possessing the full adult number of segments and grew by enlargement of segments, although Gates (1949) challenged these findings. Moment (1953a,b), working on *E. fetida* in culture, stated that newly emerged worms possessed, on average, the same number of segments as adults, and the only worms that grew by adding segments were those dissected from cocoons. He also observed individual worms from the time they emerged from the egg, and noted that although they increased in size many times, the number of segments remained the same. Probably the correct conclusion as to the way earthworms grow was given by Evans (1946), who stated that some species possess the adult number of segments on hatching, whereas other species add further segments during post-emergence growth. Supporting this conclusion, Vail (1974) reported that the young of E. *fetida* possessed between 22 and 28 distinct segments and those of *Bimastos tumidus* 69–128 distinct segments. The highest numbers are the same as those reported for adults. However, species of *Lumbricus* have fewer segments as they get older, even though they continue to increase in size.

The young worms tend to be colorless on hatching, with the circulatory system showing through the cuticle due to the presence of hemoglobin in the blood. They vary very considerably in length, and in cocoons with several young, there is a tendency for the longest worms to emerge first. The worms continue to grow, and, except for changes in pigmentation, there are few external morphological changes until maturity is reached (Vail, 1974). All the evidence is that the number of segments changes little during growth for E. *fetida*, and for B. *zeteki* (Murchie, 1960) except when there is autotomy and regeneration (Moment, 1953a,b).

The growth pattern of A. *caliginosa* in field cultures was followed by Nowak (1975). From a mean live weight of 27 mg at hatching, the worms grew to a mean live weight of 325 mg after 20 months. There was a rapid pre-reproductive phase of growth followed by a phase of steadily decreasing growth after sexual maturity was attained. Similar results, with a sigmoid growth curve were repeated by Phillipson and Bolton (1977) for A. *rosea*. Many workers have confirmed this sigmoid growth curve for E. *fetida* and other species that can grow in organic wastes (Edwards and Neuhauser, 1988).

Lumbricus terrestris bred in culture in outdoor conditions continued to increase in weight for about 3 years (Satchell, 1967). The increases in weight each year occurred almost entirely during the autumn and spring; whereas during the winter and summer months little weight was gained, and some was even lost. After 3 years, the average weight of the worms began to decrease, possibly because some individuals lost weight before dying.

Michon (1954) studied the development and growth of D. *subrubicunda* in culture at two temperatures. He reported that individuals gained weight rapidly until they reached sexual maturity, but after this there was a much slower increase in weight, until the disappearance of the clitellum indicated the onset of senescence. During this last period there was a slow decline in weight until the eventual death of the worms. Murchie (1960) studied field populations of B. *zeteki* and found that the time taken for individuals to reach full size depended on what time of year they emerged from the cocoons. Individuals that emerged in autumn and overwintered as small worms and those that emerged in the spring grew rapidly, achieving full size by August, to become reproductive the following spring, i.e. within a period of 12–24 months. Those worms that emerged during the summer months, at the beginning of a dry period, grew less rapidly, and could take 22–26 months to reach maturity.

The population of an earthworm species at any one time is made up of young immature, well-grown immature (adolescent), mature and senescent individuals, the proportions depending on the time of year (Fig. 3.5). The numbers of worms in the first three of these categories, for the species *A. nocturna* and *A. caliginosa*, were studied by Evans and Guild (1948b) in samples taken over a period of 9 months from August to May. At the beginning of September, the population consisted mainly of immature and adolescent worms but this was followed by a rapid increase in the number of sexually mature worms of both species. The numbers of adolescent worms remained low throughout the rest of the period, with only a small increase in numbers of *A. caliginosa* in spring, although the numbers of adolescent *A. nocturna* increased considerably at this time. The absolute longevity of earthworms in the field is poorly understood. *Eisenia fetida*, *L. terrestris* and *A. longa* have been kept alive for 4.5, 6.0 and 10.25 years, respectively, but it seems likely that survival is for 2 years or less in the field. In tropical conditions in West Africa, *M. anomala* populations attained sexual maturity after 8–14 months and growth continued for a further 20–21 months (Fragoso and Lavelle, 1992).

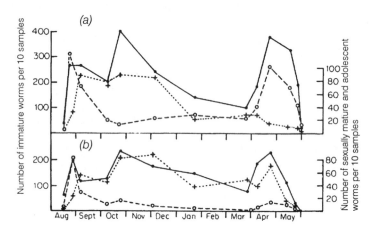

Figure 3.5   Seasonal trends in populations of immature (●), adolescent (○) and sexually mature (+) earthworms: (a) *A. nocturna*; (b) *A. caliginosa* (after Evans and Guild, 1948b).

3.5 BEHAVIOR PATTERNS

Many aspects of earthworm behavior have already been dealt with in this chapter and will be discussed further in Chapter 6, but there are other important behavioral patterns that are not mentioned elsewhere. For instance, all earthworms are very sensitive to touch, the pattern and speed of their reaction varying with both the species and circumstances. A good example is the behavior of *L. terrestris* when it scavenges for food on the soil surface. While doing this it usually keeps its tail in its burrow, and if it comes into contact with an object such as a stone, it will usually stop and move forward around the object. If the earthworm is touched, it withdraws back into its burrow, sometimes very quickly, and does not emerge again for some time. However, this withdrawal reaction is usually much less violent if the worm is almost completely in its burrow when touched, and it withdraws from the immediate vicinity of the stimulus only. If a worm is grasped while partly out of its burrow it will actively resist any attempt to pull it completely out, by extending its posterior setae into the burrow wall and expanding its posterior segments, so as to grip the walls of the burrow and completely fill its exit. It is then very difficult to pull the worm from its burrow, particularly if it is a large individual, and often it will break in two rather than relinquish its grip. After a worm has been subjected to this very violent treatment it usually becomes 'alarmed' and retreats deeply into its burrow, often remaining there for a long period. The nervous mechanism of this escape behavior was analyzed by Pallas and Drewes (1981). This behavior is particularly noticeable when the formalin method of sampling for *L. terrestris* is used. If an attempt is made to remove a worm, which has come to the soil surface under the stimulus of the formalin, before it is clear of its burrow, and the worm is allowed to escape, it does not usually emerge again, even when more formalin is poured into the burrow.

Many other kinds of stimuli applied to the soil promote activity in earthworms and cause them to come to the soil surface; these include vibrations caused by a fork inserted into the soil, stamping on the surface, and electrical and chemical stimuli; indeed birds often peck hard at stones on the soil surface to induce earthworms to emerge from their burrows.

Other species react vigorously to tactile stimuli; for instance, Murchie (1960) reported that individuals of *B. zeteki* reacted to touch in three different ways. If pricked sharply or handled, they produce a series of lashing movements from side to side, their bodies forming U-shapes with each movement. Certain species of *Pheretima* and *Eudrilus* also do this. Individuals of *B. zeteki* often eject coelomic fluid from the dorsal pores when touched and other earthworm species can eject this fluid to a considerable height, (section 1.7). The third and least common response is autotomy, or breaking off posterior segments. Either grasping or impaling the rear end of a worm may cause it to break at a point a little in front of the stimulated region, and such a break can occur at any point up to 32 segments from the posterior end. Stimuli applied to the anterior end do not cause reaction in the same way.

Stephenson (1930) considered that autotomy was an indication of senescence but this is certainly not so for *B. zeteki* which will shed posterior segments whether young or old. Evidence that autotomy was common in this species was given by Murchie (1960), who reported that many field collected individuals tended to have fewer segments than reared ones. Thus, there is little evidence of posterior regeneration in this species. *Diplocardia mississippiensis* seems to undergo autotomy readily even in healthy worms. Where this occurs spontaneously, the front one-quarter to one-third of the body, consisting of 40 or more segments, breaks off (Vail, 1972).

3.5.1 FEEDING BEHAVIOR

Individuals of *L. terrestris* feed on leaf and other plant material obtained from the soil surface. They do not feed to any great extent on the leaf material *in situ*, but first pull it into the mouth of the burrow, to a depth of 25–75 cm, so forming a plug which may protrude from the burrow. There have been several suggestions as to why they do this. It may be in order to camouflage the entrance to the burrow, although the plug often makes the burrow more obvious. Alternatively, it may be to prevent the entry of water into the burrow during heavy rain or to keep out cold air. Whatever the reason, these earthworms never leave the mouths of their burrows exposed, and will replug them very quickly if the original plug is removed. If the mouth of the burrow is in an area where there is insufficient organic material to form a plug, they will use inorganic material, commonly small piles of stones, termed 'worm cairns', which may be seen in gravel paths, blocking the mouths of burrows. Individuals of *L. terrestris* normally feed on food material only within their burrows, and since far more material is stored than can be used between normal foragings, there is adequate food when inclement conditions on the soil surface prevent them from foraging, or when the local food supply is temporarily exhausted.

Lumbricus terrestris selects its food material carefully and pulls many kinds of leaves into the burrows by the tip of the laminae, leaving the nonpalatable petioles projecting from the burrow. Darwin (1881) reported that when earthworms were offered paper triangles they always grasped them by the sharpest corner, usually without any attempt to seize any other corner, and this was confirmed by Hanel (1904), although she also found the conflicting evidence that when leaves of lime trees were cut so as to round off the pointed apex, worms still grasped them at this point. The overall general tendency seems to be for the worms to grip the pointed tips of leaves. For instance, Darwin (1881) reported that of 227 leaves of various species, 181 had been drawn in by the tips of the leaves and only 26 had been pulled in by the base of the petioles. In some unexplained way, worms can discern the best way to drag leaves into their burrows with minimum effort. Darwin also described other behavioral patterns shown by individuals of this species during their foraging activities. Baldwin (1917) demonstrated experimentally that *L. terrestris* is much more active when food is available than when it is not.

3.5.2 LEARNING

Many workers have tried to show that earthworms can learn by experience, usually by giving individuals a choice of two courses of action, one of which results in the worm receiving an unpleasant stimulus. Such experiments usually use a form of T-maze, with the bottom of the 'T' as the entrance, and the left- and right-hand arms of the cross-piece providing alternative routes, controlled by stimuli involving either electrodes or a light source in one arm of the 'T', to produce negative or positive responses. The first worker to use this type of apparatus to investigate the ability of earthworms to learn by experience was Yerkes (1912). He used a T-shaped glass box, open at the bottom, resting on a plate-glass base covered with damp blotting paper. The entrance to the bottom arm of the T-maze was joined to a wooden block through which passed a hole of diameter 14 mm, lined with blotting paper. The worms, which were individuals of the species *E. fetida*, were introduced to the maze through this hole. Just inside the entrance to one side arm, there was a strip of sandpaper on the glass plate and just beyond, two electrodes, resting on a strip of rubber. (In a previous experiment, he used a strip of blotting paper soaked in a strong salt solution instead of the electrodes.) On reaching the junction of the two arms of the cross-piece, the introduced worms could turn either into the arm containing the electrodes which was open at the end, or into the opposite arm, at the far end of which was placed another hollow wooden block, similar to that at the entrance to the T-maze.

On being introduced to the maze, the worms showed no avoidance symptoms, until they entered the arm containing the sandpaper and electrodes. After repeated trials, they learned to avoid contact with the electrodes, and eventually also the sandpaper strip (which acted as a warning of the electrical stimulus), by turning into the opposite arm of the maze. Yerkes also reported that even if up to five anterior segments (including the cerebral ganglia) were removed from individuals they still retained the imprinted lesson, but only until the cerebral ganglia were regenerated; thereafter it was lost. Heck (1920) also conducted experiments using a similar type of T-maze, and both Yerkes and he concluded that the worms learned to turn into the arm which did not contain shock-producing electrodes, as a direct consequence of avoidance of the learned negative stimulus in the arm containing the electrodes.

Robinson (1953) tested the reactions of *L. terrestris* in a similar apparatus. He found that the behavior of the worms in his maze differed from that observed by Yerkes. Before reaching the electrode area, they showed avoidance symptoms to other more generalized stimuli, such as making contact with the walls of the maze. He concluded that Yerkes had over-simplified the nervous reactions of worms to stimuli and that the process of learning to avoid adverse stimuli was more complex. However, Schmidt (1955) compared the behavior of both species under the same experimental conditions (in a maze similar to that of Yerkes and Heck), and observed that individuals of E. fetida behaved exactly as Yerkes had seen, and individuals of L. terrestris behaved as in Robinson's experiments.

Thus, L. terrestris tended to use the entire choice point of the maze as a source of discriminatory cues, whereas E. fetida seemed to discriminate mainly on the basis of the sandpaper. Perhaps such reactions relate to the different habits of the two species. Lumbricus terrestris has to seek out its food whereas E. fetida lives in manure or other media high in organic matter and has food readily available. Perhaps its increased sensitivity to stimuli has adaptive value for L. terrestris but not for E. fetida.

Certainly, the behavioral pattern and the degree of ability to learn to avoid an unpleasant stimulus differs with species, and the behavior of one species cannot necessarily be used to predict that of another. For this reason it is unfortunate that different workers have used different species in their experiments. The species tested have included: A. tuberculata (Heck, 1920; Swartz, 1929), Bimastos parvus (Swartz, 1929), A. longa (Fraser, 1958), A. chlorotica, L. castaneus (Heck, 1920) and L. rubellus and Pheretima communissima (Iwahara and Fujita, 1965).

Several other workers (Swartz, 1929; Lauer, 1929; Bharucha-Reid, 1956) used mazes to study behavioral patterns of worms. Lauer (1929) placed electrodes in both arms of a maze, so that the direction of stimulus could be reversed as required. Krivanek (1956) conditioned batches of individuals of L. terrestris to avoid high- and low-intensity light sources, electrical contacts and tactile stimuli. The conditioned state of worms exposed to the high-intensity light source was retained for a further period of 48–72 hours after 42 exposures to the stimulus. This was up to three times as long as that of those worms conditioned to react to the low-intensity light source. The conditioned state of worms in the batches exposed to electrical shock and to tactile stimuli was retained for a period similar to that of those exposed to low-intensity illumination.

Arbit (1957) also used electrodes in a T-type maze, to examine the behavior and learning response of two batches of individuals of L. terrestris in relation to their diurnal cycle of activity. One batch was tested in the maze between 8 p.m. and midnight, and the other between 8 a.m. and noon. He found that the evening batch needed significantly fewer attempts before they learned to avoid the stimulus than the morning batch, and also, that the amount of stimulus needed to start the worms moving in the maze (a camel-hair brush, and shining a light) was significantly less in the evening batch than in the morning one. He concluded that generalizations on learned behavioral patterns are not valid from one species of earthworm to another, and that the time when the pattern is learned relative to the diurnal cycle is also important. Therefore, it seems probable that the ability of an earthworm to acquire a learned response to a stimulus depends, at least partly, on its normal activity at the time of learning (Baldwin (1917) showed that *L. terrestris* is most active between 6 p.m. and midnight).

The considerable variations in results have no doubt been due to lack of standardization of experiments in terms of: species, age, length of time between tests, treatment of worms between tests, time of day for the test, amount of light or dark, design of the maze and number of trials. Last, and most important, is the cleaning of test mazes between tests; earthworms leave mucus trails which they may follow or avoid. Datta (1962) stressed the importance of washing the walls as well as the floor of the maze between trials, and Ressler *et al.* (1968) suggested that T-maze learning results could be accounted for by unlearned responses to mucus traces, but they did not test their hypothesis. Rosenkoetter and Boice (1973) believed that at least some of the learning phenomena could be accounted for in this way. Nevertheless, there seems to be adequate evidence to conclude that at least some of the behavior of some species of earthworms can be accounted for by a potential for learning. Much more work is still required on differences between species, neural mechanisms and behavior in choice situations other than those presented by T-mazes (Howell, 1974). However, even after this considerable body of work, Rosenkoetter and Boice (1973) cast considerable doubt on the theory of T-maze learning by earthworms, attributing their reactions in the numerous reported experiments to either attraction or repulsion by worm trails from earlier experiments.

3.5.3 MIGRATIONS

A phenomenon that occurs occasionally is that of mass migration of worms for a considerable distance on the soil surface. Doeksen (1967) reported a number of instances of this in The Netherlands. Individuals of *E. fetida*, which were living in the soil in greenhouses, migrated in large numbers up the sides of buildings and even on to roofs, during damp, wet, foggy weather. After heavy rain, individuals of the same species living in a dung heap migrated to a nearby farmhouse, climbed up the walls on to the roof, and were even found inside the building. The suggested cause of these migrations was that hydrogen sulfide was produced in the burrows, resulting from anaerobic conditions developing because of poor soil ventilation. Gates (1961) mentioned individuals of species of *Eutyphoeus* and *Lampito* that were seen wandering on the soil surface in daylight after rain in Burma, and that *Perionyx excavatus* and some species of *Dichogaster* were found in trees and on the roofs of high buildings. Mass migrations of *Perionyx* species were also reported by the same author. Reddy (1980) reported mass migrations of *Amynthas alexandri*, probably in response to temperature changes.

Large numbers of worms often appear on the surface of soil after rain, and many of these die, probably due to exposure to ultraviolet light or radiation. No adequate explanation of this behavior has yet been offered, although it has been suggested that low oxygen tensions or chemicals in soil solution may cause it. We have no evidence as to whether they return to their burrows if they survive.

3.5.4 LUMINESCENCE

Phosphorescence or luminescence has been reported to be produced by earthworms. This may show as luminescent blotches or spots on the body, or as a luminous trail behind the worm. It is believed that the luminescence is due to a luminous slime exuded from the anus or prostomium, but an alternative suggestion is that it is due to coelomocytes in the protective slime. The function of the luminescence may be protective because it occurs usually when the worm is irritated, or in response to vibration or some similar stimulus.

3.5.5 DIURNAL RHYTHMS OF ACTIVITY

Although is has been assumed that earthworms demonstrate diurnal rhythms of activity in response to ultraviolet light and other environmental factors (Ralph, 1957), experiments where earthworms were kept in continual dark showed that they still maintained a diurnal pattern of activity for up to 7 days after all light stimuli had been removed (Edwards and Lofty, 1977). *Millsonia anomala* was also reported to have a marked pattern of diurnal activity, coming to the surface and casting at between midnight and 9 a.m. in moist soil (Lavelle, 1978). There is evidence of a diurnal rhythm of respiration, with maximal rates occurring at about 6 a.m. and 7 p.m. in Europe (Ralph, 1957; Doekson and Couperus, 1968).

Earthworm physiology

4

Most studies of the physiology of oligochaete worms have involved terrestrial species, and more is known of the physiology of earthworms than of many other aspects of their biology. The reader is referred to a standard, if out-dated text, *The Physiology of Earthworms* by M. S. Laverack (1963), for a more detailed treatment than is possible in the present work.

4.1 RESPIRATION

Terrestrial earthworms have no specialized respiratory organs. Oxygen and carbon dioxide diffuse through the cuticle and epidermal tissues into the blood, which contains hemoglobin, a respiratory pigment. The only structural specializations are the greatly branched capillary blood vessels embedded in the body wall. In all respiratory systems, the oxygen must first dissolve in an aqueous layer on the respiratory surface; in the earthworm this is the whole body surface, from where it passes into the body by diffusion, but this is not an active process (Krüger, 1952). The cuticle is kept moist by secretions from the mucous glands of the epidermis.

The hemoglobin in the blood plasma can take up oxygen and transport it to other parts of the body, and the hemoglobin of earthworms can absorb and become saturated with oxygen at pressures as low as 19 mm of mercury (atmospheric pressure is 152 mm). It has been suggested that the coelomic fluid is at a pressure as low as 14 mm (Tembe and Dubash, 1961) so that oxygen can reach the inner tissues even when only small amounts are available. Respiration that depends on simple diffusion is inefficient, but the effectiveness of hemoglobin in the plasma as a respiratory pigment and the low internal pressure of the body fluids of the worm enable it to work very satisfactorily.

Measurements of rates of respiration of earthworms are often not comparable, due to different test methods and the lack of statistical analysis of the data. Hence, a full assessment of the influence of body size, ecological habits or taxonomic position on respiration is not yet feasible (Mendes and Almeida, 1962; Byzova, 1965). Rates of earthworm respiration are usually proportional to the surface area available for gaseous diffusion. Smaller earthworms have a greater ratio of surface area to body weight, so in terms of oxygen uptake per gram of body tissue, they respire more (Raffy, 1930). The rate of respiration for *L. terrestris* was calculated as being between 38.7 and 45.2 mm^3 of oxygen/h/g of body tissue at 10 °C, and 31–70 mm^3 at 16–17 °C (Johnson,1942), but Doeksen and Couperus (1968) recorded much higher rates than these.

Earthworms can survive for long periods in water and continue to respire, their respiratory rate in this medium depending on the partial pressure of the oxygen dissolved in the water (Raffy, 1930). Some species are adapted to living for indefinite periods in mud or waterlogged soils.

The rate of respiration of earthworms is very dependent on the soil temperature. For instance, the amount of oxygen used increased from 25 to 240 mm^3/individual/30 minutes when the temperature changed from 9 °C to 27 °C (Pomerat and Zarrow, 1936). Thus, earthworms in tropical areas tend to respire faster than those in temperate regions because of higher temperatures. There is some good evidence that earthworm respiration acclimatizes to temperature; thus when earthworms maintained at a low temperature were transferred to a higher temperature, they still respired more slowly than individuals that had been kept at a higher temperature previously (Kirberger, 1953; Saroja, 1959). There is experimental evidence of a diurnal rhythm of oxygen consumption, with maximal rates at about 6 a.m. and 7 p.m. (Ralph, 1957; Doeksen and Couperus, 1968). These times do not necessarily coincide with periods of maximal activity; hence, it has been suggested that earthworms can accumulate oxygen debts (Ralph, 1957), probably forming lactic acid which can be resynthesized into glycogen later. This may also be the mechanism by which earthworms can survive for many hours with no atmospheric oxygen. It has been shown that during anaerobic respiration, lactic acid and other compounds are formed; for instance, intestinal worms form valerionic acid and it has been suggested that earthworms can also accumulate this compound (Lesser, 1910; Stephenson, 1930).

Respiration does not usually decrease significantly until the partial pressure of oxygen in the soil falls to a very low level. At 25% of the normal pressure (38 mm mercury), respiration is depressed by 55–60% (Johnson, 1942). Such low levels of oxygen are usually associated with large concentrations of carbon dioxide, and earthworms can respire at the normal rate until the soil atmosphere contains up to 50% carbon dioxide. However, earthworms may move away from carbon dioxide concentrations greater than 25%, although such large amounts probably occur only in localized areas in soil (Shiraishi, 1954). It has also been suggested that respiration decreases on exposure to ultraviolet light (Merker and Braunig, 1927).

A comparative study of the respiration rate (oxygen consumption/gram of wet weight/hour) of a range of lumbricid species showed that typical soil-dwelling species such as *E. rosea* and *A. caliginosa* had the lowest respiration rate. Soil-litter-dwelling species such as *L. rubellus*, *L. terrestris* and *O. lacteum* and compost-dwelling species did not differ very much in respiration rate. *Lumbricus castaneus*, which is mainly a litter-dwelling species, had an intermediate rate of respiration, and the litter-dwelling species *D. octaedra* had the highest rate of respiration (Byzova, 1965).

4.2 DIGESTION

The digestive system (described in Chapter 1) is a simple one, consisting of a buccal chamber, pharynx, esophagus, crop, gizzard and intestine, although there are detailed differences between earthworms in different families. Earthworms use organic matter as a source of nutrition but depend upon protozoa, rotifers, nematodes, bacteria, fungi and other micro-organisms, for their nutrients. Nutrients are extracted from the large quantities of micro-organisms associated with decaying organic matter and soil that pass through the gut (Edwards and Fletcher, 1988). Some species, such as *L. terrestris*, feed on leaves directly, and even show preference for particular species and conditions of leaves, but they are still dependent upon the micro-organisms that grow upon them for nutrition. Gut cellulase activity was reported for a range of epigeic and endogeic earthworms (Urbasek, 1990), including *L. rubellus* which had only low cellulase activity. Cellulase activity was also reported from the gut of *E. fetida* by Whiston and Seal (1988).

The process of digestion was worked out in detail for one species of earthworm, *E. fetida*, by van Gansen (1963). The forepart of the digestive system (segments 1–14), which he termed the 'reception zone', contains the sensitive, prehensile mouth, the esophagus and the ductless pharyngeal gland which secretes an acid mucus containing an amylase but probably no proteolytic enzyme. Other workers have reported that a proteolytic enzyme is secreted by the 'salivary' glands of some species. Opening into the esophagus is the calciferous gland, which secretes amorphous calcium carbonate particles coated with mucus. The function of this secretion is unknown, but it has been suggested that it influences the pH of the intestinal fluid (Robertson, 1936), and van Gansen (1962) believed that its function was the regulation of the calcium level of the blood. It has been suggested that the main function of these glands is to excrete excess calcium, in the form of calcium carbonate, from the body tissues into the gut and thence to the exterior through the anus. Many earthworms live in soils rich in calcium and it would seem appropriate for them to have a mechanism for ridding the body of excessive amounts of this element. However, Piearce (1972) reported that *D. veneta*, which lives in calcareous woodland soil, passes considerable quantities of calcium through the alimentary canal, but that its calciferous glands are of the simplest type. It has been demonstrated that active calciferous glands, in some species, contain large quantities of carbonic anhydrase, an enzyme that figures prominently in acid–base reactions. Metabolic activities produce carbon dioxide that can increase the acidity of the coelomic fluid. Carbonic anhydrase in the calciferous glands apparently counteracts this effect by catalyzing the fixation of this carbon dioxide in the form of calcium carbonate. The glands are, in part at least, regulators of the pH of the coelomic fluid, and experimental removal of these glands results in a lowering of the coelomic pH.

Segments 15–44 were termed the 'secretory zone' by van Gansen. This contains the crop, which leads to the gizzard and intestine. The strong, muscular action of the body wall of the gizzard brings opposing surfaces of the thick internal cuticle against one another, thus grinding up the soil and organic matter which then passes on into the intestine. *Eisenia fetida* secretes two proteases and one amylase, mainly from the epithelial brush-border cells, the 'goblet' cells of the intestinal wall secreting mostly mucus. Different enzymes have been reported from this zone for other species; for instance, a lipase and a protease with a rennin-like action (Millott, 1944; Arthur, 1965) and a cellulase and chitinase (Tracey, 1951). Most of the available evidence indicates that the cellulase and chitinase are secreted by the gut wall, although it has been suggested that these enzymes might be produced by symbiotic bacteria and protozoa, as in other invertebrates. No doubt other species of earthworms produce many other enzymes, in view of the diverse nature of the food; for instance, one worker reported that an invertase was present in the intestine. The digested food passes into the bloodstream through the intestinal epithelium and is carried to the various parts of the body and tissues for use in metabolism or storage.

The last zone (segments 44 to the anus) in *E. fetida* was termed the 'absorption zone'; here the undigested matter in the intestinal contents becomes enveloped by a peritrophic membrane that lines the intestine and, when excreted, covers the casts. The composition and structure of the casts will be discussed in Chapter 10.

4.3 EXCRETION

The principal excretory organs in earthworms are the nephridia, which extract waste materials from the coelomic fluid in which they lie, and excrete them to the exterior, as urine consisting mainly of ammonia and urea (although some workers have reported uric acid and allantoin in this fluid), through the nephridiopores. There are at least two nephridia in each segment (Chapter 1), but in some species there are many more. In a few genera, such as *Pheretima*, the nephridia open into the gut.

Considerable amounts of nitrogenous matter (about half of the total nitrogen excreted per day) are excreted from the body wall as mucus. This acts as a lubricant, binds soil particles together to form the wall of the burrow and also forms a protective coat against noxious materials. There are several other excretory mechanisms in earthworms, but these are still poorly understood and, although there have been many hypotheses, their relative importance and precise mode of action still remains obscure. However, it is known that most of the excretory products of metabolism reach the blood or coelomic fluid, from which they must be excreted into the gut as mucus. Special cells, called chloragogen cells, which belong to the coelomic epithelium of the intestine, are believed to be important in removing excretory materials from the blood, although it has been suggested that they also transport nutrients between tissues and organs (van Gansen, 1956, 1957, 1958a,b; Roots, 1957, 1960). Chloragogen cells, which are always closely associated with blood vessels and capillaries, collect yellow refractive granules which have been termed chloragosomes; these contain a complex arrangement of substances (Roots, 1960). It is generally agreed that the chloragogen cells break off from the surface of the alimentary canal, fall into the coelom and form vacuoles which discharge into the coelomic fluid. Other cells disintegrate completely and their contents, believed to be mainly urea and ammonia, possibly with some fats and glycogen, are liberated into the coelomic fluid where they can be eliminated later. It seems that the chloragogen cells function very much like a mobile liver, acting as a homeostatic device to maintain required levels of substances in the blood and coelomic fluid. The substances deposited in the coelomic fluid by the chloragogen cells can be eliminated later, either directly by the nephridia, or by being taken up by amoebocytes which are then either deposited in the body wall or in bulky nodules in the coelomic fluid. Some species, such as *Pheretima* spp., have lymph glands where amoebocytes accumulate. Amoebocytes also occur in the blood, some being deposited in the intestinal wall, and later falling off into the intestine to be excreted with the feces. Other excretory cells, termed uric or bacteroidal cells, are peritoneal cells containing rod-like inclusions, believed to be small crystals of uric acid; these function very much like amoebocytes.

The most important excretory organs are the nephridia, which have several postulated modes of action. Coelomic fluid containing excretory materials in solution passes through the nephrostome and along the nephridial tube by ciliary action, the cilia acting as a sieve. It has been suggested that whole chloragogen cells may also pass into the nephridia but it is unlikely that they do so through the nephrostome, and if this does occur, it is more likely to be through the walls of the nephridia. Some earthworms have nephridia that are closed internally and thus must function differently, probably by waste materials diffusing across the nephridial wall into the lumen.

The nephridia obviously act as differential filters, because there is much more urea and ammonia, but less creatinine and protein, in the urine they produce than in the coelomic fluid. The composition of the urine changes as the fluid passes along the nephridium (Ramsay, 1949) (Fig. 4.1), but it is not known whether osmotic changes are due to resorption of salt or secretion of water. Certainly, the urine is at a much lower osmotic pressure than the coelomic fluid, but how this is achieved is not clear. There is some evidence that granules of waste material can be taken up by the ciliated tube of the middle wall and remain there, so that these parts of the nephridia may act as kidneys of accumulation. Bahl (1947) concluded that nephridia have three functions in excretion, namely filtration, resorption and chemical transformation. He believed that protein was reabsorbed through the nephridial wall against a concentration gradient, but this remains to be confirmed and was not accepted by Martin (1957). A possible mode of functioning of the nephridia is summarized in Fig. 4.2, emphasizing some of the excretion and resorption processes.

El-Duweini and Ghabbour (1971) analyzed the urine of *A. caliginosa* and *Pheretima californica* and found it contained 2.9–4.3 mg ammonia/100 ml, 31.7–39.1 mg urea and 2.0–2.2 mg protein. The urine was 6.2–7.6% of the fresh body weight and 1–2.5% of the fresh body volume per 24 hours. Nitrogen is also excreted in the feces as ammonia (Tillinghast, 1967); this was demonstrated by ligaturing worms near the posterior, which

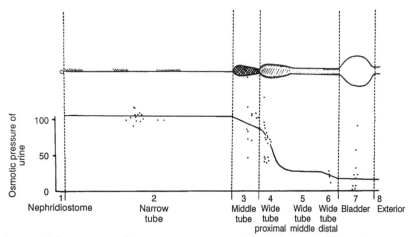

Figure 4.1 To show the osmotic pressure of the urine at different levels in the nephridium. The osmotic pressure of the Ringer's solution surrounding the nephridium has been equated to 100. Individual observations are shown as points: the line drawn through the points represents the interpretation placed upon them. (After Ramsay, 1949.)

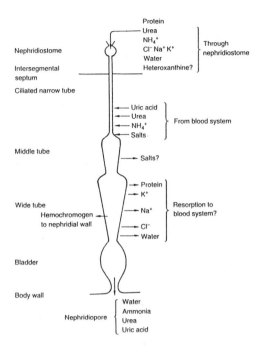

Figure 4.2 Diagram to summarize the possible mode of functioning of oligochaete nephridia (after Laverack, 1963).

restricted the output of ammonia but not of urea. The source of this ammonia was postulated as of two possible origins; first, actively feeding worms can themselves produce ammonia as an end product or, alternatively, that the ammonia is produced by the gut flora.

The same workers (Tillinghast *et al.*, 1969) reported that temperature can markedly influence both the magnitude of output of ammonia and urea and the time taken for fasting worms to become ureotilic. These results may help to explain some of the conflicting data in the literature (Cohen and Lewis, 1949a,b; Needham, 1957). The loss of carbon from ^{14}C-labeled individuals of *O. lacteum* in the form of mucus was calculated as 0.2% of the body weight daily and in casts as 0.5% daily (Scheu, 1992). Lavelle *et al.* (1983) reported that a population of *Pontoscolex corethrurus* could excrete 50 tonnes/ha of mucus; equivalent to 20% of the total organic matter of the soil they lived in.

4.4 CIRCULATION

The structure of the circulatory system of *L. terrestris* (described in Chapter 1) (Figs 1.6, 1.8) is similar to that of most terrestrial oligochaetes.

Blood flows forward along the length of the dorsal blood vessel which is the main collecting vessel and which lies in close contact with the gut all along the body. Most of the blood from the dorsal vessel passes down into wide pulsating vessels ('hearts') containing valves, although some blood is also passed anteriorly to the head. The function of the 'hearts' is to pump the blood down into the large ventral vessel in which the blood flows forward anteriorly to the head and backward posteriorly, distributing it to all parts of the body.

Blood is collected from anterior tissues and organs by branches of the subneural vessel (which lies close to the nerve cord) and it flows backwards in this vessel, being eventually returned to the dorsal vessel by the dorso-subneural vessels which run in the septa of each segment. The ventral vessel supplies blood to various organs of the body. To supply the nervous system, branches pass from the ventral vessel to ventro-parietal vessels, then into the lateral-neural vessels, which run alongside the nerve cord and branch into capillaries that supply blood to the cord. From the nerve cord, blood drains into the subneural vessel via branched capillaries.

The esophagus and intestine receive blood from the ventral vessel through ventro-esophageal and ventro-intestinal vessels. Blood returns from the esophagus in the front of the body through paired lateral-esophageal vessels which lead to the dorso-subneural vessel of the 12th segment and thence to the dorsal vessel. In the posterior part of the body the blood passes from the peri-enteric plexus of capillaries around the intestine back to the dorsal vessel via the paired dorso-intestinal vessels. Three small vessels in each segment run from the typhlosole into the dorsal vessel.

In each segment behind the 12th, blood passes from the ventral vessel to the ventro-parietal vessels, which have numerous branches that end in capillaries close to the skin; these allow the blood to become oxygenated. Blood returns from the skin via other capillaries which lead to the dorso-subneural vessels and thence to the dorsal vessel.

4.5 NERVOUS SYSTEM

The detailed functions of the components of the nervous system are still not fully understood, although they have been studied extensively. Nor is much known about the detailed functioning of the sense organs, synapses, motor fibers and innervation of muscles. Earthworm neurobiology was reviewed by Mill (1982).

4.5.1 GENERAL FUNCTIONS

The most important part of the nervous system is the large ventral nerve cord (Figs 1.7, 1.12, 1.13), which contains three longitudinal giant fibers lying in its dorsal part; one median and two smaller ones situated latero-ventrally.

The various sense organs which lie in the body wall are associated with numerous intraepidermal nerve fibers that end freely between the epidermal cells. It has been suggested (Smallwood, 1926) that the sense organ fibers synapse with the subepidermal network of fibers, which are collected together to form the segmental nerves, and the stimuli from these are then passed on to the motor nerves and fibers. Sensory stimuli in one segment may invoke a motor response either via fibers in the nerve of the same side of the same segment, the contralateral nerve of that segment, the nerve before or after that stimulated in the same segment, or even the segment in front or behind. Nerve connections extend for as many as three segments backwards or forwards via association neurons. Stimuli from one segment can also reach other segments indirectly, because the sensory fields of the nerves supplying the body wall cover more than one segment.

Strong stimuli can bypass the association neurons and be transmitted up and down the giant fibers. Giant fibers are not completely continuous along the length of the body, because in each segment there is an oblique partition across each fiber which is probably not a true synapse, but clearly is no obstacle to rapid transmission of stimuli. This arrangement of giant fibers is important in allowing the transmission of motor impulses along the nerve cord at high speed (600 m/s) and a very rapid reaction of the animal to adverse stimuli. There is some evidence that the median, dorsal, longitudinal giant fiber transmits stimuli only from the front to the back of the worm and not vice versa. There are also smaller lateral giant fibers and it is believed that these transmit stimuli in the opposite direction. The most probable explanation is that the sensory input from the front of the animal passes into the median giant fiber and in the rear of the animal into the lateral fibers. Most of the available evidence indicates that the main chemical transmitter substances in the central nervous system are acetylcholine, probably adrenalin, noradrenalin and possibly 5-hydroxytryptamine (5HT). The main function of the enlarged cerebral ganglia in the front of the body seems to be mostly inhibitory, because if they are excised, the worm moves continuously. They seem to play little part in initiating movement.

It is now well established that certain nerve cells are capable of elaborating and releasing complex organic substances which act as hormones. These can be released into the bloodstream and act upon some distant organ. Such neurosecretions were first reported for *L. terrestris* by Schmid (1947), who found that secretions were produced cyclically and that production could be initiated by adrenalin or novocaine.

Secretory cells have been reported from the cerebral, subesophageal and the first two ventral ganglia (Herlant-Meewis, 1956; Hubl, 1956; Marapao, 1959). It seems possible that main bodily functions such as reproduction, pigment migration, development of secondary sexual characters, activity cycles and diurnal respiratory cycles may be controlled by neurosecretions, but little evidence of this is available. It has been shown that the yearly reproductive cycle of *L. terrestris* seems to be controlled by neurosecretions, because immature individuals lack certain secretory cells in the cerebral ganglia (Hubl, 1953). If these ganglia are removed from mature worms, the secondary sexual characters disappear and egg-laying and cocoon production cease. It has also been demonstrated that neurosecretions are important in regenerative phenomena.

4.5.2 LIGHT REACTIONS

Earthworms do not have recognizable eyes, but possess sensory cells with a lens-like structure in regions of the epidermis and dermis, particularly on the prostomium (Fig. 1.5). The middle part of the body is rather less sensitive to light and the posterior is also slightly sensitive. Earthworms certainly respond to light stimuli, particularly if they are suddenly exposed to light after being kept in the dark. *Lumbricus terrestris* is photopositive to very weak sources of light and photonegative to strong ones, i.e. it crawls towards dim lights and away from strong ones. However, if it is kept for long periods in strong light it does not react at all to a sudden increase in intensity (Hess, 1924); it has been suggested that this is due to saturation of the light receptors. Other species, such as *Pheretima* spp., are completely photonegative and respond in proportion to the intensity of the light (Howell, 1939).

Earthworms react differently to different wavelengths of light; blue light is stimulating and red is not, so the activities of earthworms are best studied experimentally in red light. Ultraviolet light seems to be harmful, and it has been suggested that some of the earthworms lying dead on the surface of the ground after rain have been killed by the ultraviolet light of the sun (Merker and Braunig, 1927). The mode of action of the light-sensitive reaction is not yet clear, but it has been shown that some drugs depress photosensitivity, and cutting the ventral nerve cord also modifies reactions to light. Small electric potentials have been reported in the nerves following light stimulation (Prosser, 1935). All available evidence is that the reaction to light is controlled and coordinated by the cerebral ganglia. Howell (1939) suggested the following mechanism for the photonegative response in *Pheretima agrestis*: impulses from photic stimuli to light-sensitive cells travel along different nerves to the cerebral ganglia, the ventral nerve cord and the circumpharyngeal connectives. Strong impulses cross over directly in the transverse commissure of the cerebral ganglia and in the commissures of the ventral nerve cord. Weak impulses are relayed to the cerebral ganglia where, with impulses entering these ganglia from the cerebral nerves, or from the circumpharyngeal connectives, they are probably modified and cross over in the transverse commissure. After crossing over, the impulses go to the muscles of the side opposite that which was illuminated, causing them to contract, thus producing a negative response.

4.5.3 CHEMORECEPTION

Darwin (1881) showed that earthworms could distinguish readily between different food substances and it has been shown that E. *fetida* can react to a diverse range of chemical stimuli. Such reactions probably help in the selection of food, give warning of adverse environmental conditions, such as soil acidity, and assist in mating by detecting the mucous secretions of other earthworms.

Most of the sense organs, which react to chemical stimuli (Fig. 1.5), are on the prostomium or the buccal epithelium which comes into contact with substances when the buccal chamber is everted during feeding. Several workers have demonstrated the selection of, and preference for, particular forms of leaf litter by earthworms (Mangold, 1951; Wittich, 1953). Satchell (1967) reported that the palatability of leaves to earthworms depended greatly on their polyphenol content, and Edwards and Heath (1963) showed that if water-soluble phenols were washed out of oak leaves, they became much more palatable to L. *terrestris*. Mangold (1953) reported that alkaloid substances above a certain concentration were not accepted by earthworms. Earthworms reacted inconsistently to glucose and saccharose. It seems from this and other evidence that most food preferences of earthworms depend on chemoreception.

Earthworms can also detect acids. Mangold (1953) showed that acids such as phosphoric, tartaric, citric, oxalic and malic acids, found in plant materials, were accepted at low concentrations but not at high ones. Different earthworm species differ in their tolerance to soil acidity, but all have a threshold of pH below which they cannot live for long, so the ability to detect pH is essential for survival. Laverack (1961a) demonstrated that A. *longa* will not burrow into soil with a pH below 4.5 and L. *terrestris* into soil with a pH below 4.1. He showed that acid-sensitive fibers are present all over the earthworm body by taking recordings of nerve impulses which occurred when worms were dipped in buffer solutions. There was no indication that the prostomium was more sensitive to acids than the rest of the body. The threshold values for the acid-sensitive organs were pH 4.4–4.6 for A. *longa*, 4.1–4.3 for L. *terrestris* and 3.8 for L. *rubellus*, and these organs reacted only to acid stimuli and not to sodium chloride, quinine or sucrose. There is little doubt that earthworms are sensitive to chemical substances other than those named so far, because they are brought rapidly to the soil surface when dilute solutions of chemicals such as potassium permanganate, formaldehyde or dilute pesticides are poured on to the soil.

The authors of this book have found, using tests of the palatability of uniform disks of leaf tissue, that earthworms are sensitive to a wide range of chemicals. For instance, various pesticides that are commonly sprayed on to the leaves of crop plants or orchard trees make the treated leaves very unpalatable. Furthermore, in waterlogged soil many chemicals, including a wide range of pesticides and detergents, act as an irritant and cause worms to come to the soil surface.

4.5.4 THIGMOTACTIC REACTIONS

Earthworms are very sensitive to touch because of tactile receptors on discrete areas on the surface of the body. There are three nerves in each segment that are involved in touch, the anterior of these having a greater receptive area in the segment ahead, and the posterior one a greater sensory area on the segment behind. These tactile organs are involved in the thigmotactic responses of earthworms. When placed on soil or other surfaces, worms become very active until they find a suitable crevice or crack, so that the sides of the body are in contact with the substrate, when presumably the touch receptors are stimulated and the worm stops moving. This response is very strong and can override the reaction to light.

4.5.5 RESPONSE TO ELECTRICAL STIMULI

Earthworms respond rapidly to electrical stimuli, becoming U-shaped with both ends directed towards the cathode when a low voltage is applied. If the direction of the current is reversed, the worm reorientates itself (Moore, 1923). Worms also emerge on to the surface of soil to which an electric current is applied (Satchell, 1955c). The source of such a current can be either direct or alternating voltage, and the reaction occurs over quite a wide range of voltages. Currents through the soil of several amps produce the greatest response and this current is dependent on many soil factors, such as moisture content and pH.

4.6 WATER RELATIONSHIPS

The conservation of water to avoid dehydration is very important for all terrestrial animals. Earthworms have a thin, permeable cuticle overlying a mucus-secreting epidermis, neither of which can do much to prevent water losses, so that terrestrial earthworm species are exposed to considerable risks from dehydration, due to water losses through the skin. Water can also escape from the body via the mouth, anus, dorsal pores and nephridia.

Up to 85% of the fresh weight of earthworms is water, a considerable part of this being in the coelomic fluid and blood; so they must be able to prevent excessive water losses in order to survive. Earthworms living in soil are often not fully hydrated and they may increase in weight by as much as 15% if placed in water (Wolf, 1940) and lose this when replaced in soil. Thus, earthworms must have mechanisms for replacing or conserving water, although their locomotion and burrowing, which depend to some extent upon hydrostatic pressure, are not affected seriously by losses of water up to 18%. There are diurnal changes in weight of 2–3% but these are of little significance (Wolf, 1940). Amounts of moisture in soil vary greatly, and if they fall too low, earthworms eventually begin to lose moisture and weight, although as much as 70–75% of the water content can be lost without killing the worm (Roots, 1956). As the worm becomes more and more dehydrated so its behavior changes; first there is a reaction termed a 'dehydration tropism'(Parker and Parshley, 1911) which is governed by the prostomium, because if this part of the body is removed, the reaction does not occur. There is then a period when the body rolls in the soil and eventually coelomic fluid is expelled from the dorsal pores in an attempt to moisten the surface of the body. Ultimately, rigor, anabiosis and death occur. If irreversible changes have not occurred, the animals can be revived by submersion in water, even after very large water losses.

There has been disagreement on the ways in which earthworms maintain the internal fluid at a reasonably constant concentration. Greatly different estimates of the osmotic pressure of the coelomic fluid have been reported, although it is clear that the osmotic pressure of the blood of *Lumbricus* spp. is slightly below, and that of *Pheretima* spp. above, that of the coelomic fluid (Ramsay, 1949).

Probably most of the uptake and loss of water is through the body wall. There is evidence (Stephenson, 1945) that *Lumbricus* spp. can maintain a constant internal salt concentration, because if these worms are placed in dilute salt solutions, the internal chloride concentration remains above the external concentration. Worms can also keep their internal salt concentration at a level lower than that of very concentrated solutions in which they are placed. This was confirmed by Ramsay (1949), who showed that as the concentration of chloride in the medium increased, so the osmotic pressures of the body fluids also increased, keeping them always greater than those of the medium. Urine excreted from the nephridiopores is always at a lower concentration (hypotonic) to the body fluids, except in very concentrated media with a sodium chloride content greater than 1.0. Thus, although earthworms can maintain a relatively constant internal osmotic pressure in dilute solutions, they are unable to do so in concentrated ones. Although it has been clearly shown that earthworms maintain their internal fluid at a relatively constant concentration, and excrete hypotonic urine as do freshwater animals, the exact mechanisms of this process are still not clear.

Some oligochaetes are aquatic, and terrestrial earthworms differ greatly in their affinity for and ability to survive in flooded soil, ranging from semi-aquatic species to those that prefer dry soil. Nevertheless, earthworms of the species *A. chlorotica, A. longa, D. subrubicunda, L. rubellus,* and *L. terrestris,* which normally inhabit relatively dry soils, were all able to survive from 31 to 50 weeks in soil totally submerged beneath aerated water. Even then, the factor limiting survival was probably lack of food rather than submersion. Individuals of *A. chlorotica* have made burrows in flooded soil (Roots, 1956) and cocoons of this species will hatch under water.

Dendrobaena subrubicunda, L. rubellus and *L. terrestris* prefer moist rather than saturated or flooded soil, but the other species discussed above sometimes choose waterlogged soil, although most terrestrial species leave flooded soil for drier sites and earthworms are often found on the soil surface after rain. There is evidence that individuals of some earthworm species can become gradually acclimatized to flooded soils and water. There is some evidence that weight change after submergence is influenced strongly by the environmental humidity to which earthworms are exposed prior to submergence (Oka *et al.*, 1984).

Obviously the osmotic regulation and excretory systems of earthworms are such that in water they can excrete fluids as fast as these diffuse inwards, and so maintain their internal fluids at a constant pressure.

4.7 LOCOMOTION AND PERISTALSIS

When earthworms burrow through the soil, they do so by coordinated contractions of the longitudinal and circular muscle bands that lie in the body wall. These contractions are made possible by the segments being kept turgid by the coelomic fluid, although there is no movement of fluid across the septa between segments when peristaltic waves pass along the body. If coelomic fluid moved along the body, it would be difficult for one end of a worm to contract and the other to expand.

The prostomium is used to find a cavity suitable for burrowing. The earthworm protrudes the setae on its posterior segments to keep the rear end of the body fixed, then contracts the circular muscles of the front end of the body, thus causing the anterior segments to extend forwards with a thrust of between 2 and 8 g, according to the size of the worm (Fig. 4.3). This thrust depends mainly on the internal hydrostatic pressure of the worm, and the area to which the pressure is applied. Pressures of the order of 1 kg/cm^2 have been recorded at the anterior end (Newell, 1950). The contraction of the circular muscles passes from segment to segment backwards along the body, and after the wave of contraction has passed, the anterior longitudinal muscles contract in turn, drawing the posterior end of the body forwards. The rear end of the body is then held again in position in the burrow by the backward-pointing setae, and another wave of contraction passes backwards down the body. The setae are protracted

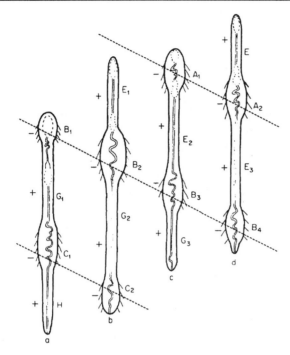

Figure 4.3   Patterns of muscular configuration of an earthworm, with concomitant pattern of the intestine and possibly of the esophagus. The wavy line represents the gut in the folded condition. +, Regions of high pressure; –, regions of low pressure. (After Arthur, 1965.)

and retracted as required, by special muscles; when the movement of the body is reversed, so is the direction of the setae.

Segmental contractions are by no means completely under control of the nerve cord, although if the cord is cut, the body muscles posterior to the cut can no longer contract. The segmental contractions seem to be caused by intersegmental stimuli, because if an earthworm is cut in half, but the two parts of the worm joined by thread and then put under tension, peristaltic waves continue to pass down the body.

If the body of the animal is cut through completely except for the nerve cord, contractions can still pass down the body, but much faster (about 25 mm/s). During normal peristalsis the nerve cord has an electrical rhythm whose frequency is identical with that of the muscular rhythm (Gray and Lissman, 1938). Thus, although locomotion can be maintained either by nervous impulses along the nerve cord or by reflexes passed from one segment to the next, normally the two methods reinforce each other. The cerebral ganglia seem to play no part in locomotion.

Contractions can be stimulated by touch, as well as by longitudinal tensions ranging from 0.1 to 1.0 g and they can be blocked by anesthesis with magnesium chloride. It is believed that the tension is detected by receptor cells lying between the ventral and dorsal setae of each side in every segment, and by other cells which lie buried in the circular muscle layer. If the tension is removed, peristalsis stops unless the ventral surface of the earthworm is in contact with the substratum. Peristalsis can also be initiated in a decapitated earthworm, by making the head electropositive to the posterior end, and inhibited by reversing the flow of the current (Gray and Lissman, 1938). Thus, the ways in which peristalsis is maintained are now understood, but we still do not know the ways in which it is initiated.

The gut also has peristaltic contractions, which seem to be mainly under nervous control. Two sets of nerves reach the gut, one from the circumesophageal nerve ring forming a plexus between the mucous membranes and the muscle layers of the intestine, and another from the ventral nerve cord running up each septum to the gut. The former nerves are believed to be excitatory and the latter inhibitory. However, it still is not clear how the gut maintains its peristaltic movements. There is some evidence, by assessing the effects of the drugs PCPCA and 5-HTP, that locomotion follows a circadian pattern. Among the four times of day – 06.00, 12.00, 18.00 or 24.00 – when the locomotion rate was determined, the responses to 5-HTP showed a circadian rhythm.

4.8 REGENERATION

The capacity for regeneration of injured or lost parts of the bodies of earthworms varies very much and ranges from a very poor capacity for regrowth in *L. terrestris* to species that seem to be able to regenerate almost any organ, such as those belonging to the genera *Criodrilus* and *Perionyx*. There is very considerable literature on the morphogenetic aspects of this phenomenon, but very little on its physiology or mechanisms. It is not proposed to deal with the detailed histology of the regenerating tissues here; the reader is referred to Stephenson (1930) and Laverack (1963) for a discussion of this subject.

Oligochaetes can regenerate either the anterior or posterior portions of their bodies, but the posterior part grows again more readily than the anterior. When the front end is replaced, wound tissue is formed over the cut end, taking about 7 days to grow, the gut being either occluded or remaining open during regrowth. The rear end of the body usually remains open during regeneration and the wound tissue forms a ring. When the front of the body regenerates, it grows with the same width as the rest of the body, but the rear usually grows as a slender appendage, which develops all the segments that it can regenerate before it begins to expand in width. Regenerated tissue takes about 2–3 months before becoming fully pigmented.

Although there have been numerous records of regeneration of different earthworm species, the extent to which this occurs is still only known for a few species. Gates (1974) stated that fewer than 10 species of earthworms were known to have regenerative capacity, but in his book (Gates, 1972a) he gives details of regeneration occurring quite commonly in 29 species from Burma.

Clearly, regeneration occurs more readily when the tail is cut off rather than the head, but even the posterior end will not grow again if too many segments are removed, e.g. anterior fragments of *E. fetida*, consisting of fewer than 13–30 segments regenerate only slowly, if at all. The full number of segments is not always regenerated; the number regrown depends on the length of the remaining fragment of the body. There is a gradual decrease in regenerative capacity along the body from front to back. Moment (1953a,b) showed that if an individual of any species was bisected at any particular segment, there was a remarkable consistency in the number of segments regenerated. No earthworm ever continues regenerating segments until there are more than the uninjured animal possessed. Moment considered that growth was controlled by electrical means. Earthworms have a voltage difference from one end to the other, and this is the same in young worms as in old ones. When a worm is cut in two, this voltage decreases sharply, but returns to its original level within 3 weeks. Moment (1953a,b) believed that the cessation of growth after a certain number of segments have been produced, could be explained on the basis that each segment has a particular voltage contribution, and when the full number of segments is attained, the total critical voltage inhibits further growth. There is some evidence to support such a theory, but none to show the mechanism of such an electrical inhibition of growth.

Some species have less ability to replace lost segments than others, e.g. *E. fetida* regenerates segments much more readily than does *L. terrestris*. The other feature essential for regeneration to occur seems to be that the nervous system in the part remaining should be intact. If the anterior end of an earthworm is cut off and the nerve cord removed from a few of the remaining segments, these segments are reabsorbed and regeneration starts at the segment that still has its nerve cord (Carter, 1940). If the nerve cord is removed completely, no regeneration occurs (Zhinkin, 1936). Nerve depressants such as lithium, acetylcholine, parathion and disulfoton also greatly inhibit regeneration. Oxygen is necessary for rapid regeneration and no regrowth occurs under anaerobic conditions; too much oxygen can also block regeneration. This has been confirmed by treatment with potassium cyanide which changes both the rate of respiration and also that of regeneration.

Some evidence exists that the chloragogenous cells are important in regeneration, because there is a mass migration of these cells to the wound after part of the worm is cut off. There is also evidence that temperature influences regeneration, all species regenerating much more readily in summer, with optimum temperatures between 18 and 20 °C; these are much higher temperatures than those which normally favor development of terrestrial earthworms. Younger worms regenerate segments much more readily and quickly than do older ones. The sexual organs are rarely regenerated if the part of the anterior end which carries them is amputated. Gras (1984) reported that the tail-flattening reflex, an escape mechanism in response to stimulation of the head, is reconstituted gradually in segments anterior to the amputation level even if new tail segments are not regenerated. It has also been suggested that one of the important factors in the onset of coordinated regenerative processes is related to neuroactive agents from the severed end of the ventral nerve cord.

Sometimes abnormal regeneration occurs; for instance, a posterior end may grow in place of the head, or vice versa, or two tails may be developed. If the anterior part of an individual *E. fetida* is removed by a cut made behind the groove between segments 18/19, either an abnormal head or a tail is grown. Thus, although the phenomenon of regeneration has attracted considerable attention, we still lack adequate knowledge of its mechanism and physiology.

4.9 TRANSPLANTATION

Some workers have experimented on transplanting portions of earthworms on to other individuals (Tembe and Dubash, 1961). When the anterior portion of one earthworm is sutured to the tail portion of another of the same species in the normal position, the intestine, blood vessels and nerve cords become continuous within 2 weeks. The parts also unite satisfactorily, even if one half is rotated 90° to the other. If two tails are joined, they remain alive for a considerable time, but if two heads are joined they do not usually unite satisfactorily, and the parts do not survive for long. Ovaries have been transplanted satisfactorily from one worm to another, even when the ovaries of an individual of *L. terrestris* were transplanted into a worm of the species *A. caliginosa*.

The earliest phase of graft rejection in earthworms, the recognition of foreign tissue antigens, may be controlled by leukocyte interactions that occur during the early phases of graft healing and agglutinin synthesis (Cooper and Roch, 1984). Second-set *L. terrestris* allografts from the same donor to the same recipient had accelerated rejection if transplanted less than 10 days after first-set allografts (Cooper and Roch, 1986). We still need more data on the factors that control the success of transplants.

Earthworm ecology: populations

5

Populations of earthworms are extremely variable in size, ranging from only a few individuals to more than 1000 per m^2 so that the assessment of the size distribution and structure of earthworm populations is difficult. A particular complication is the seasonal changes in the numbers, demography and vertical distribution of the populations, so that for comparison of communities, samples need to be taken at comparable times, using the same sampling methods.

5.1 ESTIMATION OF POPULATIONS

To estimate earthworm populations, some method of determining the number of worms in small sample areas is necessary. Sampling earthworm populations accurately is extremely difficult because their distribution is aggregated horizontally and some have deep and relatively permanent burrows. Their distribution is vertically stratified, it may be very difficult to separate them from many soils and their life cycles are complex in phenology. Most of the methods that are currently used to sample earthworms have various shortcomings and most estimates of field populations in the literature are likely to be underestimates because of the relative inefficiencies of the different methods.

Earthworm numbers in soil samples may be assessed mechanically in various ways, or the earthworms may be stimulated to move out of the soil in the field or from soil samples in the laboratory. Probably, a combination of these two approaches gives the best results. The various methods of assessing earthworm populations in the field, and extracting earthworms from soil samples have been reviewed by Bouché (1972), Bouché and Gardner (1984), Lee (1985) and Edwards (1991). They can basically be divided into passive methods, where the earthworms are separated physically from soil, and behavioral methods, where they are stimulated to emerge by physical or chemical stimuli.

5.1.1 HANDSORTING

Most of the early population studies involved digging up soil samples and sorting these by hand (Stöckli, 1928), and indeed many workers still use this method, except that they take cores or quadrats of soil of exact dimensions, to enable accurate population estimates to be made. Workers who have estimated populations by handsorting include Bretscher (1896), Bornebusch (1930), Ford (1935), Hopp (1947), Low (1955), Reynoldson (1955), Svendsen (1955), Wilcke (1955), Barley (1959a,b), van Rhee and Nathans (1961), El-Duweini and Ghabbour (1965a), Reinecke and Ljungström (1969), Rundgren (1975) and many others. Zicsi (1958a) suggested that numbers of earthworms estimated per m^2 decreased with increasing sample size, and suggested samples of 25 cm × 25 cm be used. In later work, Zicsi (1962a) compared the efficiency of estimating populations of earthworms by handsorting samples of sizes 0.06, 0.25, 0.5 and 1.0 m^2, taken with a square sampling tool. He concluded that 16 sample units/ha taken to a depth of 20 cm gave a good estimate of a population of medium-sized species. For larger worms and deeper burrowing species, he concluded that a larger area and deeper soil sample were required. Reinecke and Ljungström (1969) used four samples of 25 cm × 25 cm to sample earthworm populations in a South African pasture, and Rundgren (1975) four samples of 35 cm × 35 cm from Swedish soils.

Small earthworms may be missed by handsorting. Persson and Lohm (1977) considered that individuals less than 200 mg live weight could often be overlooked. They suggested sampling to a depth of 30 cm in summer and a 60 cm depth in winter. Axelsson et al. (1971) suggested that specimens less than 160 mg were often missed, and Reynolds (1973b) that worms less than 2 cm in length might not be counted because of their low visibility.

Lavelle (1978) sampled a mixed population of megascolecids and eudrilids in the Ivory Coast and recommended handsorting the top 50 cm of soil in 10 cm layers from 1 m^2 quadrats, supplemented by washing and sieving 20 × 20 cm samples to recover smaller individuals and cocoons. Nelson and Satchell (1962) tested how many earthworms could be recovered from soil by handsorting, by introducing known numbers of worms into soil. They found that the smaller worms and dark-colored worms were often missed, and their numbers were consequently underestimated; when 924 worms were introduced to soil, 93% of all worms were recovered by handsorting, but only 80% of immature *A. chlorotica* and 74% of immature *L. castaneus* were found. They concluded that handsorting was satisfactory only for individuals of more than 0.2 g live weight. There were considerable differences in efficiency between individual sorters. Handsorting is particularly relevant when assessing populations of earthworm species that live close to the soil surface.

5.1.2 SOIL WASHING

Cocoons and small earthworms are not easily recovered by handsorting or chemical stimulation. The only method that seems appropriate for these stages is a combination of washing and sieving soil samples, possibly accompanied by a flotation stage. For instance, Morris (1922) and Ladell (1936) used a method of washing soil samples with a jet of water through a series of sieves. Raw (1960b) handsorted samples from rather poor pasture soils, and then washed the soil away from the same samples in a sieve of mesh 2 mm within another 0.5 mm mesh sieve standing in a bowl of water. The sieves were then immersed in magnesium sulfate solution of specific gravity 1.2, and the worms that floated to the surface collected. Only 52% (84% of the weight) of the total earthworm populations collected were obtained by handsorting, and a further 48% were recovered by subsequent washing. From a heavy, poorly structured arable soil, 59% (90% by weight), and from a light, well-drained soil, 89% (95% by weight) of the total numbers of earthworms in the soil were recovered by handsorting. Obviously washing is more efficient than handsorting, and it also recovers cocoons, but the washing method takes much longer and tends to damage the earthworms mechanically. A mechanized soil washing method, which involves rotating the containers in which the sieves stand, was much faster, less laborious and suitable for most soils (Edwards *et al.*, 1970b). Flotation after washing and sieving can improve the efficiency of separating earthworms from soil (Raw, 1960b; Gerard, 1967; Martin, 1976).

Walther and Snider (1984) reported that both cocoons and earthworms could be recovered efficiently from humus and soil by a combination of washing through sieves and handsorting. In this way, 97.7% of cocoons and 96.7% of worm biomass in a soil were recovered. Bouché and Beugnot (1972) used a modified washing method which involved soaking soil samples in 2% sodium hexametaphosphate to disperse clay particles, soaking in 4% formaldehyde to kill and fix earthworms, passing the mixture through a series of sieves and handsorting the earthworms in the last sieve. Judas (1988) described an apparatus that extracted earthworms efficiently from broad-leaved litter by washing and sieving. Washing/sieving methods are clearly valuable tools in estimating earthworm populations but tend to be time-consuming and laborious; however, they are essential in estimating numbers of cocoons and very small, estivating or dormant earthworms.

5.1.3 ELECTRICAL METHODS

Fishermen have obtained earthworms from soil for bait by attaching one lead of an AC electrical mains to a copper wire attached to a nonconducting handle and inserting it into the soil. Walton (1933) and Johnstone-Wallace (1937) first reported that such a technique could be used for sampling earthworm populations, and Doeksen (1950), who experimented further, suggested that a steel rod 8–10 mm in diameter and 75 cm long with an insulated handle, was suitable as an electrode. He used 220–240 V at 3–5 A, and the strength of current was regulated either by a variable resistance, or by inserting the electrode deeper into the soil. One to three electrodes could be used simultaneously.

The conductivity of the soil, and hence the current passing, depends on its moisture content and pH, but usually the current penetrates deep into the soil, bringing earthworms up from deep burrows; however, if the surface soil is dry it may drive the earthworms downwards instead. This possibility can be minimized by insulating all of the electrode inserted into the soil except its point. Earthworms usually emerge at distances between 20 cm and 1 m from the electrodes, but there is some danger that earthworms close to the electrode may be killed by the current. Nevertheless, this method does seem effective in sampling for deep-living earthworms.

Satchell (1955d) also used an electrical method with a 2 kVA generator that led to a water-cooled electrode inserted 46 cm into the soil, with a voltage of 360 V applied. The portable generator made the method more suitable for field sampling earthworms. Satchell considered that the main defect of the method was that of defining the exact limits of the volume of soil from which earthworms were recovered.

Edwards and Lofty (1975b) described an electrical sampling method that consisted of inserting two electrodes, in the shape of forks with prongs 50 cm long, into the soil 1 m apart and passing the current from a 250 V diesel generator between the electrodes. The method was most effective when the current was kept to about 2–4 A. They found that another factor affecting the efficiency of the method was the pH of the soil through which the current was passed, more earthworms being recovered from soil with a low pH than from the same soil that was less acid.

Rushton and Luff (1984) used a circular electrode configuration. This was more efficient at extracting shallow-living than deep-burrowing species, and juveniles over adults, and extraction efficiency was correlated strongly with soil moisture content. Theilemann (1986) described an 'octet' method in which the electrodes were arranged octagonally and current was alternated between different pairs of electrodes over time.

5.1.4 CHEMICAL METHODS

The first chemical extractant used to sample earthworm populations was mercuric chloride solution (1.7–2.3 litres of solution containing 15 ml $HgCl_2$ in 18.25 litres of water) (Easton and Chandler, 1942). Evans and Guild (1947a) used a potassium permanganate solution to bring worms to the soil surface (1.5 g/litre at a rate of 6.8 litres/m^2), and later used this method in their population studies (Evans and Guild, 1947c; Guild, 1948, 1952a). Jefferson (1955) used a solution of mowrah meal (the material remaining after oil is extracted from ground seeds of the bassia tree, *Bassia longifolia*). Raw (1959) reported that a 0.55% formalin solution (25 ml of 40% formalin in 4.56 litres water applied to 0.36 m² of soil surface) was very effective in bringing most earthworms to the surface. This is because dilute formalin is less toxic to worms than potassium permanganate, which often kills worms before they reach the surface. The main disadvantage of these chemical methods is that they do not recover all species equally efficiently; those species with wide and deep burrows come to the surface much more readily than the species without such burrows (Bouché and Gardner, 1984). Baker (1983) reported that of 15 lumbricid species in a peat soil in Ireland, *L. terrestris*, *D. octaedra* and *Dendrobaena mammalis* were sampled relatively efficiently by the formalin method. However, the formalin method seems to be relatively inefficient for sampling nonlumbricid species such as *Diplocardia* and *Pheretima* (Reynolds, 1976). Satchell (1969) also used the formalin method but recommended that much more solution should be used (three applications of 9 litres of 0.165–0.55% formalin/0.5 m²). He pointed out that the efficiency of the method was seasonal, since both soil temperature and soil moisture content affect the number of worms coming to the soil surface, and he worked out a correction factor based on a regression analysis which would correct for the soil temperature at the time of sampling (Satchell, 1963), which he later modified (Lakhani and Satchell, 1970). However, numbers could not be corrected to account for worms that were estivating and hence did not respond to the chemical. Chloroacetophenone was used to sample *L. terrestris* populations by Daniel *et al.* (1992). They reported that the efficiency of the method was improved by repeated application of the chemical on three successive evenings. Walther and Snider (1984) reported that 96% of lumbricids could be recovered from litter using 1 hour of immersion in dilute (0.25%) formalin solution. Gunn (1992) suggested that mustard solution could be used to estimate earthworm populations without any of the environmental contamination that occurs with other toxic extractants such as formalin.

5.1.5 HEAT EXTRACTION

Earthworms can be stimulated to emerge from intact soil samples by application of moderate heat, and this method, which has been little used, may be useful in obtaining small surface-living species from matted turf. It involves using a container (55 × 45 cm) with a wire sieve 5 cm from its bottom. Soil quadrats cut from turf (20 × 20 × 10 cm deep) are placed on the sieve, immersed in water with fourteen 60 W light bulbs suspended above, and left for 3 hours, after which worms can be collected from the bottom of the container, where they are driven by temperature gradient.

5.1.6 VIBRATION METHODS

In the United States, mechanical stimulation by vibration has been used commonly to collect species of acanthodrilids such as *Diplocardia mississippiensis* and *Diplocardia floridana* and the megascolecid *Pheretima diffringens*. This often takes the form of vibrating with a bow a flexible rod that has been inserted into the soil. The vibrations passing through soil bring the earthworms to the soil surface. Its drawback is that it cannot be used on an area basis to assess populations accurately, and the method does not seem to work very well for lumbricids (Reynolds, 1973a).

5.1.7 COUNTING CASTS

Some species of earthworms such as *A. longa* and *Hyperiodrilus africanus* have very distinctive surface casts and many other species, such as *A. caliginosa*, have readily distinguishable surface casts (Chapter 10). Counting these casts cannot give valid assessments of earthworm populations, although Evans and Guild (1947b) reported that they obtained a close correlation between the numbers of casts deposited on the soil surface and the numbers of *A. longa* and *A. caliginosa* extracted from the soil under the casts. However, the presence of casts, which is seasonal and influenced by temperature and moisture, can be used as an excellent index of earthworm activity, which in turn is a useful parameter of earthworm populations.

5.1.8 MARK AND RECAPTURE

The tagging of soil-inhabiting invertebrates with some kind of marker, releasing them into soil and then extracting them from soil samples, has been used as a method of estimating invertebrate populations. A few workers have used this method for assessing earthworm populations. Meinhardt (1976) stained earthworms with a water-soluble nontoxic green dye which could stain the bodies of earthworms for several months before returning them to the soil. Mazaud and Bouché (1980) used the same method to study earthworm dispersal rates and mortality in a pasture soil. Other workers have used radioactive isotopes such as [198]Au to label *L. terrestris* (Joyner and Harmon, 1961), and Gerard (1963) inserted small pieces of [182]Ta wire into the coelom of individuals of *L. terrestris* to label them and trace their movements.

5.1.9 COMPARISONS OF METHODS

Several workers have compared the relative efficiency of extracting earthworms from soil by two or more methods. Svendsen (1955) reported that handsorting was much more efficient than using potassium permanganate extraction. Raw (1959) compared the use of formalin with that of potassium permanganate and handsorting. From one arable orchard he obtained 59.7 worms from 0.36 m^2 with formalin, 32.5 with potassium permanganate and 47.5 by handsorting. Comparable figures for a grass orchard were 165.1 for formalin, 83.9 for potassium permanganate and 280.0 for handsorting. Bouché (1969) compared handsorting with first applying formalin, then handsorting again to find earthworms that had not been extracted by the chemical. He reported that 55.4 worms/m^2 were extracted with formalin and a further 273.4 per m^2 by handsorting soil from the same area. This is an excellent way of improving the overall efficiency of the formalin method. Alternatively, handsorting can be performed first and formalin applied to the bottom of the handsorting pit (Bohlen et al., 1995a,b,c). Svendsen (1955) compared handsorting with potassium permanganate expulsion and obtained much better results with handsorting. Baker (1983) concluded that of 15 species in a reclaimed peat soil in Ireland, L. terrestris, D. octaedra, and Satchellius mammalis were the most efficiently sampled by the formalin method, whereas the other species, especially Aporrectodea spp., were sampled better by handsorting.

Bouché (1975a) compared the efficiency of six methods of sampling earthworm populations: digging and handsorting, soil washing, and sieving, flotation and extraction with potassium permanganate or formalin. He stressed the importance of using behavioral characteristics in improving efficiency and calculated correction factors for the various methods in the soil he sampled. He compared three methods in a later study (Bouché and Gardner, 1984).

Daniel et al. (1992) reported that populations of L. terrestris could be sampled equally efficiently by first using dilute formaldehyde or chloroacetophenone solutions and then handsorting to a depth of 110 cm. They improved the efficiency of the chemical methods by repeated application of the chemicals on three successive evenings. Gunn (1992) compared the use of mustard for estimating earthworm populations with using formalin, potassium permanganate or household detergent and reported it to be better than formalin and as good as potassium permanganate. Barnes and Ellis (1979) concluded that formalin extraction worked better in direct drilled (no-till) soils than in cultivated soils.

There seems little doubt that handsorting or washing gives the best results for most species, but these methods are very time-consuming, and do not work well for deep-burrowing species such as L. terrestris. At present, the formalin method seems the best compromise for species with deep burrows and can be combined with handsorting of surface soils for maximum efficiency.

5.1.10 NUMBER AND SIZE OF SAMPLES

Many different sizes of soil samples have been used, ranging from soil cores 20 cm in diameter to 50 × 50 cm and 1 m^2 quadrats. Zicsi (1962a) reported that the number of worms he recovered per m² by handsorting decreased with the increasing size of the sample. The minimum sample area required depends very much on the density of a population in a site. For most purposes, 0.5 or 0.25 m² quadrats seem to be a suitable size and are easy to handle. The optimum number of samples can be determined by pre-sampling and then basing the number of samples on the degree of precision needed.

5.2 SIZE OF POPULATIONS

Earthworm populations can be expressed either in terms of numbers or weight (biomass). The use of numbers is sometimes misleading because it does not differentiate between very small and large individuals, which have very different influences on soil processes. Biomass is often a preferable parameter but can also lead to misleading conclusions. It is probably best to express populations both as numbers and biomass, and most workers on earthworms report their populations in this way. Reporting biomass on the basis of ash-free dry mass, determined by ashing worms in a muffle furnace at 400–500 °C, corrects for the mass of soil in the guts of sampled worms, allowing for accurate comparisons between different investigations. This method is essential for estimating secondary productivity of earthworm tissue.

5.2.1 NUMBERS

Relatively few workers have sampled the earthworm populations in a variety of habitats at the same time of year, so it is difficult to assess precisely which habitats can support large earthworm populations. Some typical populations that have been estimated by handsorting or formalin extraction are given in Table 5.1. These estimates are all very approximate because of large variations in efficiency of extraction and seasonal changes in numbers of earthworms. For instance, populations can range from less than one earthworm to more than 2000 per m² (0.5–305 g/m²). There are almost always fewer earthworms in acid, mor soils, fallow soils and moorlands, than in good mull soils. The numbers of earthworms in regularly cultivated arable soils are usually very variable, and populations are intermediate in size between the more sterile habitats and those in pastures and natural grasslands which can support large numbers of earthworms. The populations in coniferous forests tend to be lower and those in deciduous temperate forests and tropical forests rather larger than those in arable land. Clearly, there is great variability in earthworm populations, although grassland seems able to support higher populations than most other habitats, presumably due to the availability of large quantities of organic matter.

Table 5.1   Numbers and weights of earthworms in different habitats

Habitat	No./m²	g/m²	Site	Extraction method	Reference
Fallow	18.5–33.5	4.6–8.4	USSR	Handsorting	Dzangaliev and Belousova (1969)
Fallow	22.6	79	Wales	Handsorting	Reynoldson (1966)
Fallow	210–460	16–76	S. Australia	Handsorting	Barley (1959a)
Taiga	17.4	2.8	Finland	Handsorting	Zajonc (1970)
Arable soil	146	50	N. Wales	Handsorting	Reynoldson (1955)
Arable soil	287	76	Bardsey Island	Handsorting	Reynoldson (1955)
Arable soil	220	48	Germany	Wet sieving	Krüger (1952)
Arable (with dung)	79	39.9	UK	Formalin	Edwards and Lofty (1977)
Arable	20–25	2.0–2.5	S. Australia	Handsorting	Barley (1959a)
Arable	5–100	0.5–20.0	Rumania	Handsorting	Gruia (1969)
Arable	7.4–101.8	0.23–3.64	Uganda	Handsorting	Block and Banage (1968)
Garden soil	420	153	Egypt	Handsorting	El-Duweini and Ghabbour (1965a)
Garden soil	73	–	Argentina	Handsorting	Ljungström and Emiliani (1971)
Pasture	389–470	52–110	UK	Handsorting	Svendsen (1957a)
Pasture	390	56	Bardsey	Handsorting	Reynoldson *et al.* (1955)
Pasture	481–524	112–120	N. Wales	Handsorting	Reynoldson (1955)
Pasture	260–640	51–152	Australia	Handsorting	Barley (1959a)
Pasture	742–1235	146–303	New Zealand	Handsorting	Waters (1955)
Pasture	690–2020	305	New Zealand	Handsorting	McColl and de Latour (1978)
Pasture	460–625	62–78	Australia	Handsorting	Barley (1959a)
Pasture	72–1112	–	S. Africa	Handsorting	Reinecke and Ljungström (1969)

Habitat	No./m²	g/m²	Site	Extraction method	Reference
Pasture	400–500	100–200	Ireland	Handsorting	Cotton and Curry (1980a)
Old pasture	109	59	Sweden	Handsorting	Nordström and Rundgren (1974)
Old pasture	390–470	52–110	England	Handsorting	Svendsen (1955)
Old pasture	646	149	Wales	Handsorting	Reynoldson (1966)
Old pasture	288	125	France	Washing/ sieving	Bouché (1977)
Natural grassland	250–750	–	New Zealand	Handsorting	Lee (1958)
Tropical savanna	230	49	Ivory Coast	Handsorting and wet sieving	Lavelle (1974)
Coniferous forest	14–66	–	Finland	Handsorting	Huhtà and Karpinnen (1967)
Coniferous forest	103–167	30–35	Sweden	Handsorting	Persson and Lohm (1977)
Coniferous forest	10–40	–	Norway	Handsorting	Abrahamsen (1972)
Coniferous forest	27–72	–	Japan	Handsorting	Kitazawa (1971)
Subalpine woodland	15–106	5.7–35.7	Australia	Handsorting	Wood (1974)
Pine woodland	40	17	UK	Handsorting	Edwards and Lofty (1977)
Oak woodland	106	120	Czecho- slovakia	–	Zajonc (1970)
Oak woodland	122	61	Denmark	–	Bornebusch (1930)
Beech woodlands	73–177	6–54	Denmark	–	Bornebusch (1930)
Mixed woodlands	14–142	26–280	USA	–	Reynoldson (1966)
Mixed woodlands	157	40	Wales	Handsorting	Reynoldson (1955)
Mixed woodlands	118–138	–	England	Handsorting	Phillipson et al. (1976)

Habitat	No./m²	g/m²	Site	Extraction method	Reference
Mixed woodlands	136	68.3	USSR	Handsorting	Zajonc (1970)
Mixed woodlands	106	98.1	Czecho-slovakia	Handsorting	Zajonc (1970)
Mixed woodlands	14–124	26.3–280.3	USA	Handsorting	Reynoldson (1966)
Tropical forest	34	10.2	Nigeria	–	Madge (1969)
Tropical forest	61.7	2.5	Nigeria	Handsorting	Cook et al. (1980)
Tropical rainforest	80–121	34.2–42.4	Mexico	–	Fragoso and Lavelle (1987)
Tropical rainforest	280–401	35.4–71.9	Costa Rica	Handsorting	Atkin and Proctor (1988)
Tropical rainforest	64–166	3.3–22.7	Malaysia	Handsorting	Leaky and Proctor (1987)
Tropical rainforest	6–26	0.4–0.6	Sarawak	Handsorting	Anderson et al. (1983)
Tropical rainforest	68	2.69	Peru	Handsorting	Rambke and Verhaagh (1992)
Tropical rainforest	64	21.8	Peru	Handsorting	Lavelle and Pashanasi (1989)
Tropical montane forest	40–108	1.8–2.7	Sarawak	Handsorting	Collins (1980)

5.2.2 BIOMASS

It is often difficult to obtain the live weight of large numbers of earth-worms directly, since they have to be weighed soon after collection, and Satchell (1969) described a method for calculating the live weight of earth-worms that had been kept in 10% formalin solution. He plotted a regres-sion of the live weights of worms against their weight after being kept in 10% formalin, then oven-dried them at 105 °C, and reweighed them. He obtained the expression: 1 g dry weight = 5.5 g live weight. The gut con-tents of an earthworm may be as much as 20% of its total live weight, so this must be accounted for when estimates of population biomass are made. Collins (1992) calculated a regression model which related earthworm length to dry weight for lumbricids, using specimens from northern Wisconsin forests.

5.3 POPULATION STRUCTURE: AGE DISTRIBUTION

Populations of most soil-inhabiting invertebrates tend to have a pyramidal age structure, with many more young individuals than mature ones at most times of the year; for instance, Raw (1962) reported that the proportions of *L. terrestris* individuals of different ages in his samples from an orchard, were in the ratio of 8 mature worms to 13 large immatures and 31 small immatures. Not much information is available about the seasonal changes in age structure of populations of earthworms.

Evans and Guild (1948a) sampled 12 fields at Rothamsted for earthworm populations and compared the numbers of immature and adult worms of different species (Table 5.2). Clearly, these proportions are never in a fixed ratio, and depend very much on the time of year when a population is sampled, so that after active breeding periods the proportion of immature worms will be greatest, and at all other times relatively small. However, there is some indication that the structure of populations of different species may differ considerably. Seasonal changes in proportions of adult and immature worms of *A. nocturna* and *A. caliginosa* were reported by van Rhee (1967), and by Gerard (1960) for *A. chlorotica* (Fig. 5.1). Van Rhee (1967) compared the population structure of *L. terrestris*, *L. castaneus*, *A. rosea*, *A. caliginosa* and *A. chlorotica* in four orchard soils in five successive years, and for all species, except *L. castaneus*, the immatures greatly out-

Table 5.2 Ratios of immature worms to adults during October (from Evans and Guild, 1948a)

Field	*A. nocturna*	*A. caliginosa*	*A. chlorotica*	*A. rosea*	*L. terrestris*
Parklands	4.5	3.0	0.5	2.2	10.3
Great Field I	1.9	1.4	0.5	1.9	12.3
Great Field III	2.5	1.5	0.3	1.8	11.1
Pastures	3.1	0.7	0.2	1.2	7.3
Longcroft	2.3	1.6	0.6	3.4	6.9
Claycroft	1.5	1.1	–	1.8	7.5
Appletree	5.3	1.0	0.5	2.1	5.0
Great Field II	4.5	1.2	0.3	2.5	19.3
Parklands Wedge	3.5	1.5	–	5.0	12.9
New Zealand	2.0	1.7	0.3	1.7	–
Stackyard	12.8	2.3	0.6	3.0	0.7
Delharding	3.3	2.5	1.6	1.9	1.5

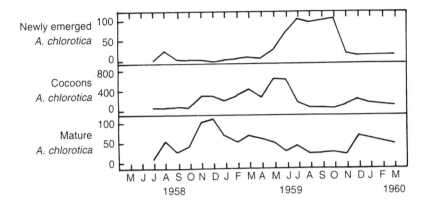

Figure 5.1 Seasonal abundance of *A. chlorotica* in 'Pastures' at Rothamsted 1958-60 (Gerard, 1960).

numbered the adults. For the species *Pheretima hupeiensis*, in the north-western United States, mature individuals were abundant in August but many died after reproduction, and immatures were more numerous in September, although by November, the entire population became imma-ture and remained so throughout the winter months (Grant, 1956).

McCredie *et al.* (1992) reported from Australia that the abundance of *A. trapezoides* increased from 58 per m² at the time of the opening rains to 170 per m² (88.6 g live weight/m²) after 10.5 weeks. Near the end of the wet season (in October) the density was 37 per m². At the time when the rains began, the population consisted of juvenile and semi-mature individuals. Clitellate earthworms were found 1 month later and predominated from August (10.5 weeks) to the end of the season. Cocoons occurred from August through October, and those incubated in the laboratory at 16 °C hatched on average after 42 days and produced about two juveniles each. Juvenile and immature earthworms collected from a quiescent state at the end of summer, matured within 1 month when reared in moist soil in the laboratory.

The only data on the age structure of nonlumbricid earthworms are those from studies by Lavelle (1971, 1974, 1978) for species from western Africa. A clear pyramid of age classes was reported for *Millsonia anomala*, with 25% being 1–2 months old; 48.3%, 6–8 months; 18.3%, 11–12 months; 6.7%, 16–19 months; and 1.7% older than 24 months (Lavelle, 1971). Analyses of the age structure of other species are also reported in Lavelle (1978).

5.4 POPULATION STRUCTURE: SPATIAL DISTRIBUTION

Earthworms are by no means distributed randomly in soil. Guild (1952a) and Murchie (1958a) classified the possible factors that were likely to be responsible for variability in horizontal distributions as:

1. physico-chemical (soil temperature, moisture, pH, inorganic salts, aeration and texture);
2. availability of food (herbage, leaf litter, dung, consolidated organic matter); and
3. reproductive potential and dispersive powers of the species. To which we add:
4. historical factors (including disturbance and colonization of new habitats).

5.4.1 HORIZONTAL DISTRIBUTIONS

Murchie (1958a) concluded that no single one of these factors was likely to be solely responsible for the horizontal distribution, but rather the interaction of several or all of the factors, thereby providing suitable soil conditions for earthworms in some areas.

There is not much experimental evidence that physico-chemical factors other than soil moisture cause aggregations of earthworms, but several workers have correlated such aggregations with the available food supplies. For instance, in pasture, populations of *D. octaedra* and *L. rubellus* were aggregated significantly beneath dung-pats in spring (Boyd, 1957b, 1958), and pigmented species such as *Lumbricus festivus, L. rubellus, L. castaneus, D. rubida, D. octaedra* and *B. eiseni* were shown to aggregate under dung-pats more than nonpigmented *Allolobophora* and *Aporrectodea* species (Svendsen, 1957a). If they react so readily to dung, then it seems probable that variations in the distribution of other forms of organic matter may also influence earthworm distribution. However, aggregations cannot always be explained on the basis of heterogeneity of the habitat; Satchell (1955a) showed that *L. castaneus* and *A. rosea* were greatly aggregated in a relatively uniform pasture which had not been grazed. He suggested that aggregations might occur when earthworms are reproducing more rapidly than the offspring can disperse from the breeding site.

When all the species of earthworms in a habitat are considered, a pattern of overlapping aggregations is usually found. When Satchell (1955a) calculated indices of dispersion for adults and immatures, he found that the adults were nearly randomly dispersed but the immatures were aggregated significantly. Hence, on this assumption, a species with distinct seasonal changes in abundance can be expected to pass from a very aggregated phase in the breeding season in early summer to a more randomly distributed phase in winter (Satchell, 1955a).

Hamblyn and Dingwall (1945) claimed that the rate of advance of the margin of populations of *A. caliginosa*, from inoculation points in recently limed grasslands, was of the order of 10 m/yr, and they suggested that any more rapid horizontal dispersal was probably by cocoons carried in soil on agricultural implements, hooves of animals, feet of birds or in streams. This has been supported by later studies.

Many workers have introduced earthworms to new habitats and studied their rate of multiplication and spread. For instance, van Rhee (1969a,b) inoculated a new polder in The Netherlands with 4664 individuals of *A. caliginosa* and reported that they had multiplied to 384 740 individuals 1 year later. At the same site 2558 individuals of *A. chlorotica* increased to 121 660 in the same period. He calculated a horizontal rate of spread of the population of *A. caliginosa* of 6 m/yr and 4 m/yr for *A. chlorotica*. Hoogerkamp *et al.* (1983) reported an annual dispersal rate of about 9 m for *A. caliginosa* and 4 m for *L. terrestris*. The spread of earthworms in inoculated polder soils was calculated by Stein *et al.* (1992) to be from 10 to 13 m/yr. Stockdill (1982) inoculated *A. caliginosa* into New Zealand pastures at a spacing of 10 m and reported that the whole area became colonized after 8–10 years.

5.4.2 VERTICAL DISTRIBUTIONS

Different species of lumbricids inhabit different depth zones in the soil (Fig. 5.2), but the vertical distribution of each species changes considerably with the time of year. The seasonal vertical distribution of the common British lumbricids has been studied by several workers. Species

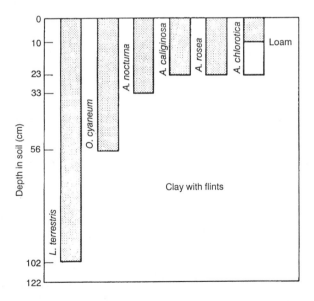

Figure 5.2 Vertical distributions of earthworms in a Rothamsted pasture (adapted from Satchell, 1955a).

such as *D. octaedra* and *B. eiseni* live in the surface organic horizon of soil for most of the year. *Allolobophora caliginosa, A. chlorotica, A. rosea, L. castaneus* and *L. rubellus* occur commonly within 8 cm of the soil surface, as do immature individuals of *O. lacteum, O. cyaneum, A. longa, A. nocturna* and *L. terrestris*. Most adult and nearly mature individuals of *O. cyaneum* are in the top 15 cm, and although they have distinct burrows, these are usually temporary. *Aporrectodea longa* and *A. nocturna* have fairly permanent burrows which usually penetrate as deep as about 45 cm, but the deeper vertical burrows of *L. terrestris* commonly go down to a depth of 1 m and can penetrate as deep as 2.5 m. Gerard (1967) studied the changes in vertical distribution of common earthworms at different times of the year in England (Fig. 5.3). Most earthworms in his samples were below 7.5 cm deep in January and February, when the soil temperature was about 0 °C, but by March when the soil temperature had risen to 5 °C at a depth of 10

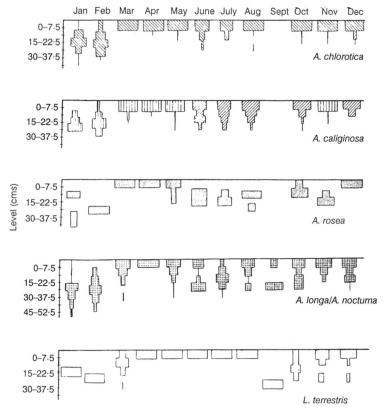

Figure 5.3 The depth of six species of earthworms in monthly soil samples from January to December, 1959 (expressed as percentages for each species in each sample). Samples were taken in 7.5 cm layers to a depth of 30.5 cm, and sometimes deeper (up to 53 cm). (After Gerard, 1967.)

cm, most individuals of *A. chlorotica*, *A. caliginosa* and *A. rosea*, and small and medium-sized individuals of *A. longa*, *A. nocturna* and *L. terrestris*, had moved into the top 7.5 cm of soil, although most of the larger worms were still deeper in the soil. From June to October, earthworms of most species were below the top 7.5 cm again, except for newly hatched individuals. In November, December and the following April, most worms had returned to the top 7.5 cm. The two factors influencing movement to deeper soil, seemed to be very cold or very dry surface soil. All species of worms (except *L. terrestris*) seemed to be quiescent in summer and mid-winter, and at both these times were deeper than 7.5 cm below the surface. More worms were quiescent in summer than in winter. Nearly all cocoons were found in the top 15 cm of soil, most being in the top 7.5 cm.

Seasonal changes in vertical distributions were also studied in Sweden (Rundgren, 1975). *Dendrobaena octaedra* and *D. rubida* occurred close to the surface throughout the year, but *A. longa*, *A. caliginosa*, *A. rosea* and *L. terrestris* all penetrated deeper into the soil at some times in the year (Rundgren, 1975). Earthworms of the genus *Diplocardia* in south-east United States move from a depth of about 10 cm in October, to about 40 cm deep in January, and return to the surface soil in spring, with a critical temperature for downward movement of approximately 6 °C (Dowdy, 1944). In the north-west United States, earthworms of the species *P. hupeiensis* were active in the 15–20 cm soil level and were near the surface only in September and March, and from November to February they were deeper than 55 cm (Grant, 1956).

In Australia, *A. trapezoides* and *A. caliginosa* were most abundant in the surface layers of the soil (0–10 cm depth) for the 3–7 months from autumn to spring when the soils were most moist. During summer, most individuals were deeper than 20 cm below the surface, were inactive and coiled tightly within spherical chambers (Baker *et al.*, 1992a,b). In other studies (Baker *et al.*, 1993a,b), *A. rosea*, *A. trapezoides*, *Microscolex dubius* and *Microscolex phosphoreus* were most abundant when the soils were wettest, and during this period the majority of earthworms were in the top 10 cm of soil. During the other months of the year, most earthworms were located at a depth below 10 cm and mature adults with clitella were found only in winter and spring. Similarly, *Gemascolex walkeri* and *O. cyaneum* occurred predominantly in the top 10 cm of soil in Australian pastures, for 4–5 months each year, in autumn to spring, when the soils were wettest; during the dry season, they were much deeper (Baker *et al.*, 1993a,b).

It seems that seasonal vertical migrations of earthworms occur in most parts of the world, and these are initiated by the environmental factors in the upper soil levels becoming unsuitable for earthworms to feed and grow satisfactorily. A positive correlation between the length of the lumbricid body and the depth of burrowing into soil was suggested by Piearce (1983).

Reddy and Pasha (1993) investigated the influence of rainfall and temperature on the vertical distribution of earthworms in two semi-arid grassland soils in India. The species were *Octochaetona philloti* (Michaelsen) (16–96 per m²) and *Barogaster annandalei* (Stephenson) (3.2–58.3 per m²). These earthworms migrated to deeper soil layers during winter and summer, *O. philloti* reaching a maximum depth of 40 cm and *B. annandalei* going down to 45 cm. They concluded that physical soil factors were more important collectively in influencing seasonal variations in populations and distributions than chemical factors. Usually those species that feed on or near the surface were dark-colored, whereas the subterranean species were predominantly paler in color.

5.5 SEASONAL POPULATIONS AND ACTIVITY

The numbers of earthworms and their degree of activity vary greatly during the annual cycle, but not all earthworm ecologists have differentiated adequately between seasonal changes in numbers and changes in activity. This applies particularly to those workers who used chemical extractant methods for determining populations; such techniques depend on the activity of the earthworms and quiescent individuals do not respond, so the numbers collected reflect both size of populations and their activity.

Evans and Guild (1947c) followed changes in numbers of earthworms in an old pasture field in England for more than a year, using a chemical sampling method, and they concluded that the two soil conditions that affected earthworm activity most were temperature and moisture (Fig. 5.4), although another important factor was the obligatory diapause that occurred from May to October, for the two species *A. nocturna* and *A. longa*, or the periods of quiescence or facultative diapause of *A. chlorotica*, *A. caliginosa* and *A. rosea* during adverse conditions. Tiwari *et al.* (1992) also found a significant correlation between earthworm populations and temperature and moisture in a pineapple field.

The amounts of seasonal activity of surface-casting earthworm species can also be assessed by the numbers of worm casts produced. Evans and Guild (1947c) found that the numbers of casts deposited, and the numbers of *A. longa* and *A. nocturna* that were obtained by potassium permanganate sampling, were correlated closely. Other workers have attempted to estimate seasonal activity by counting the numbers of worms on the surface at night, but this is very inaccurate because many species of worms only come to the surface when it is wet.

Gerard (1967) reported that in pasture soil in England, *A. chlorotica, A. caliginosa* and *A. rosea* usually occurred within 10 cm of the soil surface, but when the soil temperature fell below 5 °C, or the soil became very dry, individuals of these species moved to deeper soil. In hot, dry periods in

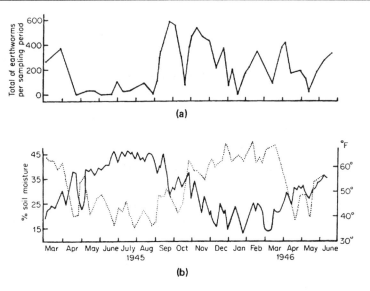

Figure 5.4   (a)The seasonal changes in activity of all earthworms; 'Total of earth-worms' = total number of earthworms in 10 samples. (b) The corresponding changes in soil moisture at 5 cm (....), and temperature at 10 cm (____), from March 1945 to June 1946. (From Evans and Guild, 1947c.)

summer, most species became inactive, and were deeper in the soil. Hopp (1947) believed that the most important factor influencing seasonal changes in numbers of worms in arable soils in the United States was the death of worms when the unprotected surface soil became frozen in winter. Seasonal changes in earthworm populations have also been ascribed to other causes; for instance, Waters (1955) suggested that flushes in the availability of dead root material and herbage debris were a main cause of increased numbers of worms, but this hypothesis is not generally accepted.

The five species *L. terrestris*, *A. rosea*, *A. chlorotica*, *A. nocturna* and *A. caliginosa* were most active in an English pasture between August and December, and April to May (Evans and Guild, 1947c) (Fig. 5.5). The prin-cipal period of activity of *Hyperiodrilus africanus* in Nigeria occurred during May and June, which is the beginning of the wet season; there-after, numbers lessened gradually until November, when very few were found (Madge, 1969). The numbers of cocoons produced also varied sea-sonally, and Gerard (1967) showed that in an English pasture, most cocoons were produced in late spring and early summer (Fig. 5.6).

Gates (1961) reported that earthworm activity in the tropics is also lim-ited to certain seasons; in the monsoon tropical climate of Burma and the humid subtropical climate of India, earthworms are active mainly in the

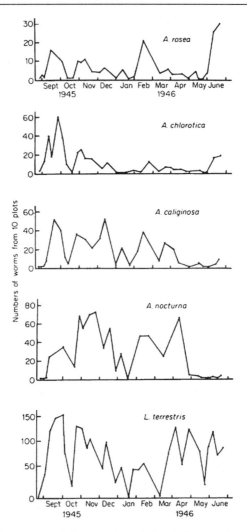

Figure 5.5 The seasonal changes in activity of five species of earthworms (after Evans and Guild, 1947c).

4–6 months of the rainy season between May and October. *Millsonia anomala* has a marked seasonal activity in West Africa, with phases of inactivity and quiescence occurring when soil water content falls below 7% (Lavelle, 1971). By contrast, in the humid continental climate of the south-eastern United States this species is most active in the spring and autumn months. Hopp (1947) presented data on the seasonal earthworm population changes in Maryland, USA, which confirmed this, but since he only took samples to a depth of 7.5 cm, the proportion of the total population he extracted varied considerably with the time of year. Grant (1956)

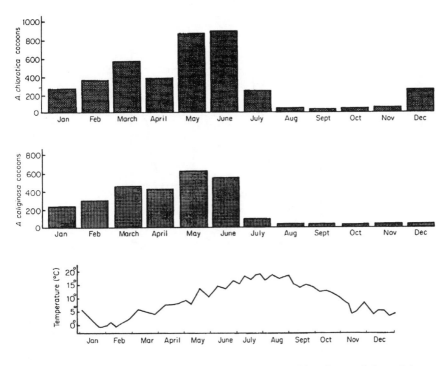

Figure 5.6 Seasonal production of cocoons by *A. chlorotica* and *A. caliginosa* (after Gerard, 1967).

reported that *P. hupeiensis* was most active in the north-western United States during the summer months, and retreated to soil levels below 55 cm deep from November to February.

In grassland in Japan (Nakamura, 1968a,b), the greatest numbers of earthworms occurred in autumn, especially in October, and the numbers were very low in winter, particularly during January and February (Fig. 5.7). In Australia, populations of *A. rosea*, *A. caliginosa* and *O. cyaneum* in pasture increased from May to July and decreased from July to October (Baker *et al.*, 1993a). Soil temperatures were not low enough during the wet season (10 °C at 15 cm depth) to prevent breeding from occurring. The vertical distribution of the worms in these soils also changed during the year, so that in winter most earthworms were in the top 15 cm and few were below 30 cm, whereas when soils began to dry out in spring, few worms remained in the top 15 cm. During summer, 60% of all worms were between 15 and 30 cm deep. The abundance of *A. rosea*, *A. trape-zoides*, *Microscolex dubius* and *Microscolex phosphoreus* was monitored monthly in lucerne and cereal fields in Australia (Baker *et al.*, 1993b), and overall populations of about 303 per m² were recorded. The highest numbers occurred in winter and spring. In a New Zealand pasture between

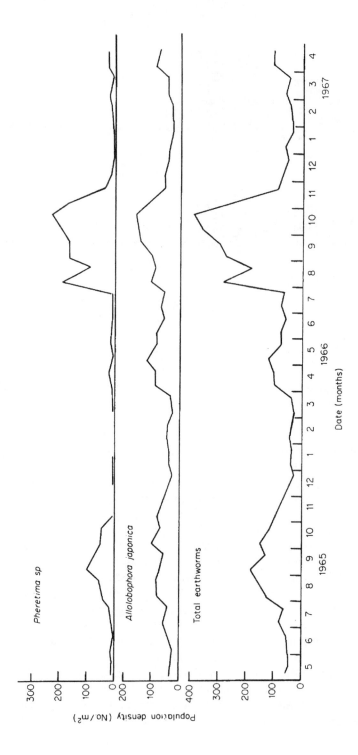

Figure 5.7 Seasonal variation in the population density of earthworms in alluvial soil grassland in Japan during the period May 1965 to April 1967 (after Nakamura, 1968a).

1951 and 1954, peak populations occurred in mid-winter (Waters, 1955; Barley, 1959a,b) (Fig. 5.8).

In addition to such seasonal changes in activity, many species of earthworms are distinctly diurnal in their activity. Experiments by the authors demonstrated clearly that *L. terrestris* tended to be active from 6 p.m. to 6 a.m. although this varied with season. This diurnal activity is intrinsic and seems to be at least partially independent of temperature and light. Patterns of activity differ between species; for instance *Millsonia anomala* has a marked diurnal rhythm of activity, with two daily maxima of casting at about midnight and 9 a.m. in moist soil (Lavelle, 1971).

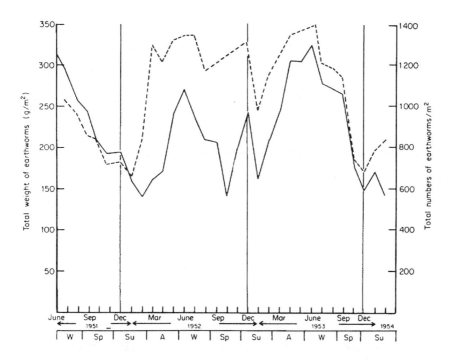

Figure 5.8   Seasonal fluctuations in the abundance of earthworms found in pasture land at Palmerston North, New Zealand. W, Winter; Sp, spring; Su, summer, A, autumn; ——, weight; – – –, number. (From Waters, 1955.)

Earthworm ecology: communities

6

Earthworm communities may be simple or more complex but species tend to be complementary in their activities. There are many species associations and individual species that have relationships with many other organisms including micro-organisms, invertebrates and vertebrates.

6.1 THE STRUCTURE OF EARTHWORM COMMUNITIES

6.1.1 MAJOR ECOLOGICAL GROUPS

Several schemes have been proposed to classify earthworm species into major ecological categories, which are based mainly on differences among species in the burrowing and feeding activities, and vertical stratification in soil. These major ecological groups represent functional adaptations to the soil environment that allow different species of earthworms to coexist by exploiting different food resources and habitat space. Earthworm communities nearly always include species that pursue different ecological strategies, and a familiarity with these strategies is essential to an understanding of the structure of earthworm communities.

Early research on British Lumbricidae distinguished between surface-dwelling species that produce no burrows or casts, and soil-dwelling species that live within a defined burrow system (Evans and Guild, 1948a; Satchell, 1980). Graff (1953a) defined these two groups further, with the surface-dwelling species being more pigmented, producing more cocoons and maturing faster. The soil-dwelling species are less pigmented, have low rates of cocoon production and mature more slowly. Piearce (1972) divided woodland species of earthworms into ecophysiological groups based on their feeding habits and ability to mobilize calcium. He recognized three types of earthworm:

1. pigmented litter feeders, which include *L. castaneus*, *L. rubellus* and *D. rubida* f. *subrubicunda*, all of which have active calciferous glands;

2. unpigmented topsoil feeders, including *A. caliginosa*, *A. chlorotica* and
    O. lacteum, with calciferous glands that are intermediate in complexity
    and activity;
3. unpigmented humus or compost feeders, of which only one species,
    D. veneta var. *hybernica* f. *typica*, was defined (Piearce, 1972).

Lee (1959) recognized three main ecological groups of earthworms
among New Zealand Megascolecidae, based on the soil horizons in which
the earthworms were commonly found: i.e. litter, topsoil or subsoil. His
litter species form no burrows, are generally heavily pigmented dorsally
and ventrally and feed on decomposing litter. The **topsoil** species he
described live in permanent burrows that descend into the mineral horizon. They have medium pigmentation dorsally, are unpigmented ventrally and feed on decomposing litter on the soil surface and some soil.
His **subsoil** species have constantly extending burrow systems and are
unpigmented or lightly pigmented. They feed on soil and organic matter
in the soil. They are often called 'geophagous' because of their habit of
eating large quantities of soil.

Bouché (1971, 1977), working with European lumbricids, also recognized three major ecological groups, which he termed: (1) epigeics, (2)
anecics and (3) endogeics. **Epigeic** earthworms are in a category similar to
Lee's litter species. These worms typically live on the soil surface or in the
upper reaches of the mineral soil, beneath a litter layer, have relatively
high reproductive rates and grow rapidly. This group is represented by *L.
rubellus*, *D. octaedra* and *L. castaneus*, among others. **Anecic** earthworm
species fall into the same general category as the topsoil species recognized by Lee (1959). They form permanent or semi-permanent vertical
burrows in the soil, which descend into the mineral horizon and open at
the surface, where the earthworm emerges to feed, primarily on dead
leaves and other decaying organic materials. This group includes *L. terrestris*, *A. longa* and *Nicodrilus longus* which are the major European representatives of this group. Anecic earthworms are more predominant
among European lumbricids than among New Zealand megascolecids.
Endogeic earthworm species, like Lee's subsoil species, inhabit the mineral soil horizons. They consume more soil than do either epigeic or
anecic species and derive their nourishment from more humified organic
matter, although some species will occasionally come to the surface to
feed beneath the litter layer. Bouché (1977) described these three major
groups as being evolutionary extremes on three corners of a triangle, with
many species occupying intermediary positions with respect to these
extremes. His terminology has been widely adopted and is commonly
encountered in the scientific literature on earthworms. The general characteristics of the three major groups of European lumbricid earthworms
described by Bouché are summarized in Table 6.1.

Table 6.1 General diagnostic features of the major ecological groups of European lumbricid earthworms as described by Bouché (1977)

Diagnostic characters	Epigeic species	Anecic species	Endogeic species
Food	Decomposing litter on the soil surface; little or no soil ingested	Decomposing litter on soil surface some of which is pulled into burrows; some soil ingested	Mineral soil with preference for material rich in organic matter
Pigmentation	Heavy, usually both ventrally and dorsally	Medium–heavy, usually only dorsally	Unpigmented or lightly pigmented
Size of adults	Small–medium	Large	Medium
Burrows	None; some burrowing in upper few centimeters of soil by intermediate species	Large, permanent, vertical burrows extending into mineral soil horizon	Continuous, extensive, subhorizontal burrows, usually in the upper 10–15 cm of soil
Mobility	Rapid movement in response to disturbance	Rapid withdrawal into burrow but more sluggish than epigeics	Generally sluggish
Longevity	Relatively short-lived	Relatively long-lived	Intermediate
Generation time	Shorter	Longer	Shorter
Drought survival	Survives drought in the cocoon stage	Becomes quiescent during drought	Enters diapause in response to drought
Predation	Very high, particularly from birds, mammals and predatory arthropods	High, especially when they are at the surface; somewhat protected in their burrows	Low; some predation by ground-dwelling birds and predatory arthropods

Bouché (1977) and Lee (1959) based their three major ecological cate-gories on similar morpho-ecological characteristics. Discrepancies between the two classification schemes are due to differences in the earth-worm fauna being described. For example, among the New Zealand Megascolecidae, some endogeic species can attain very large size and tend to be larger than the anecic species; whereas anecic species are the largest earthworms among the European lumbricids described by Bouché (1977). However, most of the characteristics of the corresponding groups of the two classification systems are basically the same.

In addition to the three major ecological groups and intermediary species, a relatively small number of earthworm species inhabit special-ized habitats and are rarely encountered in the soil/litter environment. Lee (1985) recognized four such groups of earthworm species, including those that live under the bark of trees (Lee, 1981), in rotting logs, and at the base of epiphytes and leaf axils of understory forest trees in tropical forests (Lee, 1959). However, not all species inhabiting these specialized habitats are found exclusively in these habitats. Earthworm species that are commonly thought of as soil- or litter-dwelling species are sometimes encountered in these habitats. For example, of the four megascolecid earthworms Lee (1959) found in rotting logs, only one, *Megascolides suteri*, was considered to be primarily a log-inhabiting species. Common European lumbricids, such as *L. terrestris* and *L. rubellus*, are sometimes found in logs. Other earthworm species, such as those in the North American lumbricid genus, *Bimastos*, are usually associated with rotting logs, but are also found in the litter layer or on the soil surface as they migrate to new habitats (Reynolds *et al.*, 1974).

Lavelle (1979) has questioned the adequacy of Bouché's ecological cate-gories to describe relationships among earthworms living in the tropical savannas of the Ivory Coast. The vast majority of earthworms in these savannas are endogeic species or species that fall somewhere between the epigeic and endogeic category. Anecic species may account for up to 50–75% of lumbricid biomass in temperate regions but make up less than 10% of the total earthworm biomass of tropical savannas and dry forests (Lavelle, 1983). Lavelle (1979) proposed that endogeic earthworm species at Lamto could be divided into those that live in the upper soil horizons (epiendogeics) and those that live deeper in the soil profile (hypoendoge-ics). He also divided endogeic species into three subcategories which are based on the food material ingested by particular species: **polyhumic** endogeic earthworm species ingest soil with a high organic matter con-tent; **mesohumic** ones feed indiscriminately on both mineral and organic particles in the upper 10–15 cm of soil; and **oligohumic** species, which are found in tropical ecosystems, feed on soil of the deep horizons (30–40 cm deep) that are poor in organic matter (Lavelle, 1988).

Perel (1977) separated lumbricids into two major morpho-ecological groups: **humus formers** and **humus feeders**. As their names suggest, these groups are characterized, in part, by their feeding habits and by differences in their general morphological, behavioral and physiological traits. Humus formers feed on coarse particulate organic matter at various stages of decomposition, and are represented by litter-feeding species such as the soil-dwelling species *L. terrestris* and *L. rubellus* and the litter-dwelling *D. octaedra*. Humus feeders use more humified and fine particulate organic matter and consume more soil than humus formers. This group is represented by endogeic species in the genera *Aporrectodea*, *Octolasion* and *Diplocardia*.

These groups are further distinguished by traits related to their diet and feeding behavior. The humus formers have a maneuverable prostomium that enables them to manipulate plant debris. They have a relatively simple intestine with a small typhlosole that lacks secondary folds. They include anecic and epigeic species and tend to be strongly pigmented. Humus feeders have less flexible mouthparts. They have a more complex intestine with a large typhlosole that often has secondary folds. Humus feeders are endogeic and have little, if any, pigmentation.

All of these various approaches to classifying ecological groups of earthworms have validity but it is difficult, if not impossible, to develop categories into which all of the world's earthworm species will fall neatly. The generally good agreement between the two ecological classification systems of Bouché (1977) and Lee (1959) and the general applicability of these categories to earthworms from other parts of the world, such as the genus *Diplocardia* in the US, indicate that their three major ecological groupings have broad applicability. Most of the discussion on this subject has been limited to the European Lumbricidae and New Zealand Megascolecidae, with the further distinctions drawn by Lavelle (1979), for earthworms of tropical savannas. It is reasonable to expect that earthworm communities in each region of the world evolved in response to the unique vegetation, climate and geography of that region. The degree to which peregrine species of the European lumbricids and Asian megascolecids have succeeded in colonizing, and even coming to predominate in, soils around the world to which they are not endemic, suggests the wide adaptability of the strategies developed by these species.

6.1.2 SPECIES DIVERSITY

The number of species in a given earthworm community, which is the simplest measure of species diversity, ranges from 1 to 15 species. Most earthworm communities contain around 3–6 species, with a remarkable degree of consistency among different habitats and different geographic regions. The diversity of the earthworm community at a given locality is influenced by the characteristics of the soil, climate and organic resources of the locality, as well as its history of land use and soil disturbance.

Edwards and Lofty (1977) and Lee (1985) summarized data for the number of species in a wide variety of habitats from around the world. Lee reported that communities of the lumbricid earthworms in Europe, for which there are the greatest amount of data, are relatively rich in species in deciduous woodlands and permanent pastures or meadows, and relatively poor in species in coniferous forests, peatlands, heathlands and cultivated fields. The species-poor communities are characterized by extreme soil conditions, such as low pH or poor fertility, low-quality litter or a high degree of soil disturbance.

Lavelle (1983) made a factorial analysis of 42 earthworm communities from around the world. He described eight distinct communities, falling into a latitudinal-vegetational sequence, that correspond closely to the different biomes in which they are found:

1. coniferous forest;
2. heath;
3. cold grassland;
4. cold deciduous forest;
5. temperate deciduous forest;
6. temperate grassland and Mediterranean woodland;
7. moist savanna;
8. dry savanna.

These communities contain much different proportions of the major ecological groups of earthworms. Epigeic species account for nearly all of the biomass in cold coniferous forests of Sweden, less of the biomass of cold deciduous forest and a very small percentage of the earthworm biomass in heathlands. They contribute from 2 to 13% of the biomass of earthworms in the remaining temperate and tropical environments, being less abundant in habitats with lower quantities of litter production and lower amounts of rainfall. Anecic species are relatively more abundant in temperate than in tropical environments, comprising 64% and 52% of total earthworm biomass in cold and temperate deciduous forest, respectively, 72% of earthworm biomass in temperate grasslands and Mediterranean forest but less than 10% of earthworm biomass in tropical savanna and dry tropical forest.

Based on the results of his analysis, Lavelle (1983) proposed that there was a latitudinal sequence in the dominance of different ecological groups of earthworms. From the northern latitudes to the equator, the successively dominant groups were epigeic, anecic, meso- and oligohumic endogeic worms. Earthworm communities in tropical regions tend to have a greater degree of vertical stratification than the communities in temperate regions, because the higher temperatures of tropical regions allow oligohumic endogeic earthworm species to subsist on the poor-quality organic matter of deeper (30–40 cm) soil horizons.

More recently, Fragoso and Lavelle (1992) showed that some earthworm communities in tropical rainforests are dominated by epigeic or anecic earthworms, contrary to Lavelle's earlier theory that epigeic and anecic earthworms did not dominate earthworm communities at tropical latitudes. Lavelle's earlier ideas were based on data from tropical savannas and tropical dry forests and did not consider the earthworm fauna of tropical rainforests. This was probably because relatively little research has been done on earthworm communities in tropical rainforests. Research there has been limited by the general view that earthworms are not abundant in tropical rainforests, and therefore do not play an important role in the soil dynamics of these ecosystems.

However, Fragoso and Lavelle (1992) showed that average biomass and population density of earthworms in tropical rainforests does not differ much from that in temperate woodlands (Table 5.1). They analyzed data comprising 31 earthworm communities from 14 different rainforest localities, including sites in Central America, South America, Africa and Asia. Comparing 12 communities for which data were available, they reported that earthworms accounted for 51% of the total biomass of the soil macrofauna, followed by termites which accounted for 13% of total biomass. The earthworm communities of the tropical rainforests contained from 4 to 14 species, and had a mean species richness (6.5 ± 1.3 species) that was not significantly different from that of temperate forests (5.7 ± 2.0 species, calculated from data in Lee, 1985).

6.1.3 SPECIES ASSOCIATIONS

Certain earthworm species tend to be associated with one another. Usually, such associations result from some characteristic of the habitat, for instance, *L. terrestris, A. longa, A. caliginosa* and *O. cyaneum* are characteristic pasture species in England, although they are not the only species that occur in pasture. These species occur together commonly, associated with *L. rubellus, A. chlorotica* and other related species, in arable fields that are dominated in temperate regions by European lumbricids, such as those in central Europe and eastern North America (MacKay and Kladivko, 1985; Bauchhenss, 1991; Bohlen *et al.*, 1995a). Similarly, there are rarely more than four species in peaty soils, and these are usually small worms. Calluna heath and coniferous mor soils often contain only two species, *D. octaedra* and *B. eiseni*, which reside in the litter layer. In woodland mull soils *A. rosea, A. longa, A. caliginosa, A. chlorotica, O. cyaneum, L. castaneus, L. terrestris* and *L. rubellus* are commonly found together. *Dendrobaena rubida* was consistently associated with *Bimastos zeteki* in

Michigan woodlands (Murchie, 1960). There are many such casual associations of species.

In pastures in Scotland, Guild (1951b) showed that no more than 7–10 species occurred in any one habitat, and *A. caliginosa* and *A. longa* usually seemed to be the dominant species, with as many as 50% of the total populations of worms belonging to these species. There seemed to be little relationship between the age of a pasture and the number of species it contained. Nevertheless, the previous agricultural history of a grass or arable field is an important factor in determining which species are present. For instance, an old permanent pasture had many *A. nocturna* and rather fewer *A. caliginosa*, but ploughing and reseeding to grass, after 1 or 2 years in arable cultivation, favored increases in numbers of *A. rosea* (Evans and Guild, 1947c). *Allolobophora chlorotica* is often the dominant species in arable fields, and remains so for several years after arable land is reseeded with grass, eventually being replaced by more common pasture species.

Four associations of earthworm species were defined in a study in Sweden of 20 habitats (Nordström and Rundgren, 1973). These were:

1. *D. octaedra–D. rubida–L. terrestris*;
2. *A. caliginosa–A. longa–L. terrestris*; and
3. *L. castaneus–A. rosea–A. chlorotica*.

Abrahamsen (1972) studied associations in Norwegian coniferous forest soils and reported considerable overlap between species associations in different communities.

Phillipson *et al.* (1976) studied species associations in earthworm communities of an English beechwood. They used four indices of similarity to investigate the relationships among the 10 earthworm species encountered in the beechwood. These indices were: Sørensen's quotient of similarity, Cole's coefficient of similarity, Fager's recurrent grouping and Mountford's ordination, all of which are described in Southwood (1966). The index values given by Sørensen's quotient and Cole's coefficient distinguished the same two significant species associations: (1) *A. rosea–L. terrestris–L. castaneus–D. mammalis*, and (2) *A. rosea–A. caliginosa–L. terrestris*. Analysis by Fager's recurrent grouping also gave rise to two species associations which were similar to the associations described above. Phillipson *et al.* (1976) pointed out that the co-occurrence of species does not necessarily mean that those species have similar ecological requirements. Similarity of requirements can be inferred by ordination methods. Using the ordination method of Mountford (1962), Phillipson *et al.* (1976) noted the close similarity of requirements for *D. rubida* and *L. terrestris*, *D. mammalis* and *L. castaneus*, and *A. rosea* and *A.caliginosa*.

Baker (1983) used Fager's recurrent grouping and Mountford's index of similarity to demonstrate the degree of association and ecological similarity among 15 species of earthworms in a reclaimed peat soil in Ireland. Fager's method indicated that there was one recurrent group, that of *A. chlorotica–A. tuberculata–E. rosea*. Mountford's indices showed a significant association between *A. chlorotica*, *A. turgida*, and *A. chlorotica* and *A. tuberculata*, and *A. longa* and *D. mammalis*. Such associations between two or more species are likely to be commonly encountered in earthworm communities. Some of these associations are casual, whereas others are dependent upon similarity between species in ecological requirements.

Some associations of species are almost a form of commensalism and independent of the habitat. For instance, Lukose (1960) reported an association between the giant earthworm, *Drawida grandis* (Moniligastridae) and worms of a megascolecid species (11–32 mm long), which were found crawling over individuals of the larger species. When separated, the smaller worms always returned to the host worm. Baylis (1914) reported a similar association between an enchytraeid worm *Aspidodrilus* and a large species of earthworm, and Cernosvitov (1928) between an enchytraeid worm *Fridericia parasitica* and the earthworm *Allolobophora robusta*. A small lumbricid, *D. mammalis*, sometimes lives commensally in the burrows of *L. terrestris* and *A. longa* (Saussey, 1957).

6.2 DISPERSAL

Although earthworms are confined to the soil habitat, they can be readily dispersed in a wide variety of ways.

6.2.1 ACTIVE AND PASSIVE DISPERSAL

Earthworms can move from one location to another either by active or passive dispersal. Active dispersal of earthworms across the soil surface, which proceeds relatively slowly, can occur for several reasons. Earthworms may migrate to evade unfavorable environmental conditions or to seek new habitats, but the behavioral cues leading to such dispersal are poorly understood. Earthworms are often seen in huge numbers moving across the soil surface, sidewalks or pavements, after heavy rains, a phenomenon that was noted by Darwin (1881). These migrations may be in response to flooded soil conditions and the low oxygen tensions in flooded soils, but the number of earthworms migrating is often particularly large at certain times of year, suggesting that other cues may be important in influencing the mass emergence and migration of these populations. Schwert (1980) pointed out that, for most lumbricid earthworm species, mass emergence of populations occurs during cool, moist weather, and does not necessarily depend on heavy rainfall. Although many individuals die during these occasions of mass emergence, others survive, and their dispersal to new habitats may decrease competition in areas of high earthworm density and enhance genetic exchange among different localized populations (Schwert, 1980). The distances traversed by individual worms have rarely been quantified, although individuals of *L. terrestris* have been reported to migrate up to 20 m during a single night (Mather and Christensen, 1988), although they often return to the same burrow.

Gates (1972a) reported mass migrations of huge numbers of *Perionyx* sp. in the Chin Hills in Burma. These migrations took place in the early morning at the beginning of the cold season, when earthworms emerged in large numbers and then crawled or tumbled down the hillsides, disappearing into the soil by nightfall. Gates (1972a) observed similar mass migrations of earthworms uphill in the early part of the rainy season in India, and suggested that these mass movements, as well as those he observed in Burma, were due to the earthworms seeking a more favorable environment. Reddy (1980) observed a mass downhill migration of *Amynthas alexandri* in uncultivated grassland, in early morning at the beginning of the cold season, which was the same time and season that Gates (1972a) reported for the mass migrations of *Perionyx* sp. in Burma. Similar sorts of migrations between different habitats at the onset of the rainy season have been noted among certain earthworm species in tropical rainforests. Mather and Christensen (1992) concluded from field studies of 857 earthworms from 10 species that surface migration was an important component of dispersal for both juvenile and mature earthworms.

Passive transport of earthworms occurs as a result of incidental anthropochorous transport by humans, by being carried in streams or surface flow of water during heavy rains, or through cocoons being carried on the feet of birds and other animals. Incidental transport by humans may have an important influence on the local dispersal of earthworm populations. Marinissen and van den Bosch (1992) used a simple mathematical model to describe the dispersal of earthworms in newly colonized agricultural fields on Dutch polders. They found that passive dispersal by tractor tires can have a major influence on the rate of population expansion into surrounding agricultural fields.

Another potentially important form of passive dispersal is by streams and surface water that flows during heavy rainfall. During the analysis of a small stream in southern Ontario, Canada, Schwert and Dance (1979) recovered over 300 lumbricid cocoons in drift samples. Ninety-two per cent of these cocoons were apparently viable, suggesting that large numbers of viable cocoons are transported in this way to new habitats downstream. Not only cocoons, but also immature and adult earthworms can be transported by streams or surface runoff. Atlavinyte and Payarskaite (1962) reported that large numbers of earthworms were consistently washed away with eroded soil following heavy rains in Russia, with a greater number being washed away from fallow and arable land than from natural grassland.

The simple model developed by Marinissen and van den Bosch (1992) to study dispersal in newly colonized habitats showed that passive transport can have a major influence on the rate of population expansion. There is little evidence that other animals, such as birds or livestock, are important in the passive dispersal of earthworms, although the importance of this form of dispersal has been suggested frequently. Several authors have suggested that birds may have a role in dispersing earthworms to new habitats, especially by dropping captured worms from their beaks as they return to their nest to feed their young. Earthworms apparently have been introduced to Dutch polders and New Zealand islands in this way (Lee, 1959; Meijer, 1972).

6.2.2 INTRODUCTION AND SPREAD OF EARTHWORM POPULATIONS

The primary modes of introduction of earthworm species into new environments have usually been the incidental transport by humans in the soil of potted plants and in the ballast of ships. Such incidental transport is responsible for the bulk of the accidental introductions of alien earthworm species to new regions, and has contributed greatly to the world-wide spread of peregrine earthworm species. Smith (1893) concluded that lumbricid earthworms were introduced to New Zealand mainly from England in plants imported for nurseries, gardens and landscaping, or in the ballast of ships that was dumped overboard on to shore. Gates (1976) examined earthworms intercepted by the US Bureau of Quarantine over a period of 25 years. Among the earthworms that had been recovered were 24 species of Lumbricidae and 26 species from other families, the majority of which were Megascolecidae belonging to the genus *Pheretima*. Most of these earthworms were found in the soil of potted plants. Twenty-one of the species found were already known from the United States, but most of these are believed to have been introduced to North America from Europe.

Earthworms have often been introduced purposely to new environments to improve soil fertility and plant production and ameliorate soil (Chapter 12). In several instances, the rate of spread of the introduced populations into the new environment and development of the earthworm populations from an inoculum point has been determined. In The Netherlands, van Rhee (1969a,b) introduced *A. caliginosa* and *A. chlorotica* into polder soils recently reclaimed from the sea. He introduced the earthworms during 1964–67 and followed their rate of spread and population growth. Earthworms that were introduced at eight equidistant points in a 60 m × 50 m field had spread throughout the field within 3 years. The calculated rate of spread for *A. caliginosa* was 6 m/yr and that for *A. chlorotica* was about 4 m/yr. Estimates for the rates of spread of various earthworm species in grassland and arable fields are summarized in Table 6.2. Stockdill (1982) introduced European lumbricid earthworms, primarily *A.*

Table 6.2 The rate of spread (in m/yr) of populations of various species of earthworms introduced into different habitats (adapted from Marinissen and van den Bosch, 1992)

Earthworm species	Land use	Rate of spread (m/yr)	Reference
A. caliginosa	Grass strips in orchards	6	van Rhee (1969a,b)
	Grazed polder grassland	9	Hoogerkamp, Rogaar and Eijsackers (1983)
	Grazed cut-over peat	2.5–10	Curry and Boyle (1987)
	Pasture	8	Hamblyn and Dingwall (1945)
	Arable polder soil	7	Marinissen and van den Bosch (1992)
A. chlorotica	Grass strips in orchards	4	van Rhee (1969a,b)
L. rubellus	Grazed cut-over peat	>10	Curry and Boyle (1987)
	Arable polder soil	14	Marinissen and van den Bosch (1992)
L. terrestris	Grazed polder grassland	4	Hoogerkamp, Rogaar and Eijsackers (1983)

caliginosa, into pastures in New Zealand and monitored their rate of spread over a long period of time. The earthworms were introduced at 10 m intervals in blocks of turf that had been cut from pastures with established earthworm populations. The initial rate of spread (during the first 4–5 years) of the earthworms was slow but increased rapidly thereafter, reaching a rate of about 10 m/yr, and they had colonized the whole field in 8–10 years (Table 6.2). Marinissen and van den Bosch (1992) developed a simple model to describe the rate of colonization of new habitats by earthworms, and concluded that optimizing growth conditions for the earthworms was the best way to promote population expansion.

6.3 PREDATORS, PARASITES AND PATHOGENS

Earthworms can be the target of a wide range of predators and are also attacked by diverse kinds of parasites and pathogens.

6.3.1 PREDATORS

Earthworms are an important component of the diet of many vertebrate predators (reviewed by MacDonald, 1983). They are preyed upon by very many species of birds, such as blackbirds (*Turdus merula*), starlings (*Sturnus vulgaris*), thrushes (*Turdus ericetorum*), gulls (*Larus canus*), crows (*Corvus corone*), robins (*Erithacus rubecula* in England, *Turdus migratorius* in North America) and storks (*Ciconia ciconia*). Earthworms were a major component of the diet of white storks (*Ciconia ciconia*) in Spain (Alonso *et al.*, 1991). Bengston *et al.* (1975) reported that caged golden plovers (*Pluvialis apricaria*) could consume 4.5 lumbricids/m²/day in an Icelandic hayfield. Tucker (1992) considered that the use of cultivated fields by invertebrate-feeding birds was greatest in those fields that received regular applications of farmyard manure. Earthworm populations were correlated positively with the input of manure to these fields. Woodcock (*Scolopax* spp.) feed preferentially on earthworms, and Granval and Muys (1992) proposed the concept of controlled biostimulation of earthworm communities to create more favorable conditions for woodcock in western Europe. Similarly, Reynolds (1977) demonstrated that the habitat preference of the American woodcock (*Philohela minor*) was correlated positively with the biomass of earthworm populations in the habitat and that woodcock may eat their own body weight of earthworms per day. He suggested that secondary woodland could be used to maximize woodcock production by managing for tree species that produced litter that was palatable to earthworms. Berg (1993) reported that earthworms were the most important component of the diet of curlews (*Numenius arquata*) in the pre-breeding period in Norway.

Some species of birds that feed on earthworms are endangered, and the preservation of appropriate habitats with abundant earthworm populations may be the key to preserving these species. For example, Lofaldli *et al.* (1992) showed that earthworms constitute more than 90% of the food of the endangered great snipe (*Gallinago media*) in central Norway. Earthworms also comprise an important part of the diet of sandhill cranes (*Grus canadensis*) in the Great Plains of North America. Davis and Vohs (1993) argued that management of native grasslands for this threatened species needs to focus on creating favorable conditions for abundant earthworm populations in the spring.

Mammals such as hedgehogs (*Erinaceus europeus*), badgers (*Meles meles*), shrews (*Sorex* spp., *Blarina brevicauda*) and especially moles (*Talpa europaea, Parascalaps breweri, Condylura cristata*) eat large numbers of earthworms. Hofer (1988) showed that earthworms contributed the bulk of the diet of badgers in Wytham Woods, Oxfordshire in England. The proportion of earthworm-rich habitat in the group territories of the badgers was a good predictor of the total group consumption of earthworms. Kruuk and Parish (1985) considered that earthworms were a major part of the diet of badgers. In contrast to the situation in England, earthworms were not found to be an important part of the diet of European badgers in a dry Mediterranean coastal habitat in central Italy (Pigozzi, 1991). Moles tunnel in search of earthworms, catch them and sometimes store them in caches until required as food; they incapacitate the worms, by biting 3–5 segments from their anterior ends (Evans, 1948a). Skoczen (1970) found a mole's cache containing 470 worms weighing a total of 820 g.

Red foxes (*Vulpes vulpes*) also feed on earthworms, which may comprise the bulk of their diet in some areas during certain seasons. MacDonald (1980) studied the foraging behavior of foxes hunting for earthworms in grasslands near Oxford, England. He determined that foxes caught as many as 10 earthworms per minute, with an average of 2–5 earthworms per minute. The number of hours of more than 90% relative humidity prior to foraging was the most important predictor of capture rate (MacDonald, 1983). Pigs were reported to eat many earthworms in New Guinea (Rose and Wood, 1980), and the long-beaked echidna (*Zaglossus bruijnii*) feeds almost exclusively on earthworms (Griffiths, 1978).

It was reported that shrews used earthworms as the most important prey in The Netherlands (Denneman, 1991), and many other vertebrates also feed on earthworms. For instance, Ljungström and Reinecke (1969) reported that in South Africa, giant microchaetid worms are attacked by night adders (*Causus rhombeatus*). Many other species of snakes readily eat earthworms, including the garter snakes, *Thamnophis sirtalis* (Gregory and Nelson, 1991), and the Australian blindsnake, *Ramphotyphlops subocularis*, which apparently feeds exclusively on earthworms (Webb and Shine, 1993). Butler's garter snake, *Thamnophis butleri*, was shown by Catling and

Freedman (1980) to feed almost exclusively on earthworms in Ontario, Canada. Earthworms apparently are a minor component of the diet of many species of frogs and toads, but certain species of frog are specialized earthworm eaters. Blum and Menzies (1988) examined the gut contents of 13 different species of frogs from New Guinea, belonging to the genera *Xenobatrachus* and *Xenorhina*, and found that their diet consisted almost exclusively of earthworms. Certain newts (*Trituris* spp.) (Smith, 1951) and salamanders (*Salamandra salamandra*) (Bas Lopez *et al.*, 1979) also feed extensively on earthworms.

The effect of vertebrate predators on populations and biomass of earthworms has rarely been quantified. Bengston *et al.* (1976) reported that golden plovers (*Pluvialis apricaria*) captured around 4.5 earthworms/m^2/day in a hayfield in Iceland. The birds were excluded from field plots by nets, and after 22 days the earthworm population was 238 per m^2 in the protected plots and 107 per m^2 in adjacent, unprotected plots. Thus, the plovers appear to have reduced earthworm populations by nearly 50% in 22 days. Others have reported similar rates of predation of earthworms by birds (Moeed, 1976). Cuendet (1977) estimated that black-headed gulls (*Larus ridibundus*) consumed nearly 30–100 kg/ha of earthworms after autumn plowing in agricultural fields in Switzerland, but this represented only 1–5% of the total annual production of earthworm biomass in these fields. In an experimental test of the effects of predation on earthworm populations, Judas (1989) fenced out predators (birds, shrews, other rodents and carabid beetles) from an experimental plot for a 7.5 month period from October to May. He reported that there were no significant differences between the fenced off plot and a control plot in the abundance and distribution of earthworms. The available evidence suggests that predation does not always have a major influence on the structure of earthworm communities or the size of earthworm populations, but may have an important influence on earthworm populations in some situations or at certain times of year. Judas (1989) calculated that predation by birds, shrews, other rodents and carabid beetles increased earthworm populations markedly, from 31% to 64% per annum in the beech forest.

A variety of invertebrates also prey on earthworms, although the importance of earthworms in the diet of most soil-dwelling predatory invertebrates has not been assessed. Many species of carabid and staphylinid beetles and their larvae attack earthworms. For instance, the staphylinid, *Quedius* (*Microsaurus*) *mesomelinus* fed on immature and mature individuals of *Eisenia fetida* under experimental conditions (McLeod, 1954). Centipedes also attack earthworms, and three species of carnivorous slugs, *Testacella scutulum*, *T. haliotidea* and *T. maugei* prey chiefly on earthworms, each individual eating about one worm per week. Larvae of the anthomyiid fly, *Coenosia tigrina*, attack earthworms; earthworm mucus stimulates egg laying in the adult females of this species

(Morris and Pivnick, 1991). The diet of driver ants (*Dorylus gersteackeri* and *D. nigricans*) was shown to consist almost entirely of earthworms in cocoa farms in the high forests of Ghana (Gotwald, 1974). A few species of leeches, including the British species *Trocheta subviridis*, also feed on earthworms. Several planarians (flatworms), including one species in England (*Bipalium kewensis*) and two in the USA (*Bipalium adventitum* and *Bipalium pennsylvanicum*) feed preferentially on earthworms (Reynolds, 1973a; Ogren and Sheldon, 1991). The terrestrial planarian, *Artioposthia triangulata*, feeds mainly on earthworms and may contribute significantly to earthworm mortality in some regions. This planarian originates from New Zealand and can grow to a length of 15 cm and a weight in excess of 2 g. Blackshaw (1990) contended that these planarians were responsible for the total elimination of earthworm populations from a grass field near Belfast, Ireland. In controlled feeding studies, Blackshaw (1991) showed that planarians were capable of eating 1.4 earthworms per week. At this feeding rate, a planarian population of 6.5 per m^2 could nearly eliminate an earthworm population of 450 per m^2. These planarians were first reported from Ireland in 1961 and are now found throughout Northern Ireland. They have the potential to have an extremely important influence on the population dynamics of earthworms in this area, although they do not seem to have a major impact on earthworm populations in their native New Zealand. Boag *et al.* (1993) examined data on the known distribution of *A. triangulata* in Northern Ireland, Scotland and New Zealand and suggested that this species has the potential to become established in the north of France, Belgium, The Netherlands, Germany and west of Norway, with possibly disastrous results for indigenous earthworm populations.

Lavelle (1983) identified a large carnivorous earthworm (*Agastrodrilus dominicae*) from central Ivory Coast that is apparently adapted to feed on smaller earthworms. He placed two individuals of *A. dominicae* in a laboratory culture with 36 small eudrilid earthworms (*Chuniodrilus zielae* and *Stuhlmannia porifera*) and after 24 days only six of the smaller earthworms remained. Some of the smaller earthworms were missing parts of their tails and the large earthworms increased their weight during the brief incubation. In a control culture, only two of 18 smaller worms were lost during 24 days. Lavelle (1983) discusses several adaptations of *Agastrodrilus* that enable it to feed on smaller worms, such as their ability to move rapidly through soil without ingesting it and modification of their digestive tract (e.g. reduced gizzard and no typhlosole).

6.3.2 PARASITES AND PATHOGENS

Earthworms have many internal parasites, including Protozoa, Platyhelminthes, Rotatoria, nematodes and dipterous larvae. Bacteria that have been reported in earthworms include *Spirochaeta* and *Bacillus botulinus*, but little is known of their effects. No viruses have yet been recorded from earthworms. A fungal pathogen, *Exophiala jeanselmei*, has been isolated from naturally infected cocoons of *O. tyrtaeum* and from the cocoon albumen of *E. fetida* (Vakili, 1993). The most common, and probably the most important, protozoan parasites of earthworms belong to the Gregarina. These protozoa have been found in many different parts of the bodies of earthworms, including the alimentary canal, coelom, blood system, testes, spermathecae, seminal vesicles and even in the cocoons. The records of species found in earthworms are too extensive to be included in full, but the genera of gregarines that have been found in earthworms include: *Distichopus, Monocystis, Rhyncocystis, Nematocystis, Echinocystis, Aikinetocystis, Grayallia, Nellocystis, Craterocystis* and *Pleurocystis* (Stephenson, 1930; Stolte, 1962).

Little is known about the dynamics of the gregarine parasites or their effects on earthworm survival, growth or fecundity. Pižl and Sterzynska (1991) reported that rates of infection of earthworms by monocystid gregarines was significantly greater in field plots adjacent to roads than in plots in centers of parks. They found a significant correlation between the degree of earthworm infection by gregarines and heavy metal concentrations in soil, indicating that pollution stress may increase the susceptibility of earthworms to infections by these pathogens. Pižl (1985) noted that earthworms in an orchard treated intensively with herbicides were extensively infected with gregarines. He exposed earthworms to a herbicide (Zeazin 50) under controlled laboratory conditions and reported that earthworms exposed to the herbicide had significantly higher infection rates than controls that were not exposed to the herbicide. The mechanism for the greater infection rate of earthworms exposed to the herbicide is not known, but Pižl suggested that the natural defense of earthworms against the monocystid parasites was weakened by the herbicide treatment.

A number of ciliate protozoa also infest the bodies of earthworms, although few cause any serious harm to the worms. The genera recorded include *Anoplophrya, Maupasella, Parabursaria, Hoplitophrya, Plagiotoma* and *Metaradiophyra* (Stolte, 1962). Other protozoa that have been found in the bodies of terrestrial earthworms include *Myxocystis, Sphaeractinomyzon* and *Thelohania*.

There are several instances of platyhelminth worms being found in the bodies of earthworms. The cysticercoid stage of *Taenia cuneata* has been found in *E. fetida* and in the wall of the alimentary canal of a *Pheretima* sp. (Stephenson, 1930). The larvae of *Polycercus* have been found in the tissues of *L. terrestris* (Haswell and Hill, 1894).

Many nematodes occur in the tissues of earthworms; few seem to cause serious damage, and often the worm is merely acting as an intermediate host for them (Chapter 8, section 4). Nematode genera that have been found in worms include: *Rhabditis*, *Heterakis*, *Syngamus*, *Dicelis*, *Dionyx*, *Stephanurus*, *Megastrongylus* (Stolte, 1962), *Spiroptera* and *Synoecnema* (Stephenson, 1930). Some nematodes are carried passively by earthworms. Others develop within the earthworm, and some are true parasites, earthworms being the sole host (Poinar, 1978).

An anoetid mite, *Histiostoma murchiei* Hughes and Jackson, has been reported to parasitize the cocoons of earthworms. In one habitat in Michigan, up to 40% of cocoons of *Allolobophora chlorotica* were parasitized (Oliver, 1962). The mites feed on and destroy the developing earthworm cocoon. Formerly known only from North America, *H. murchiei* has since been reported in a nature reserve in Denmark, where it was found to parasitize about 20% of the cocoons of *A. caliginosa* and 7% of the cocoons of *L. terrestris* (Gjelstrup and Hendriksen, 1991). Another species of mite, *Uropoda agitans*, also attacks earthworm cocoons (Stone and Ogles, 1953).

The larvae of muscoid flies parasitize earthworms. For example, the cluster fly, *Pollenia rudis*, is a major parasite of earthworms in the US. A single worm may harbor 1–4 larvae but only one larva becomes fully grown, and as it grows it progressively destroys the worm and eventually kills it. This fly was introduced into the US from Europe many years ago, and its numbers have so increased that the adult flies are a considerable nuisance to human beings in many areas (Cockerell, 1924). In Europe, the *Pollenia viatica* species group includes six species, all of which have one generation per year, with some species surviving the winter as first-instar larvae within earthworms (Rognes, 1991). Other flies that parasitize earthworms are *Onesia subalpina* (Takano and Nakamura, 1968), *Onesia sepulchralis*, *Sarcophaga haemorrhoidalis* (Keilin, 1915), *Sarcophaga striata* and *Sarcophaga carnaria* (Eberhardt, 1954).

6.4 ECOLOGICAL ENERGETICS

6.4.1 PHYSIOLOGICAL ENERGETICS

The ecological energetics of earthworms is concerned with the consumption of energy by earthworms, the allocation of this energy to growth, reproduction and oxidative metabolism, and the loss of energy through egestion, excretion and death. Earthworms obtain their energy by consuming reduced carbon compounds in the form of dead plant or animal tissues, animal feces or soil organic matter. The simplest way of expressing the energy (or carbon) metabolism of earthworms is in a simple equation that applies to all heterotrophic organisms:

Consumption (C) = Production (P) + Respiration (R) + Egestion (E)

where C is the total energy ingested, P the amount of ingested energy that is allocated to growth, storage or reproduction, R is the energy assimilated but lost in respiration and E is the energy egested in feces. Assimilation is defined as the sum of P plus R.

Few energy budgets have been made for earthworms. Among the most detailed budgets are those of Bolton and Phillipson (1976), for the different life stages of *A. rosea* from an English beechwood, and of Lavelle (1974), for a mixed community of megascolecid and eudrilid worms from a tropical savanna in Lamto, Ivory Coast (Table 6.3).

The data for *A. rosea* were determined at 10 °C, and those for the Lamto earthworms were determined at 26 °C, temperatures at which the earthworms are likely to be active in their respective temperate and tropical environments.

The data from Bolton and Phillipson (1976) show that the energy demand of small immature earthworms is greater than that of large immature and adult earthworms. Small immatures consume much more energy and devote a greater portion of consumed energy to production of new tissue than do larger earthworms. Small immature individuals allocate nearly as much energy to respiration as they do to growth ($R/P = 1$), whereas mature individuals allocate a larger portion of their assimilated energy to maintenance and respiration than to production of new tissues ($R/P = 3.8$). The assimilation efficiencies are greater in the smaller life stages than in the more mature life stages. The earthworms of the Lampto savanna allocate a much higher proportion of their assimilated energy to respiration ($R/P = 13$) than does *A. rosea*, which may be due to the fact that the savanna worms live at higher temperatures and must expend larger amounts of energy to burrow through large amounts of soil. Other data for tropical earthworms support the idea that they allocate a larger portion of assimilated energy to respiration than do earthworms in temperate regions. Mishra and Dash (1984) reported that an earthworm population in a subtropical dry woodland in India allocated about 27% of its assimilated energy to growth and 73% of its assimilated energy to respiration ($R/P = 2.7$), and Dash and Patra (1977) reported that earthworms in a tropical grassland devoted 15% of assimilated energy to growth and 85% to respiration ($R/P = 5.7$).

The total amount of soil consumed by the Lamto earthworms contained only about one-third to one-quarter as much energy as did the soil consumed by *A. rosea*, although the Lamto earthworms consumed nearly six times as much total soil as did *A. rosea* (Table 8). This means that the Lamto earthworms had to extract the energy needed for growth and respiration from a comparatively energy-poor substrate. The Lamto earthworms have apparently adapted to the energy-poor soil of the savanna by developing high assimilation efficiencies; the assimilation efficiencies of the Lamto earthworms are about eight times greater than those of *A. rosea*, an earthworm that is adapted to live in relatively energy-rich soil (i.e. soil with large amounts of organic matter).

Table 6.3 Daily energy budgets (J/g/day) for three life stages of *Allolobophora rosea* (at 10°C) from an English beechwood (from Bolton and Phillipson, 1976) and a mixed community of Megascolecidae and Eudrilidae (at 26°C) from a savanna in Lamto, Ivory Coast (from Lavelle, 1974)

Earthworm	Consumption (C)	Production (P)	Respiration (R)	Egestion (E)	Assimilation (R+P)	R/P	Assimilation efficiency (A/C%)
A. rosea							
Small immatures	5730	35	37	5660	72	1.0	1.3
Large immatures	4910	19	34	4860	53	1.6	1.1
Adults	4260	8	30	4220	38	3.8	0.9
Lamto savanna worms	1570	10	130	1430	140	13.0	8.9

Much higher assimilation efficiencies than those reported for geophagous earthworms have been reported for litter-feeding earthworms, suggesting that the litter feeders are more efficient than the geophages at assimilating ingested energy sources. Daniel (1991) estimated that juvenile *L. terrestris* assimilated 55% and 43% of ingested dandelion leaves when the amounts of leaves made available were at rates of 0.25 and 1.0 g/g earthworm/week, respectively. Using data from Dickschen and Topp (1987), Daniel (1991) estimated that *L. rubellus* assimilated from 55% to 61% of ingested alder leaves. Although the estimates of assimilation efficiencies from the various experiments cited above are not entirely comparable, it seems clear that the assimilation efficiencies of litter-feeding earthworms are much higher than those of geophagous earthworms. Litter-feeding earthworms also allocate a higher proportion of assimilated energy to respiration than do endogeic species (Phillipson and Bolton, 1976; Bouché, 1977).

These differences in assimilation efficiencies between litter-feeding and geophagous earthworms relate mainly to differences in the feeding habits and digestive capabilities of the two different groups of earthworms. Irrespective of their different types of feeding ecology, earthworms assimilate mostly easily decomposable organic matter, with a mean turnover time in soil of a few years (Martin *et al.*, 1992). Older, more humified organic matter is not readily assimilated by earthworms. This may explain why geophagous earthworms have a lower assimilation efficiency than do litter-feeding earthworms, because geophages consume a larger proportion of energy in the form of resistant organic compounds that are not easily assimilated.

Under field conditions, the assimilation efficiencies of earthworms probably vary greatly depending on environmental conditions and on the amounts and types of food that are available. For example, Daniel (1991) demonstrated that the assimilation efficiency of immature *L. terrestris* fed on dandelion leaves was greater when less food was available and when the earthworms were exposed to higher temperatures and low soil water potential. Furthermore, Cortez and Hameed (1988) reported that the assimilation efficiency of *L. terrestris* fed with leaves of ryegrass (*Lolium perenne* L.) was greater when they were fed leaves that were partially decomposed than when they were fed fresh leaves.

6.4.2 ECOSYSTEM ENERGETICS

Energy flow through earthworms rarely accounts for more than about 5–6% of the total energy flow through the decomposer subsystem in temperate ecosystems (Lee, 1985). Coleman and Sasson (1978) estimated that of the 90% of the total ecosystem energy flow that enters the decomposer subsystem in temperate grasslands, the portion that flows through earthworms accounts for only about 3% in dry grasslands and about 6% in moist pastures or meadows. Persson and Lohm (1977) estimated that earthworms consumed about 3% of the total net primary production in a Swedish grassland. Similarly, Nowak (1975) concluded that a population of *A. caliginosa* consumed between 0.4 and 2.9% of the total available energy in a pasture in Poland.

The contribution of earthworms to total energy flow is greater in tropical ecosystems than in temperate ecosystems. Lavelle (1974) showed that earthworms assimilate around 10–15% of the gross annual primary production in the tropical savannas of Lamto, Ivory Coast, which means that they must ingest around 12% more than the total gross primary production annually. Dash *et al.* (1974) and Dash and Patra (1977) showed that the oligochaetes (earthworms plus Enchytraeidae) in a pasture in India assimilated from 14% to 16% of the total annual primary production of the ecosystem. However, most estimates of the total energy flow through earthworms are based on incomplete accounting. Bouché (1982) pointed out the inadequacy of basing energy budgets solely on estimates of consumption, respiration and egestion. Other important flows of energy that are not included in the basic equation shown at the beginning of section 6.4.1 include:

1. excretion of urea and other urinary products;
2. excreted $CaCO_3$;
3. mucus production; and
4. cocoon production.

Of these, secretion of mucus is likely to represent the greatest energy flow. Of the total energy utilized by earthworms ($4820 \ kJ/m^2/yr$) in pastures in Berhampur, India, nearly 56% was allocated to the production of mucus (Dash and Patra, 1977). Scheu (1991) estimated that mucus production by *O. lacteum* accounted for up to 63% of the total loss of carbon (including mucus excretion and respiration) from the earthworms. These results indicate that energy budgets based solely on energy losses resulting from respiration have underestimated the loss of assimilated energy by at least 50%.

The production of cocoons may be another important factor in energy flow, although it has rarely been estimated for earthworm communities. Lavelle (1974) calculated that 0.3% of the energy assimilated by earthworms in the Lamto savannas was devoted to reproduction. This amount is negligible compared to the overall energy budget of the earthworms.

The influence of environmental factors on earthworms

7

Earthworms are thin-skinned invertebrates with little protection against the changes in moisture and temperature that occur in soils or the physical and chemical characteristics – such as soil type, pH, porosity and organic matter content – of the soils in which they live.

7.1 MOISTURE

Water constitutes 75–90% of the body weight of earthworms (Grant, 1955a), so prevention of water loss is a major factor in earthworm survival. Nevertheless, they have considerable ability to survive adverse moisture conditions, either by moving to an area with more moisture or by estivating. If they cannot avoid dry soil they can still survive the loss of a large part of the total water content of their bodies. For instance, *L. terrestris* can lose 70% and *A. chlorotica* 75% of their total body water and still survive (Roots, 1956). Grant (1955a) considered that most lumbricid worms could sustain a water loss of at least 50%. Earthworms apparently lack a mechanism to maintain a constant internal water content, so that their water content is influenced greatly by the water potential of the soil (Kretzschmar and Bruchou, 1991).

Earthworm activity also depends upon adequate availability of soil moisture, but not all species have the same moisture requirements and, within a species, the moisture requirements for earthworm populations from different regions of the world can be quite different. For example, in Europe, *A. caliginosa* goes into diapause at a soil moisture content below 25–30% and does not survive well below 20% soil moisture (Baltzer, 1956; Zicsi, 1958b), whereas in seasonally arid regions of Argentina, *A. caliginosa* and *A. rosea* are active in soils with a moisture content as low as 15% (Ljungström and Emiliani, 1971; Ljungström *et al.*, 1973). Buckerfield (1992) reported that populations of *A. trapezoides* and *A. rosea* were still active at soil moisture contents as low as 10%, in cereal fields of the semi-arid region of southern Australia. Such results indicate that certain peregrine species are capable of adapting to a wide range of moisture conditions.

The moisture preferences of only a few earthworm species have been determined experimentally. Madge (1969) placed earthworms of the species *H. africanus* in moisture gradients, and reported that they preferred soil between 12.5 and 17.2% moisture content. Soil with a moisture content of about 23.3% appeared to be optimal for them to produce casts. *Allolobophora caliginosa* and *Metaphire californica* seemed to prefer soil moisture contents of 20–45% and 35–55%, respectively, in a clay soil (El-Duweini and Ghabbour, 1968). The preference of *M. californica* for wetter soils may be due to the fact that it is not as tolerant of desiccation as *A. caliginosa* (Ghabbour, 1975) and the same may be true for *L. rubellus*. Earthworms that live in compost or dung heaps tend to prefer moister conditions than most species of soil-dwelling earthworms. For instance, juvenile and adult *P. excavatus* preferred a moisture content of around 80% in cattle manure at 25 °C (Hallett *et al.*, 1992) and 80–85% in other organic materials (Edwards, 1988).

Prolonged droughts decrease numbers of earthworms markedly, and it may take as long as 2 years for populations to recover once conditions become favorable again. One important factor seems to be that the fecundity of earthworms such as *A. chlorotica* is influenced greatly by moisture (Table 7.1). Gerard (1960) showed that some species can withstand dry conditions much better than others; for instance, *L. terrestris* survived as

Table 7.1 The effect of soil moisture on cocoon production by *A. chlorotica* (from Evans and Guild, 1948a)

Site	Bones Close						
Moisture content of soil (%)	11	13.5	21	28	35.5	42.5	SE of difference
Mean no. of cocoons produced by five worms	0	0	8.6	13.6	8.8	6.6	0.94

Site	Westfield					
Moisture content of soil (%)		1.6	24.5	33	42	50	SE of difference
Mean no. of cocoons produced by five worms		0	0.6	8.4	9.4	3.0	0.93

well in nonirrigated plots as in irrigated ones, whereas *A. chlorotica*, *A. caliginosa* and *A. rosea* did not survive in nonirrigated plots.

Different earthworm species have adopted different strategies for coping with dry soil conditions. Earthworms that migrate to deeper soil when the surface soil is too dry include *L. terrestris*, *A. longa*, *E. fetida* and *P. hupeiensis*. Despite not migrating to deeper soil, *A. caliginosa* is usually considered able to survive very dry conditions, in spite of Gerard's conclusions (1960), but *Octolasion lacteum* is not (Michon, 1949; Grant, 1955a). Lack of moisture can cause some earthworm species to become quiescent or go into diapause. For instance, *Aporrectodea* spp. are active in the upper 10 cm of soil when the soil is moist, but when the soil is dry they are usually found below 20 cm, where they estivate, tightly coiled within spherical, mucus-lined cells (Lee, 1985; Baker *et al.*, 1992a). When individuals of *A. caliginosa* were kept in soil that was dried slowly in the laboratory, they went into diapause, but when kept in moist soil, they remained active for 18 months (Gerard, 1960). Kretzschmar and Bruchou (1991) subjected *A. longa* to constant levels of soil water suction and reported that a water suction level below –35.5 kPa caused water loss from the earthworms, and levels below –620 kPa led to diapause.

Parmelee and Crossley (1988) and Edwards *et al.* (1995) suggested that cocoons may act as the main survival stage during drought for some earthworm species, such as *L. rubellus*. Parmelee and Crossley reported that cocoon biomass increased as drought severity increased during a severe drought in a no-tillage agroecosystem on the Georgia Piedmont, USA, and that there were many small, immature earthworms of this species immediately following the drought.

Soil moisture can influence the numbers and biomass of earthworms at any given location. Olson (1928) surveyed areas of Ohio, USA, for earthworms, and reported that the largest numbers of earthworms occurred in soils containing between 12 and 30% moisture. El-Duweini and Ghabbour (1965a), who investigated the survival of *A. caliginosa* in relation to soil moisture content in Egypt, reported that in soils with 5–85% gravel and sand, an increase in moisture content of from 15 to 34% was associated with an increase in numbers of *A. caliginosa*, but above 34% extra moisture had no effect. Wood (1974) showed that there was a strong positive correlation between earthworm biomass and increased soil moisture content for topsoil-inhabiting earthworm species surveyed at 18 different sites on Mt. Kosciusko in south-eastern Australia.

Most earthworms are more active in moist soils than dry ones, and during periods of considerable rain individuals of some species, such as *L. terrestris*, come out on to the soil surface at night. The activity of a West African species, *Millsonia anomala*, which normally lives in the top 10 cm of soil, depends mainly on soil moisture. When moisture is inadequate, these earthworms go into dormancy and even die under extreme drought conditions. When the soil moisture content rises to 8–10%, surviving worms become active again, with 10–17% soil moisture being optimal; above this moisture content conditions are often unsatisfactory (Lavelle *et al.*, 1974).

Soil moisture and temperature can act synergistically to influence earthworm activity. For example, Scheu (1987b) showed that an increase in temperature from 10 to 15 °C, at 60% water content, doubled the numbers of casts produced by *A. caliginosa*, whereas at 48% water content, increasing the temperature to 15 °C increased cast production by only 20%. At 15 °C, an increase in water content from 48 to 60% caused a doubling of cast production.

El-Duweini and Ghabbour (1968) compared two species, *A. caliginosa* and *P. californica*, for their ability to resist desiccation, and showed that *A. caliginosa* could resist desiccation, but *P. californica* could not, preferring soil moisture levels just below the waterlogging levels. These differences could be related to the relative efficiency of water discharge via the earthworms' nephridia. Ayres and Guerra (1981) compared the ability of different earthworms from the central Amazon to withstand desiccation. They showed that *Pontoscolex corethrurus*, a species that inhabits soils with a wide range of moisture content, was much more resistant to dehydration than were two other species, *Androrrhinus caudatus* and *Meroscolex marcusiz*, which inhabit only moist soils.

Many species of earthworms can survive long periods submerged in water. For instance, *A. chlorotica*, *A. longa*, *D. subrubicunda*, *L. rubellus* and *L. terrestris* were all able to survive from 31 to 50 weeks in soil totally submerged below aerated water (Madge, 1969). *Allolobophora caliginosa* was also able to survive for 10–20 weeks in aerated water without food; *H. africanus* could survive for more than 9 weeks submerged. The main problem during submergence is the uptake of water into the worm, but most species have adequate means of overcoming this, although they depend on maintenance of oxygenation in the water (Chapter 4). Cocoons can hatch underwater, and the young worms can feed and grow for a time although totally immersed (Roots, 1956). Some earthworms, such as the North American species *Sparganophilus eiseni*, are semi-aquatic and spend most of their life cycle submerged (Harman, 1965).

7.2 TEMPERATURE

The activity, metabolism, growth, respiration and reproduction of earthworms are all influenced greatly by temperature. Temperature and moisture are usually inversely related and high surface temperatures and dry soils are much more limiting to earthworms than low temperatures and waterlogged soils (Nordström and Rundgren, 1974).

Fecundity is affected very much by different temperatures. For instance, the numbers of cocoons produced by *A. caliginosa* and certain other lumbricid species were reported to quadruple over the range from 6 to 16 °C (Evans and Guild, 1948a). The optimum temperature for cocoon production by *L. terrestris* was 15 °C with 25.3 cocoons produced per annum (Butt, 1991). Cocoons also tend to hatch sooner at higher temperatures; for example, cocoons of *A. chlorotica* hatched in 36 days at 20 °C, 49 days at 15 °C and 112 days at 10 °C when there was adequate moisture (Gerard, 1960). Cocoons of *L. terrestris* hatched more rapidly at 20 °C than at other temperatures (Butt, 1991). However, there may be an inverse relationship between percentage hatch of cocoons and temperature. Cocoon production by *E. fetida* increased linearly with increases in temperature from 10 to 25 °C, although the number of hatchlings per cocoon was lower at 25 than at 20 °C (Reinecke and Kriel, 1981).

The duration of incubation time in relation to ambient temperature was measured for cocoons of four lumbricid earthworm species (Holmstrup *et al.*, 1991). Cocoons were obtained from laboratory cultures of *A. caliginosa*, *A. rosea*, *A. chlorotica* and *D. octaedra*. The cocoons were incubated at 5, 10, 15 and 20 °C at optimum moisture conditions. Incubation time increased greatly with decreasing temperature but embryonic development took place at all four experimental temperatures. Cocoons of *A. caliginosa*, *A. rosea* and *A. chlorotica* developed within 34–38 days at 20 °C and within 59–66 days at 15 °C. *Dendrobaena octaedra* cocoons developed within 47 days at 20 °C and 92 days at 15 °C. At 10 °C, *A. caliginosa* had the shortest cocoon incubation time, followed by those of *A. rosea*, *A. chlorotica* and *D. octaedra*. At 5 °C, *A. chlorotica* and *D. octaedra* cocoons developed for about 400 days, whereas those of *A. caliginosa* needed only about 230 days. It is suggested that a lower threshold temperature for cocoon hatching exists for most earthworm species, but that embryonic development can still take place even though temperature is below this threshold. The ability of the embryo to develop at low temperatures should be regarded as an adaptation to the particular habitat in which the species is living, making it possible for the juveniles to emerge as soon as environmental conditions are favorable.

The growth period from hatching to sexual maturity is also dependent on temperature; for instance, *A. chlorotica* took 29–42 weeks to mature in an unheated cellar (Evans and Guild, 1948a), 17–19 weeks at 15 °C (Graff, 1953) and only 13 weeks at 18 °C (Michon, 1954). *Eisenia fetida* took 9½ weeks to mature at 18 °C and only 6½ weeks at 28 °C (Michon, 1954). Viljoen *et al.* (1992) reported that *D. veneta* completed its life cycle in 107 days at 15 °C and in 151 days at 25 °C. However, the hatching success and the number of hatchlings emerging per cocoon were greater at the lower temperature, which was more characteristic of the earthworm's natural habitat in Europe. Experiments in England on the intensive culturing of

L. terrestris at various temperatures showed that cocoon production was greatest at 15 °C, and the length of cocoon incubation was shortest at 20 °C, with the ideal temperature for maximum production falling within the range of 15–20 °C (Butt, 1991). The temperature at which earthworms thrive best and which they prefer is not necessarily the same as that at which they grow fastest or are most active.

Lee (1985) reported that the optimal temperature for growth of indigenous populations of Lumbricidae in Europe ranges from 10 to 15 °C. Daughberger (1988) verified these temperature optima in laboratory experiments, in which he showed that the temperature preference of *L. terrestris* was 10 °C, and that of both *A. caliginosa* and *A. longa* was between 10 and 15 °C. *Dendrobaena rubida*, which inhabits litter, and *L. rubellus*, which lives close to the soil surface, have temperature optima in the range of 15–20 °C. Some species of earthworms that inhabit compost or dung heaps have temperature optima closer to 25 °C (Edwards, 1988). Grant (1955b) showed that the temperature preferendum was 15–23 °C for *P. hupeiensis*, 15.7–23.2 °C for *E. fetida*, and 10–23.2 °C for *A. caliginosa*. *Eisenia rosea* preferred temperatures of between 24.1 and 25.6 °C (Reinecke, 1974).

Kollmannsperger (1955) reported that the number of lumbricid earthworms found on the soil surface at night was correlated positively with temperature, and that the optimum temperature for activity was 10.5 °C. Satchell (1967) concluded that the most suitable conditions for activity of earthworms on the surface were nights when soil temperatures did not exceed 10.5 °C, grass-air temperatures were above 2 °C, and there had been some rain during the previous 4 days. These optima have been used to identify the best nights to collect *L. terrestris* for fish bait in Canada (Tomlin, 1983). Graff (1953a) listed the optimum temperatures for development of lumbricids (Table 7.2).

European lumbricids introduced and established in warmer climates can have different temperature optima than representatives of the same species in Europe. For example, the optimum temperature for *A. rosea* was

Table 7.2 Optimum temperatures (°C) for the development of earthworms (from Graff, 1953a)

A. rosea	12
A. caliginosa	12
A. chlorotica	15
O. cyaneum	15
L. rubellus	15–18
D. attemsi	18–20
D. rubida	18–20
E. fetida	25

reported to be around 25 °C in laboratory experiments in South Africa (Reinecke, 1975), but was reported to be 12 °C for field populations of earthworms in Germany (Graff, 1953a). Tropical earthworm species tend to have higher temperature optima than species from temperate regions (Lee, 1985).

Earthworms can be killed by temperatures outside their survival limits. For instance, it has been suggested that earthworm populations in arable soils in the United States can be destroyed by frost in the absence of ground cover (Hopp, 1947), but in pasture or woodlands it is unlikely that the soil would freeze deeply enough to affect populations of most species.

The upper lethal temperature for earthworms is lower than for many other invertebrates, although there is considerable variation in estimates of these temperatures by different workers. For instance, El-Duweini and Ghabbour (1965b) reported median upper lethal temperatures of 37.0–37.75 °C for *Pheretima californica* and 39.55–40.75 °C for *A. caliginosa* after only 30 minutes' exposure. Other workers have reported much lower upper lethal temperatures after 48 hours' exposure, such as 28 °C for *L. terrestris*, 26 °C for *A. caliginosa*, 25 °C for *E. fetida* and 25 °C for *P. hupeiensis* (Grant, 1955b) and 29.7 °C for *A. rosea* (Reinecke, 1974). Such temperatures can occur in field soils in many areas in which these species live. El-Duweini and Ghabbour (1965b) determined the preferred and lethal temperatures for *P. californica* and a species of *Alma* (Fig. 7.1). The preferred temperatures for these species were 26–35 °C and 24–26 °C, respectively, and the upper lethal temperatures 37 °C and 38 °C, respectively. Worms could survive the higher temperatures much better in air than in soil.

The lower lethal temperatures for earthworms from temperate regions are apparently close to the freezing point. Earthworms are known to survive in surface soils that are frozen in winter and it is likely that species from extremely cold habitats possess some capacity to tolerate cold or resist freezing of their tissues. However, neither *A. caliginosa* nor *E. fetida* survived exposure to 0 °C for 48 hours in the laboratory (Grant, 1955b). Schmidt (1918) reported that *E. fetida* had a lower lethal temperature limit of –1.3–2.0 °C, indicating only a limited capacity for this species to endure freezing conditions. Bodenheimer (1935) reported a similar lower limit (–1.3 °C) for *Bimastos samerigera* in Israel. It has been claimed that *Eisenia nordenskioeldi* in Russia can survive long periods of being entirely frozen (Lee, 1985). The lower thermal death point of the tropical earthworm *H. africanus* was reported to be 7.5 °C, which is well below temperatures that this species encounters in its natural tropical environment. Cocoons of earthworms that inhabit extremely cold environments have been shown to survive freezing temperatures (Holmstrup *et al.*, 1990; Holmstrup, 1994). Holmstrup (1994) investigated the cold hardiness of five earthworm taxa (*D. octaedra*, *D. rubidus tenuis*, *D. rubidus norvegicus*, *A. caliginosa* and *A. chlorotica*) and showed that *D. octaedra* was the most cold hardy, being

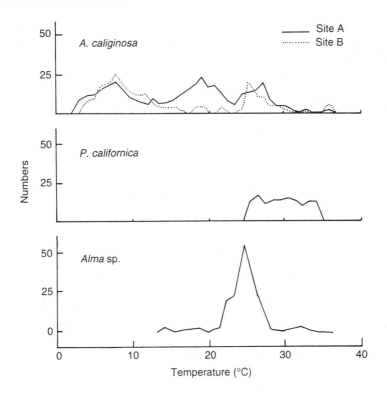

Figure 7.1   Temperature preferences of *A. caliginosa*, *P. californica* and *Alma* sp. (after El-Duweini and Ghabbour, 1965b).

able to survive at –8 °C for 3 months and –13.5 °C for 2 weeks in frozen soil. Earthworm cocoons are apparently not freeze-tolerant, but rather their cold-hardiness strategy is to become extremely desiccated to prevent freezing of their tissues in frozen environments.

Earthworms are capable of acclimatizing to different temperatures. Grant (1955b) studied the effect of conditioning temperatures on the upper lethal levels of temperature for *P. hupeiensis*. Earthworms of this species conditioned at 4 °C, 9 °C and 15 °C had thermal death points of 19.4 °C, 20.9 °C and 22.7 °C, with an average gain in heat tolerance of 0.3 °C for every 1 °C rise in conditioning temperature, up to an upper limit of about 25 °C. Earthworms kept at 15 °C took 12 days to acclimatize to a new environmental temperature of 22 °C. Such mechanisms allow gradual adjustment to seasonal changes in temperature. Earthworms may have some limited capacity to cool themselves by evaporating water from the surface of their bodies, but this is effective only at high soil water potentials when the risk of desiccation is low (Hogben and Kirk, 1944).

Earthworms can migrate away from soil at unsuitable temperatures; in experiments in which individuals of *A. caliginosa* could choose between soils at different temperatures, they preferred soil from 10 to 23 °C, and individuals of *E. fetida* preferred soils from 16 to 23 °C (Grant, 1955b). Dowdy (1944) reported that *Diplocardia* sp. moved down to lower soil strata at temperatures below 6 °C. Madge (1969), who tested species of *H. africanus* in a temperature gradient, found that they aggregated between 23.9 and 31.5 °C, and avoided temperatures above 34 °C. These temperatures are higher than those at which most other worms can survive, but this is a tropical species, which must tolerate high temperatures to be able to survive in its environment.

Temperature also affects the numbers of leaves buried by *L. terrestris*, which can be used as an index of earthworm activity in relation to temperature (Edwards and Lofty, 1977). Daniel (1991) expressed the consumption of dandelion leaves by *L. terrestris* as a function of temperature and moisture. He reported that food consumption of this species increased linearly with temperature up to 20 °C but declined above 22 °C, and declined linearly with decreases in soil water potential below −7.0 kPa.

7.3 pH

It has been demonstrated that earthworms are very sensitive to the hydrogen ion concentration (pH) of aqueous solutions (Chapter 4), so it is not surprising that soil pH is sometimes a factor that limits the species, numbers and distribution of earthworms that live in any particular soil. Some species are intolerant of acid soil conditions, whereas others thrive under acid conditions, and many species can tolerate a wide range of pH.

Several workers have stated that most species of earthworms prefer soils with a neutral pH (pH = 7.0) (Arrhenius, 1921; Moore, 1922; Phillips, 1923; Salisbury, 1925; Allee *et al.*, 1930; Bodenheimer, 1935; Petrov, 1946). However, *L. terrestris* occurs in soils with a pH of 5.4 in Ohio, USA (Olson, 1928), and *A. caliginosa* in soils with a pH of 5.2–5.4 in Denmark, although there are few earthworms in soils of pH lower than 4.3, except for one species, *D. octaedra*, which seems to be very acid-tolerant (Bornebusch, 1930). Further evidence for the acid tolerance of *L. terrestris* comes from Vimmerstedt (1983) who introduced *L. terrestris* into a revegetated mine spoil with a pH of 3.5 in Ohio, USA, and reported that the earthworms became established on the spoil. *Eisenia fetida* has been reported to prefer soils with a pH between 7.0 and 8.0 (Rivero-Hernandez, 1991) but Edwards (1988) reported that it could tolerate a pH range from 4.0 to 7.0. Certain tropical species of *Megascolex* thrive in acid soils from pH 4.5 to 4.7 (Bachelier, 1963) and *Bimastos lonnbergi* is numerous in soils between pH 4.7 and 5.1 (Wherry, 1924).

Spiers *et al.* (1986) discussed an acid-tolerant earthworm species, *Arctiostrotus* sp., endemic to the organic horizons of coniferous podzols of Vancouver Island, Canada. These earthworms occurred in organic soil horizons with pH ranges from 2.6 to 6.2 and were most abundant at a site with a pH of 2.9. The authors concluded that these earthworms have a major role in the decomposer subsystem of these ecosystems, despite the extremely low pH prevailing at some sites.

Satchell (1955a) considered that *Bimastos eiseni*, *D. octaedra* and *D. rubida* were acid-tolerant species, and *A. caliginosa*, *A. nocturna*, *A. chlorotica*, *A. longa* and *A. rosea* were acid intolerant (Fig. 7.2). He considered that *L. terrestris* was not very sensitive to pH, and Guild (1951b) agreed with this conclusion, although Richardson (1938) considered that it was. Guild (1951b) also confirmed that the relative abundance of *A. longa* and *A. caliginosa* was less in more acid soils. Madge (1969) reported that the optimum pH for *H. africanus* was between 5.6 and 9.2, so this species is somewhat acid tolerant. *Bimastos lonnbergi* and *B. beddardi* also seem to prefer acid soils (Wherry, 1924). An increase of pH from 7.25 to 8.25 was associated with a decrease in numbers of earthworms in 14 Egyptian soils studied by El-Duweini and Ghabbour (1965a), demonstrating that soils can also be too alkaline to favor earthworms. Jeanson-Luusinang (1961) observed in laboratory experiments that individuals of *Eophila icterica* that occurred in a field soil with a pH of 7.0, could tolerate soils with pH from 4.2 to 8.0,

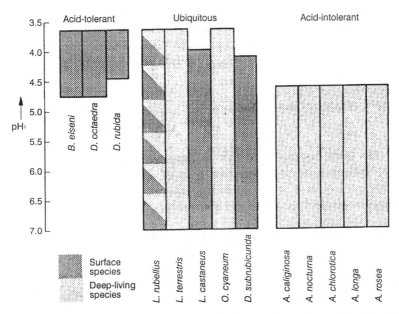

Figure 7.2 Classification of earthworms as a function of the pH of litter (after Satchell, 1955a).

but were much more active at pH 8.0 than at pH 4.2; she speculated whether this might limit their penetration into deeper soil, which is normally more acid than the surface soil.

Bouché (1972) related the distribution of 67 taxa of Lumbricidae in France in relation to the pH of soil. Most species occurred in soils with a pH range of 7.0–7.4. Twenty-six acid-tolerant species were found in soils with a pH below 4.0 and four species were found only in soils with a pH above 6.6. The conclusion for European lumbricids is that most species have a wide range of tolerance for varying pH, some species being more acidophilic and others basiphilic.

Satchell (1955a) took soil samples from plots with long-term different fertilizer treatments, on a pasture experiment on Park Grass at Rothamsted. The pH values of the soils were: 4.0, 4.1, 4.4, 5.0, 5.1, 5.6, 5.8, 6.9 and 7.0. He placed mature individuals of *A. chlorotica* on the surface of these soil samples, and studied their reactions and the time they took to bury themselves. In the three most acid soils, worms at first showed a violent avoiding reaction, twisting, jerking convulsively, and exuding coelomic fluid from their dorsal pores. They then extended to their full length and crawled about the soil surface, intermittently raising and waving the anterior segments. Activity gradually became sporadic and after 1–2 hours they lay motionless and became flaccid. After 21 hours, 58 out of 60 worms exposed to a pH below 4.4 were dead.

Edwards and Lofty (1975a) studied the populations of earthworms in the same Park Grass long-term fertilizer experiment in plots that had a range of pH from 3.7 to 7. They reported that *L. terrestris* became increasingly numerous as the pH increased, but most of the other species present, *A. nocturna, A. caliginosa, A. rosea, O. cyaneum*, tended to have an optimum pH range of 5.0–6.0. This agrees with the results of a study by Piearce (1972) who reported that, in a wide range of soils in North Wales, the number of species present and populations were greatest in soils with a pH of about 6.0 and least in soils with a pH below 5.0. Because soil pH is related to other soil factors that have an important influence on earthworm populations (e.g. clay content, cation exchange capacity), it is often difficult to establish a direct cause and effect relationship between soil pH and the size of earthworm populations, except in cases where earthworm populations are clearly inhibited by extremely high or extremely low hydrogen ion concentrations.

Reddy and Pasha (1993) reported a significant positive correlation between pH and the seasonal abundance of juvenile and young adult *Octochaetona phylloti* in semi-arid tropical grassland in India, but the population of adult earthworms was affected more by rainfall than by pH. Staaf (1987) suggested that pH and factors related to pH were very important influences on the distribution and abundance of earthworms in acid beech forest soils in Sweden.

Soil pH may also influence the numbers of worms that go into diapause. Doeksen and van Wingerden (1964) reported that when they put individuals of *A. caliginosa* into soils with pH of 4.9, 5.4, 6.4, 7.6 or 8.6, the more acid the soil, the sooner worms went into diapause, and they remained in diapause longest in soil with a pH of 6.4.

It has been reported by several workers that earthworm casts are usually more neutral than the soil in which the worms live (Salisbury, 1925; Dotterweich, 1933; Puh, 1941; Stöckli, 1949; Finck, 1952; Nye, 1955; Sharpley and Syers, 1976; Reddy, 1983). One explanation suggested for this is that earthworms neutralize soil as it passes through their guts, by secretions of the calciferous glands. A more probable explanation is that the soil is neutralized by secretions from the intestine and by excretion of ammonia, or that it is a combination of the two.

7.4 AERATION AND CARBON DIOXIDE

There is little experimental evidence that the soil oxygen tension affects the distribution of earthworms in soil, although Satchell (1967) stated that the distribution of *B. eiseni* and *D. octaedra* appeared to be limited in some sites by the minimum oxygen tensions occurring at certain seasons; however, this was confused by factors such as pH, soil moisture content, amount of raw humus, plant cover and soil microbial status. Satchell showed that there was some correlation between numbers of *B. eiseni* and oxidation–reduction potentials (E_h) of the soil. *Eisenia fetida* retreats from layers of organic matter when they become anaerobic (Edwards, 1988). In heathlands of Yorkshire, England, Satchell (1980) related extremely low numbers of earthworms in saturated depressions in the soil to the low E_h (174 mV) at these locations. Kaplan *et al.* (1980) showed that *E. fetida* could not survive in aged sewage sludge when the E_h was below 250 mV.

However, some earthworm species can survive for long periods at very low oxygen tensions. Not much is known of typical oxygen tensions that occur in soils, but Boynton and Compton (1944) reported that oxygen tensions in an orchard soil were below 10% for 11 weeks in the year at a depth of 90 cm, and for as long as 6 months at depths below 150 cm. Deep-burrowing species such as *L. terrestris* do not survive well in poorly-drained soils that have low redox potentials below the surface horizons.

There is little evidence of the vertical distribution of earthworms being affected by CO_2 concentrations in soil, and the worms do not seem to migrate in response to high CO_2 concentrations. For example, *E. fetida* did not respond to soil concentrations of CO_2 up to 25% (Shiraishi, 1954). The limits of CO_2 concentration in soil are normally between 0.01 and 11.5%, and earthworms can survive much greater concentrations than this, even up to 50% (Russell, 1950; Chapter 4).

7.5 SOIL TYPE

Earthworms are influenced by soil type and texture, although there have been relatively few studies of the direct influence of soil type on earthworm populations. Guild (1948, 1951b) made a survey of the main soil types in Scotland, and reported that there were differences both in total numbers and relative numbers of each species in soils of different textural composition (Fig. 7.3, Table 7.3). Light and medium loams had greater total populations of worms than heavier clays or more open gravelly sands and alluvial soils. *Allolobophora caliginosa* was the dominant species in all soil types, and *A. longa* was less important in open soils, gravelly sands and alluvial soils. In a survey of the distribution of earthworm species in the Hebrides, Boyd (1957b) compared the relative abundance of earthworm species in light soils with those in calcareous sand and dark peaty soils. Six species were more abundant on the light soils and six on the dark ones. In particular, *A. caliginosa* and *L. castaneus* were much more numerous in the light soils, and *B. eiseni* and *D. octaedra* in the dark soils. A few small species of earthworms can survive in deserts and semi-deserts (Kubiena, 1953; Kollmannsperger, 1956), and some worms can inhabit the arid, cold soil of north-eastern Russia. Deep-burrowing anecic species such as *L. terrestris* require adequate soil depth to build their burrows and do not thrive in shallow soils (Piearce, 1978; Muys *et al.*, 1992).

El-Duweini and Ghabbour (1965a) showed that populations of *A. caliginosa* in Egypt decreased with increasing proportions of gravel and sand in soils. This relationship did not hold in soils that had higher water contents, indicating that some of the effects of texture were due to its influence on soil moisture. In forest, pasture and heathlands in Sweden, Nordström and Rundgren (1974) found a significant positive correlation between earthworm abundance and soil clay content in soils that were categorized as having 0–5%, 5–15% or 15–25% clay. In particular, popula-

Table 7.3   Relations of soil type to earthworm populations (from Guild, 1951b)

Soil type	Population		No. of species
	Thousands/ha	No./m^2	
Light sandy	232.2	57	10
Gravelly loam	146.8	36	9
Light loam	256.8	63	8
Medium loam	226.1	56	9
Clay	163.8	40	9
Alluvium	179.8	44	9
Peaty acid soil	56.6	14	6
Shallow acid peat	24.6	6	5

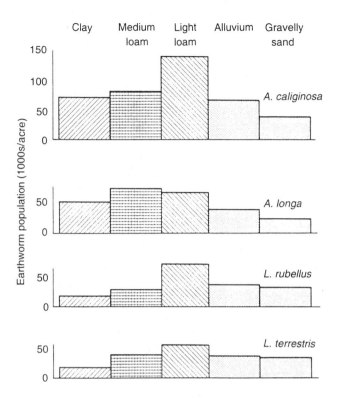

Figure 7.3 Density of earthworm populations (thousands/ha) in various soil types in Scotland (adapted from Guild, 1948).

tions of *A. caliginosa*, *A. longa*, *A. rosea* and *L. terrestris* were correlated positively with clay content at 0–60 cm.

Hendrix *et al*. (1992) studied the distribution of earthworms at sites representing various ecosystem types, management practices, landscape positions, soil texture and soil erosion status on the Georgia Piedmont in south-eastern USA. Moderately and severely eroded sandy clay loams supported significantly more earthworms and greater biomass than slightly eroded soil with a higher sand content. Of the soil texture variables, the silt content of the soil was correlated most with earthworm abundance. These results may have been due to the lower organic matter content and water-holding capacity of the sandy soils.

In their investigation of the diversity and abundance of earthworms in pasture soils at 113 sites in South Australia, Baker *et al*. (1992a) found weak positive correlations between clay content of the soil and the abundance of *A. trapezoides*, *A. rosea* and *A. caliginosa*. Of these three species, *A. caliginosa* exhibited the strongest positive correlation with clay content.

The numbers and weights of *A. caliginosa* and *A. trapezoides* were correlated negatively with the sand content of the soil.

Soil texture can influence earthworm populations because of its effects on other soil properties such as soil moisture relationships, nutrient status and cation exchange capacity, all of which can have important influences on earthworm populations. In an analysis of earthworm communities of tropical rainforests, Fragoso and Lavelle (1992) showed that earthworm communities dominated by geophagous species are characteristic of nutrient-rich soils, whereas those dominated by litter-feeding, epigeic species are normally present in nutrient-poor soils. They showed that the percentage dominance of epigeic species in the earthworm community was negatively correlated with the amounts of calcium, magnesium and nitrogen in the soil.

7.6 ORGANIC MATTER

The distribution of organic matter in soil influences the distribution of earthworms greatly. Soils that are poor in organic matter do not usually support large numbers of earthworms. Conversely, if there are few earthworms, the decaying organic matter usually lies in a thick mat on the soil surface. Such mats of undisturbed organic matter occur in both woodlands (Richardson, 1938) and grassland (Raw, 1962). In irrigated pastures in New South Wales, Australia, that contained no earthworms, such mats of organic matter were up to 4 cm thick until earthworms were introduced experimentally (Stockdill, 1966).

Some species of earthworms are attracted readily to animal droppings and dung on the soil surface. In experiments at Rothamsted, which compared plots to which dung was added annually with plots which were left unmanured, large differences in earthworm populations were noted (Fig. 7.4). Plots that grew wheat regularly for many years had earthworm populations that were 3–4 times greater in plots receiving 35 tonnes/ha of dung than in unmanured plots (Table 7.4). The numbers in Park Grass at Rothamsted, a long-term pasture fertilizer experiment, were three times greater in plots receiving 3 tonnes/ha of dung than in unmanured plots after 140 years. In Barnfield, in a fertilizer experiment in an arable field growing continual mangolds, there were about 15 times more earthworms in plots receiving dung annually than in unmanured plots after 130 years (Table 7.4). In Sweden, Nordström and Rundgren (1974) found significant correlations between the organic matter content and clay content of soils and the abundance of deep-burrowing species of earthworms. Large amounts of dead roots and other organic matter in pasture usually coincide with large numbers of earthworms, and it is probably the gradual decrease in soil organic matter, when pasture is ploughed and used for arable crops, that leads to a corresponding decrease in earthworm populations.

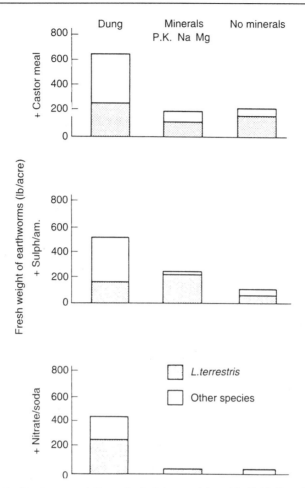

Figure 7.4 Earthworm populations (kg/ha) in an old arable field (Barnfield, mangolds) (Edwards and Lofty, 1977).

Decaying leaves in woodlands are also a source of organic matter that usually favors earthworm multiplication. Earthworms can remove a large part of the annual leaf fall in a woodland or orchard if populations are large and the leaf litter is of a species that is palatable to the worms present.

Several workers have reported strong positive correlations between earthworm numbers and biomass and the organic matter content of the soil. Increases in the organic carbon content of semi-arid agricultural soils in Egypt were associated with increased numbers and biomass of earthworms (El-Duweini and Ghabbour, 1965a; Ghabbour and Shakir, 1982). Hendrix *et al.* (1992) reported a strong positive correlation between earthworm population density and soil organic matter content across 10 sites, which included conventional and no-tillage agroecosystems, grass

Table 7.4   Earthworm populations in plots with and without dung

| Species | 1. Grassland Park Grass, Rothamsted (Satchell, 1955a) | | 2. Arable land Barnfield, Rothamsted (Edwards and Lofty, 1977) | |
	Unmanured	Dung	Unmanured	Dung
L. terrestris	13.1	22.5	0.23	10.8
L. castaneus	16.0	59.6	–	–
A. caliginosa	2.9	8.0	0.8	15.4
A. chlorotica	1.6	–	3.2	44.6
A. rosea	10.0	21.3	–	0.23
A. longa	–	–	0.46	1.8
A. nocturna	1.3	18.9	–	–
O. cyaneum	6.9	24.5	–	–
Total	51.8	154.8	4.69	72.83

meadows and a mixed deciduous forest, in the south-eastern USA. There is little doubt that the availability of organic matter is one of the most important factors influencing earthworm abundance.

7.7 FOOD SUPPLY

Earthworms can use a wide variety of organic materials for food, and even in adverse conditions can extract sufficient nourishment from organic matter and micro-organisms in soil to survive. The kind and amount of food available influences not only the size of earthworm populations but also the species present, and their rate of growth and fecundity. Evans and Guild (1948a) investigated the influence of food on earthworm cocoon production (Table 7.5), and showed clearly that more cocoons were produced by worms that were provided with decaying animal organic matter, than by those provided with fresh plant material. They also showed that earthworms fed on any nitrogen-rich diets grew faster and produced more cocoons than those with little nitrogen available. Barley (1959a) provided individuals of A. caliginosa with different diets and reported that they grew at very different rates on these diets, growing fastest when provided with animal dung (Table 7.6).

Boström and Lofs-Holmin (1986) reported that the growth of A. caliginosa depended, not only on the type of food available, but also on the particle size of the food material. Earthworms grew best with shoots of meadow fescue, followed by provision of alfalfa shoots and roots. Earthworms provided with barley straw fragments smaller than 0.2 mm gained more than twice as much weight during 150 days as earthworms provided with fragments ranging from 0.2 to 1.0 mm in size, indicating the importance of food particle size to the nutritional quality of food material.

Table 7.5   Mean number of cocoons produced by five earthworms in 3 months (from Evans and Guild, 1948a)

Food	A. chlorotica	L. castaneus
Fodder	0.8	9.4
Oat straw	1.4	12.0
Bullock droppings	12.4	73.2
Sheep droppings	14.0	76.0

Table 7.6   Changes in body weight of A. caliginosa when fed for 40 days on various diets (from Barley, 1959a)

Food	% change in weight
Angaston soil	−53
Phalaris roots	−26
Phalaris leaves	−26
Clover roots	−2
Clover leaves	+18
Dung, on surface	+71
Dung, incorporated	+111

Guild (1955) stated that most species of earthworms prefer dung or succulent herbage to tree leaves, and that pine needles are preferred least of all; Barley (1959a) corroborated this conclusion for *A. caliginosa*. Guild (1955) estimated that mature individuals of *A. longa* can ingest 35–40 g dry weight of dung per annum, *A. caliginosa* 20–24 g and *L. rubellus* 16–20 g.

Svendsen (1957b) reported that, whereas individuals of the pigmented species *L. festivus*, *L. rubellus*, *L. castaneus*, *D. octaedra*, *D. rubida* and *B. eiseni* became aggregated in dung, those of the unpigmented, geophagous species *O. cyaneum*, *O. lacteum*, *A. caliginosa*, *A. chlorotica*, *A. longa*, *A. rosea* and *E. tetraedra* f. *typica* did not. Earthworms of the species *A. rosea* and *A. caliginosa* are not attracted to leaf litter (Lindquist, 1941) but readily eat dung, and *A. caliginosa* and *L. rubellus* will also feed on dead root tissues in pasture (Piearce, 1978). Müller (1950) considered that *A. caliginosa* fed extensively on fungal mycelia but it seems that this species tends to be omnivorous. Individuals of *A. caliginosa* do not eat fallen leaves until these have become moist and brown (Barley, 1959a), and they cannot maintain their body weight on dead roots, which tends to discount Waters' (1955) claim that dead roots are the main diet of this species.

An examination of the diets and gut contents of *A. caliginosa* and *L. rubellus* in a deciduous woodland suggested that *L. rubellus* is primarily a litter-feeder, whereas *A. caliginosa* consumes organic matter in a more advanced state of decomposition. The material ingested by *L. rubellus* was richer in calcium than that consumed by *A. caliginosa*, and the former species has a mechanism for excreting calcium (Piearce, 1972). Piearce (1978) examined the gut contents of six species of Lumbricidae in a pasture in North Wales, UK, and identified different ecological groups of worms based on gut contents and feeding behavior. *Lumbricus castaneus* and *L. rubellus* consumed material rich in relatively undecomposed plant remains, whereas *A. caliginosa* and *A. chlorotica* fed mainly on well-decomposed organic detritus; *A. longa* and *D. mammalis* were intermediate in their dietary reqirements.

Most species of earthworms can distinguish between different kinds of forest litter. Darwin (1881) claimed that earthworms showed preference for leaves of particular shapes, but Satchell (1967) found that there was an order of preference for certain leaf species, if uniform disks of a range of species of leaves were offered. Gast (1937) ascribed this preference to the mineral content of certain species of leaves, but this has not been confirmed. Mangold (1951) believed that some species of leaves were unattractive to earthworms because of their bitter alkaloid or noxious aromatic content. Litter rich in protein is accepted more readily than that deficient in protein (Wittich, 1953), but protein content is often correlated with sugar content (Mangold, 1953; Laverack, 1960a), which may also be more important. Edwards and Heath (1963) buried leaf disks in nylon bags made from mesh with 7.0 mm apertures, so that *L. terrestris* could feed on them, and found that oak disks were preferred to beech, and in another experiment (Heath *et al.*, 1966) stated that the order of preference for a range of different species of leaves buried in similar mesh bags was: lettuce, kale, beet, elm, maize, lime, birch, oak and beech.

Leaves of larch, spruce, oak and beech, which are all comparatively unpalatable to earthworms, contain condensed tannins that are not found in dog's mercury, nettle, elderberry, ash and wych elm, all of which are more palatable (Brown *et al.*, 1963). King and Heath (1967) reported that the amount of water-soluble polyphenols in litter was inversely proportional to the rate at which it was consumed, and that litter became much more palatable after a few weeks of weathering. In a study of the palatability of litter to individuals of *L. terrestris*, Satchell (1967) showed that there was an inverse correlation between the palatability of litter and its total polyhydric phenol content, and a positive correlation with the amount of soluble carbohydrates. Edwards and Lofty (1977) showed that the palatability of oak and beech leaves could be increased by washing out water-soluble polyphenols. Such observations help to explain why the leaves of certain species of trees disappear from the soil surface faster than others and become broken down and incorporated into soil more readily.

Hendriksen (1990) used litter-bags in experiments with 10 different types of litter in a pasture and observed that detritivore earthworm species (*Lumbricus* spp.) preferred ash, basswood and predecomposed elm and beech litter to undecomposed beech litter. The numbers of litter-feeding earthworms found beneath the litter-bags were correlated negatively with the C:N ratio and final polyphenol concentration of the litter. In mulching experiments with five types of plant residues in the humid tropics, Tian *et al.* (1993) reported that earthworm populations were negatively correlated with the ratio of lignin to nitrogen of the plant residues.

Other factors also influence the palatability of leaves to earthworms. Starved worms will consume leaves of species that they would normally reject. Addition of bacterial cells to leaves increased their acceptability (Wright, 1972). Many chemicals, such as pesticides, can render normally palatable leaves unpalatable.

Vegetation can affect earthworm populations by altering the quantity and quality of their food supply. Phillipson *et al.* (1976) showed a strong positive association between numbers of *A. caliginosa* and eight species of ground flora in a beechwood in England. They suggested that the ground flora may have provided important food resources to the worms in the form of dead roots and micro-organisms associated with the decaying roots. Populations of *Pontoscolex corethrurus* and *Amynthas gracilis* were influenced strongly by tree species in tropical tree plantations in Hawaii (Zou, 1993). The population density of worms ranged from 92 per m^2 in pure *Eucalyptus* stands, 281 per m^2 in mixed *Eucalyptus–Albizia* stands (25% *Albizia* and 75% *Eucalyptus*) and 469 per m^2 in pure *Albizia* stands. Although *Eucalyptus* trees produced more litter, *Albizia* trees produced more fine litterfall and earthworm densities were correlated positively with the nitrogen content and concentration of fine litterfall. Boettcher and Kalisz (1991), working in forests in eastern Kentucky, USA, showed that earthworm populations were significantly lower under yellow poplar (*Liriodendron tulipifera*) and eastern hemlock (*Tsuga canadensis*) when rhododendron (*Rhododendron maxima*) was present in the understory than when it was not present. The effect of rhododendrons on the mineral soil in which they grow affected both the abundance and species composition of earthworms in the soil.

Other evidence for the influences of different tree species on earthworm communities comes from Muys *et al.* (1992), who showed that earthworm communities differed considerably on grassland sites in Belgium that were afforested with different tree species. Earthworm biomass diminished under oak *(Quercus palustris)* stands due to poor quality of the oak litter and acidification of the soil by oak leaves.

There have been several different estimates of the amounts of food taken by earthworms and passed through their guts, but there is good agreement between different workers. Guild (1955) calculated that worms of 0.1 g body weight eat as much as 80 mg of food/day/g of body weight of worm. Satchell (1967) calculated that individuals of *L. terrestris* pass between 100 and 120 mg/g/day through their guts, and Crossley *et al.* (1971) calculated that a species of *Octolasion* had a soil intake of 1.2% of its live body weight per hour, which equals about 29% of its body weight per day or 290 mg/g/day. Barley (1959a) calculated that *A. caliginosa* consumed 200–300 mg/g/day.

7.8 INTERACTIONS BETWEEN ENVIRONMENTAL FACTORS

There have been relatively few studies which assessed how interactions between environmental factors influence earthworms. Clearly, temperature and moisture are usually inversely correlated and it is often difficult to separate their effects. Similarly, it is difficult to separate the effects of soil organic matter content from the availability of decaying plant material as food. However, Briones *et al.* (1992) reported that a group of species, *A. caliginosa, D. mammalis, L. rubellus, E. tetraedra* and *D. rubida,* could be separated from *A. rosea* and *A. chlorotica* by combined preferences for organic matter, soil texture and moisture content.

The role of earthworms in organic matter and nutrient cycles

8

Earthworms have a major role in the breakdown of organic matter and the release and recycling of the nutrients that it contains. They remove partially decomposed plant litter and crop residues from the soil surface, ingest it, fragment it and transport it to the subsurface layers. Their fecal material is in the form of casts which can vary greatly in size and form, and are deposited on the soil surface, in their burrows or in spaces below the soil surface, thereby having a major role in the development of soil horizons. The casts tend to be much more microbially active than the surrounding soil and have plant nutrients in a form that can be readily utilized. Through these various interactions, earthworms are often key organisms in the overall breakdown of organic matter and transformation of major and minor mineral nutrients.

8.1 FRAGMENTATION, BREAKDOWN AND INCORPORATION OF ORGANIC MATTER

Plant organic material that reaches the soil is subject to many agents that promote decomposition, including both micro-organisms and animals. Some plant and animal residues are decomposed rapidly by the micro-organisms, but much organic matter, particularly the tougher plant leaves, stems and root material, breaks down more readily after being eaten by soil-inhabiting invertebrates and acted upon by enzymes in their intestines. Earthworms are probably the most important invertebrates in many soils in this initial stage of the recycling of organic matter.

This was demonstrated by Edwards and Heath (1963), who placed disks, cut from freshly fallen oak and beech leaves, in nylon bags of four different mesh sizes, which were then buried in woodland or old pasture soil. Only the bags with the largest mesh (7 mm) would allow the entry of earthworms as well as smaller animals. After 1 year, none of the 50 oak disks originally placed in each of the 7 mm mesh bags remained intact, and 92% of the total oak-leaf material and 70% of the beech had been removed (Fig. 8.1). Much less had disappeared in bags that allowed access only to microarthropods. Earthworms ate not only the softer parts of the leaves but also veins and ribs (Edwards and Heath, 1963). Curry and Byrne (1992) in a similar experiment in which wheat litter was confined by mesh of different sizes in a winter wheat field in Ireland, found that the decomposition rate of straw that was accessible to the earthworms was increased by 26–47% compared with straw from which earthworms were excluded. Other workers have compared the disappearance of isolated samples of litter in 1 mm mesh nylon bags with samples pinned under a nylon net with the lower surface open to access by all soil-inhabiting invertebrates. The leaves under the nylon net decomposed 2–3 times faster than those in bags, and this difference was due mainly to earthworms (Perel *et al.*, 1966). MacKay and Kladivko (1985) placed maize and soybean residues on the soil surface in pots with and without earthworms in a greenhouse. After 36 days, pots with no earthworms had retained 60% of the soybean residues and 85% of the maize residues, whereas pots with earthworms had only 34% of the original soybean residues and 52% the original maize residues.

Organic matter that passes through the earthworm gut and is egested in their casts is broken down into much finer particles, so that a greater surface area of the organic matter is exposed to microbial decomposition. Martin (1991) reported that casts of the tropical earthworm *M. anomala* had much less coarse organic matter than the surrounding soil, indicating that the larger particles of organic matter were fragmented during pas-

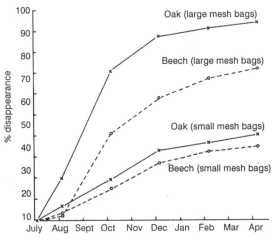

Figure 8.1   Decomposition of leaf disks by soil animals (after Edwards and Heath, 1963).

sage through the earthworm gut. Parmelee *et al.* (1990) used the pesticide, carbofuran, to decrease earthworm populations in no-tillage agroecosystems by more than 90%, and reported that the amounts of fine, coarse and total particulate organic matter in the treated plots increased by 43%, 30% and 32%, respectively, compared to that in control plots after 292 days. The significant increases in the amount of particulate organic matter in plots with decreased earthworm populations illustrates the extremely important role of earthworms in the fragmentation and breakdown of organic matter as well as in the release of nutrients that it contains.

The feeding habits of different earthworm species influence their effects on litter fragmentation and incorporation into soil (Chapter 6). Anecic species, such as *L. terrestris*, incorporate large amounts of organic matter into soil, and are capable of breaking down and feeding on large litter fragments by stripping off smaller particles with their mouthparts. Epigeic species, that reside in surface litter, consume large amounts of litter, but do not incorporate much of it into the mineral soil layers. Endogeic species feed mainly on fragmented organic matter, mixing it thoroughly with mineral soil. Ferrière (1980) examined the gut contents of 10 species of lumbricid earthworm in a pasture and observed distinct differences in the types of food consumed by the various species. Epigeic species fed primarily on relatively undecomposed fragments of leaves and roots, anecic species fed on partially decomposed, but identifiable, fragments of above-ground plant litter, and endogeic species fed mainly on unidentifiable organic matter together with roots and leaves in a more advanced stage of decomposition. Judas (1992) reported similar materials in the guts of lumbricids in a beechwood in Germany, although he found that the feeding behavior of individual species was quite variable, even within a given habitat. Anecic and endogeic species of earthworms occur together in many soils and probably have a synergistic effect on the redistribution of organic matter throughout the soil profile. Shaw and Pawluk (1986a,b) reported that when the anecic species, *L. terrestris*, and the endogeic species, *O. cyaneum*, were kept in soil microcosms together, they distributed the crop residues from the soil surface more evenly throughout the soil matrix, than when either species was present alone.

Earthworm species such as *L. terrestris* are responsible for a large proportion of the overall fragmentation and incorporation of litter in many woodlands of the temperate zone, and are primarily responsible for the formation of mulls, which are forest soils in which the surface litter and organic layers are mixed thoroughly with the mineral soil (Müller, 1878; Scheu and Wolters, 1991a). Soils with no earthworms or only a few earthworms often have a well-developed layer of undecomposed litter and organic matter on the soil surface, separated from the underlying mineral soil by a sharp boundary. These are known as mor soils, which represent the opposite extreme to mulls, along a continuum of forest soil types.

Earthworms can convert mor soils to mulls rapidly, after they colonize a site previously lacking earthworms. Langmaid (1964) reported that earthworms had thoroughly mixed the surface humic horizons with the underlying mineral soil in virgin podzols in New Brunswick, Canada, completely modifying the soil profile within only 3–4 years of first colonizing the soils. The mixing and fragmentation of forest litter by earthworms has been identified as being of fundamental importance to the renewal of spruce forest ecosystems in the French Alps (Bernier and Ponge, 1994). Anecic species, such as *L. terrestris*, play a particularly important role in mixing the surface humus horizons with mineral soil in these ecosystems, forming a favorable environment for the germination and growth of spruce seedlings. When earthworms are eliminated from forest soils, such as by changes in food quality or a decrease in soil pH, the result is a decreased bioturbation, a slowing of organic matter decomposition, and development of distinct litter and organic layers. Beyer *et al.* (1991) observed such changes in oak forests in Germany, which they attributed to a decline in earthworm populations resulting from decreasing soil pH due to air pollution and acid precipitation.

The effectiveness of *L. terrestris* in initiating the fragmentation and incorporation of fallen leaves in an apple orchard was illustrated vividly by Raw (1962) who compared the soil profile and structure of an orchard with a large *L. terrestris* population, with one in which earthworms were almost totally absent (due to frequent and heavy spraying with a copper-based fungicide). The orchard with few earthworms had an accumulated surface mat, 1–4 cm thick, made up of leaf material in various stages of a very slow decomposition, and demarcated sharply from the underlying soil, which had a poor crumb structure.

Earthworms in agricultural grasslands and turf ecosystems also have an important role in incorporating surface organic matter into soil. In New South Wales, pastures containing no earthworms accumulated surface mats or thatches up to 4 cm thick, but these disappeared progressively after earthworms were introduced experimentally (Barley and Kleinig, 1964). Such mats are common on poor upland grasslands in New Zealand in areas with no earthworms (Stockdill, 1966). Similar mats occurred in plots on Park Grass, Rothamsted, which had no earthworms due to regular treatment with ammonium sulfate as a fertilizer. Potter *et al.* (1990) reported that the rate of thatch breakdown in plots of Kentucky blue grass (*Poa pratense* L.) was slowed greatly in plots from which earthworms had been eliminated with insecticides. Clements *et al.* (1991) examined plots of perennial ryegrass (*Lolium perenne*) from which earthworms had been absent for 20 years, due to regular applications of the pesticide phorate. By the end of the 20 year period, they observed a dramatic increase in the depth of leaf litter, and a great reduction in soil organic matter content in plots from which earthworm populations had been eliminated. Stout and Goh (1980), working in grassland in New Zealand, analyzed the distribution in litter and soil of ^{14}C produced during atmospheric testing of atomic bombs. In grasslands without earthworms, the surface litter and upper 10 cm of soil was greatly enriched with ^{14}C. In grasslands with earthworms, there was no surface litter accumulation and the ^{14}C was mixed thoroughly throughout the upper 18 cm of soil.

Many kinds of leaf litter are not acceptable to earthworms when they first fall to the soil surface, and they require a period of weathering before they become palatable to the earthworms. It is believed that this weathering leaches water-soluble polyphenols and other unpalatable substances from the leaves (Edwards and Heath, 1963; Heath and King, 1964; Satchell, 1967). Zicsi (1983) fed four different litter-feeding species of earthworms, including *L. terrestris*, with litter of five different tree species, in soil columns in the laboratory. Earthworms began feeding immediately on rapidly decomposing, higher-quality litter, such as that of maple (*Acer platanoides* L.), but did not feed on the lower-quality litter of beech (*Fagus sylvatica* L.) and oak (*Quercus* spp.) until it had weathered for several months.

The rate of breakdown depends also upon the type of litter, so that beech leaves disappear much more slowly than oak leaves (Edwards and Heath, 1963) (Fig. 8.1), which in turn are more resistant to attack by earthworms than are apple leaves (Raw, 1959). Elm, lime and birch disappear more rapidly than beech (Heath *et al.*, 1966). Earthworms are much more attracted to moist litter material than to dry, and they are much more active in moist soil and litter. Earthworms can accelerate the decomposition of pine litter; when earthworms of the species *D. octaedra, D. attemsi, D. rubida, L. rubellus, B. eiseni* and *A. chlorotica* were put into cultures containing pine litter, they fragmented and decomposed the pine needles (Heungens, 1969b). Haimi and Huhta (1990) showed that *L. rubellus* increased the mass loss of coniferous forest humus by a factor of 1.4 in a 48 week laboratory incubation. Earthworms apparently do not participate in the primary stages of decomposition of pine needles but have a progressively more important role during later stages of decomposition (Ponge, 1991; see Chapter 7 for further discussion on the effects of food supply and litter type on feeding by earthworms).

A final stage in processing of organic matter is known as humification, which is basically the breaking down of large particles of organic matter into complex amorphous colloids containing phenolic materials. Only about one-quarter of the fresh organic matter becomes converted to humus in this way. Much of the humification process is caused by soil micro-organisms, and some is due to small soil-inhabiting invertebrates, such as mites, springtails and other arthropods. However humification is also accelerated by the passage of the organic material through the guts of earthworms when they feed on decomposed organic matter together with mineral soil. Probably some of the final stages of humification are due to the intestinal microflora in the earthworms' gut, because most of the evidence indicates that the chemical processes of humification are caused more by the microflora than by the fauna. Nevertheless, earthworms have been shown to accelerate the humification of straw as well as leaf litter. In pot experiments, earthworms accelerated the rate of straw humification by 17–24% (Atlavinyte, 1975) and in field experiments by 15–42%. Neuhauser and Hartenstein (1978) suggested that earthworms may enhance the polymerization of aromatic organic compounds, possibly enhancing the formation of humus as an end product. The gut of earthworms has a high specific peroxidase activity, which is a key enzyme in these polymerization reactions (Hartenstein, 1982).

8.2 AMOUNTS OF ORGANIC MATTER CONSUMED

Earthworms can consume very large amounts of litter, and the amount they ingest seems to depend more on the total amount of suitable organic matter available than on other factors. If physical soil conditions are suitable, the numbers of earthworms usually increase until food becomes a limiting factor.

Earthworms pass a mixture of organic and inorganic matter through their guts when feeding or burrowing, and surface-feeding worms in particular, such as *L. terrestris*, consume large amounts of organic matter. The smaller, epigeic species of earthworms that feed on litter in woodlands, such as *L. castaneus* and *E. fetida*, produce casts that are almost entirely fragmented litter, whereas endogeic species such as *A. caliginosa* consume a large proportion of soil, and there is much less organic matter in their casts (Piearce, 1978).

Some workers have calculated the amounts of leaf litter of different plant species eaten by earthworms. *Lumbricus rubellus* consumed 20.4 mg dry weight of hazel litter/worm/day (Franz and Leitenberger, 1948); six other species of worms consumed an average of 27 mg of alder leaves/g fresh weight of worm/day (van Rhee, 1963); and *L. terrestris* consumed about 80 mg of elm leaves/g fresh weight of worm/day (Needham, 1957). The consumption of beech litter during laboratory incubations lasting 24 weeks was estimated to be 19 mg/g wet weight of earthworms/day for *L. rubellus* and 26 mg/g wet weight/day for *D. octaedra* (Haimi and Huhta, 1990). *Lumbricus terrestris* was shown to consume 10–15 mg litter/g fresh weight/day in reclaimed peat soils in Ireland (Curry and Bolger, 1984). Kaushal et al. (1994) fed a variety of leaves (corn, wheat and mixed grasses) to *Amynthas alexandri* and reported a food consumption rate that varied from 36 to 69 mg/g live worm/day. Daniel (1991) showed that leaf litter consumption by juvenile *L. terrestris* could be described by a nonlinear function of soil temperature, soil water potential and food availability.

These three factors probably govern the amounts and rates of food consumed by most litter-feeding earthworm species.

Earthworm populations are capable of consuming a large portion of the entire annual litter fall in some ecosystems. In an apple orchard, *L. terrestris* consumed the equivalent of 2000 kg/ha of leaf litter between leaf fall and the end of February (98.6% of the total leaf fall) (Raw, 1962). Based on an estimate of litter consumption of 27 mg dry litter/g wet weight of earthworms/day, Satchell (1967) estimated that a population of *L. terrestris* in a mixed forest in England could consume the entire annual leaf fall, of 300 g/m², in about 3 months. Nielson and Hole (1964) reported that earthworm populations in mixed forests in Wisconsin, USA, could consume the entire annual leaf fall of the forest. Knollenberg *et al.* (1985) calculated that a population of *L. terrestris* present in a woodland floodplain in Michigan, USA could consume 94% of the annual leaf fall in 4 weeks during spring. Sugi and Tanaka (1978a) calculated that a population of earthworms composed of six species of *Pheretima* and one species of *Allolobophora*, could ingest 1071 g/m²/yr of litter from the soil surface in evergreen oak forests in Japan. This amount of litter was 1.4 times the annual litter fall in these forests, suggesting that the earthworms could only obtain adequate food by reingesting their casts or feeding on other fractions of organic matter in the soil. At a nearby site, with smaller earthworm populations, Sugi and Tanaka (1978b) estimated that earthworms consumed about 56% of the total annual leaf fall. Lavelle (1978), working in the Lamto region of Ivory Coast, calculated that a mixed population of eudrilid and megascolecid earthworms ingested about 30% of the litter decomposed in a grass savanna and 27% of that decomposed in shrub savanna annually. Madge (1966) calculated that in tropical forests in Nigeria, the litter fall was three or four times as much as in a temperate forest, and suggested that earthworms were the most important animals in fragmenting and incorporating it into soil.

When individuals of *A. longa*, *A. caliginosa* and *L. rubellus* were fed on cow dung in cultures for 2 years, the average dry weight of dung each individual consumed during this time was 35–40 g by *A. longa*, 20–24 g by *A. caliginosa*, and 16–20 g by *L. rubellus* (Guild, 1955). On this basis, the annual consumption of dung in the field by these species, at a population density of 120 000 adults/ha, can be calculated as 17–20 tonnes/ha, with a total estimated consumption for the whole population of about 25–30 tonnes/ha. Immature individuals of *A. caliginosa* consumed dung in culture at a rate of 80 mg oven-dry matter (400 mg wet weight)/g fresh weight of worm/day (Barley, 1959a), which was about twice the amount reported by Guild for this species. The amount of dung produced by dairy cattle (675 tonnes/ha) has been estimated as only one-quarter of the amount that a typical earthworm population could consume (Satchell, 1967). Hendriksen (1991) estimated that a field population of *L. festivus* and *L. castaneus* in a pasture in Denmark could consume 10–15 tonnes of dung/ha in 180 days. This corresponds to the dung produced by two or three dairy cows, which is slightly above the normal stocking rate per hectare.

Crossley *et al.* (1971) calculated the rate of throughput of soil by *Octolasion* sp. in cultures of soil tagged with radiocesium (^{137}Cs), and reported that soil passed through the worms' guts at a rate of about 86 mg/day/worm, equivalent to 28.8% of the live weight of the earthworms. This agrees well with calculations by Satchell (1967), who multiplied the weight of soil in dissected earthworms by an estimate of how rapidly food passes through their guts, and calculated that individuals of *L. terrestris* consumed 100–120 mg or 10–30% of their live body weight per day and individuals of *A. longa*, 20% of their live body weight per day.

Even when suitable organic material such as dung or litter is freely available to earthworms, many species also ingest large quantities of mineral soil. When individuals of *A. caliginosa* had unlimited quantities of litter available, they still ingested 200–300 mg of soil/g body weight/day, and the ingested mineral soil passed through the gut in about 20 hours (Barley, 1961). Lumbricids in old pasture land at Rothamsted consumed between 50 and 90 tonnes of oven-dry soil/ha, according to calculations by Evans (1948a), but this was almost certainly an underestimate because the sampling method he used was inefficient. Scheu (1987b) estimated that a population of *A. caliginosa* in a beechwood in Germany consumed up to 6 kg/m^2 of soil/yr. James (1991) studied organic matter processing by a mixed earthworm community containing several species of the native North American genus *Diplocardia* and the European lumbricids, *A. caliginosa* and *O. cyaneum*. He estimated that the earthworms consumed from 4 to 10% of the soil, and 10% of the total organic matter, in the top 15 cm of soil annually.

8.3 NUTRIENT CYCLING

Earthworms can have major influences on nutrient cycling processes in many ecosystems. By turning over large amounts of soil and organic matter, they can increase the rates of mineralization of organic matter, converting organic forms of nutrients into inorganic forms that can be taken up by plants. Understanding the influence of earthworms on nutrient transformations in soil requires knowledge of their effects on a variety of spatial and temporal scales, because different results are often observed at different scales of observation (Blair *et al.*, 1994b). Earthworms influence organic matter and nutrient cycles on four scales:

1. during transit through the earthworm gut;
2. in freshly deposited earthworm casts;
3. in aging casts; and
4. during the long-term genesis of the whole soil profile (Lavelle and Martin, 1992).

Their effects at each of these scales are influenced by soil type, climate, vegetation and the availability and quality of organic matter. Integrating across scales and understanding the interrelationships among multiple factors is essential to understanding the overall influence of earthworms on nutrient cycling processes. Many of the influences of earthworms on nutrient cycling processes and the mineralization of organic matter are mediated by the interactions that occur between earthworms and micro-organisms. These interactions and their importance to biogeochemical cycles are discussed more thoroughly in Chapter 9.

8.3.1 CARBON

Although earthworms consume and turn over a large amount of organic matter, their contribution to total heterotrophic soil respiration is small, accounting usually for only 5–6% of the total energy flow in terrestrial ecosystems (Chapter 7). It has been calculated for a population of the species *A. caliginosa* in Australia, that earthworms were responsible for only 4% of the total carbon consumption (Barley and Kleinig, 1964), and in two English woodlands, *L. terrestris* was responsible for only 8% of the total carbon consumption (Satchell, 1967). This was assuming that the consumption of 22.9 liters of oxygen/m² was equivalent to a carbon consumption of 118.6 kg/ha, and that 3000 kg of litter that was 50% carbon fell on to the soil surface per hectare.

The small contribution of earthworms to overall CO_2 output from ecosystems is due to their low assimilation efficiencies. Carbon assimilation efficiencies ranging from 2 to 18% have been reported for several species of endogeic earthworm (Bolton and Phillipson, 1976; Barois *et al.*, 1987; Scheu, 1991; Martin *et al.*, 1992). Assimilation efficiencies of litter-feeding earthworms tend to be greater than those of endogeic species. For example, Dickschen and Topp (1987) reported assimilation efficiencies of 30–70% for *L. rubellus*, depending on the quality of the litter ingested by the worms and the temperature at which they were incubated. Daniel (1991) reported efficiencies of 43–55% for *L. terrestris* fed on fresh dandelion leaves, although actual assimilation efficiencies for *L. terrestris* feeding on decaying plant litter, under natural conditions, are probably much lower.

Earthworms may make a substantial contribution to the total heterotrophic soil respiration when their populations are very large and active. Hendrix *et al.* (1987) estimated that earthworms were responsible for about 30% of the total heterotrophic soil respiration, during late winter and early spring, in a no-tillage agroecosystem in the south-eastern US; population densities at their site reached a maximum of nearly 1000 individuals/m².

Earthworms apparently assimilate carbon from recently deposited fractions of soil organic matter, composed of more readily decomposable substances, such as those utilized by the overall decomposer community. This was shown by Martin *et al.* (1992), who used $^{13}C/^{12}C$ ratios of carbon in earthworms to determine the source of their assimiliated carbon. The ratio of ^{13}C to ^{12}C in an earthworm should reflect the ratio of these isotopes in the organic matter that the worms are assimilating. Martin *et al.* (1992) incubated earthworms in soils where recent changes in vegetation had led to distinctive patterns of $^{13}C/^{12}C$ in the pool of recently deposited organic matter. The $^{13}C/^{12}C$ ratios of the earthworms matched those of the recently deposited organic matter in the soil, indicating that the worms assimilated carbon primarily from recent organic matter pools, rather than from older, more humified and recalcitrant pools, although their diet contained carbon from both these pools.

A large amount of water-soluble organic compounds are added to the gut contents as food passes through the earthworm gut (Barois and Lavelle, 1986). These high-energy mucous compounds stimulate microbial activity in the earthworm gut and may enable the intestinal microflora to digest the more complex organic compounds of the soil to the benefit of the earthworm. A large proportion of these high-energy, water-soluble compounds are resorbed in the posterior portion of the gut, but some are egested in earthworm casts (Scheu, 1991), where they continue to serve as energy substrates for micro-organisms. The influence of earthworms on microbial respiration is examined further in Chapter 9.

The types and amounts of carbon in earthworm casts differ from those of the surrounding soil. There is a considerable increase in the polysaccharide content of casts relative to uningested soil (Parle, 1963b; Bhandari *et al.*, 1967). Shaw and Pawluk (1986a,b) reported a greater amount of clay-associated carbon in earthworm casts than in surrounding soil, which they suggested may promote the stabilization of soil carbon through binding with clays. The carbon contents of casts tend to be higher than in the surrounding soil, in part due to the addition of intestinal mucus, but also because earthworms may select soil fractions enriched in organic compounds (Lee, 1985; Blair *et al.* 1994b). The turnover of carbon by earthworms is quite rapid. Ferrière and Bouché (1985) labelled the earthworm *Nicodrilus longus* by feeding it algae labelled with ^{14}C and ^{15}N. They reported that the entire carbon content of the earthworm could turnover in 40 days and a considerable portion of this turnover was due to mucus excretion. Scheu (1991) reported that secretion of mucus in casts and from the body wall accounted for 63% of total carbon losses (mucus excretion plus respiration) from the geophagous earthworm, *Octolasion lacteum*; this corresponded to a daily loss of 0.7% of total carbon in this species. Respiration, by contrast, accounted for only 37% of total carbon losses from the earthworms. Lavelle (1988) estimated that populations of *Pontoscolex corethrurus*, in tropical pastures of Mexico, secrete up to 50 mg of mucus/ha in a single year. This is 20% of the total carbon in the soil.

An understanding of the influence of earthworms on the mineralization of carbon in decomposing plant litter has been aided by the use of plant materials labeled with the radioactive isotope ^{14}C. Cheshire and Griffiths (1989) added uniformly ^{14}C-labeled grass to soils with earthworms or with no earthworms, and monitored the loss of ^{14}C over time. They showed that total loss of ^{14}C, after 1 year, was 63% in the presence of earthworms but only 53% in their absence. Cortez *et al.* (1989) added ^{14}C-labeled wheat straw to soil with or without the earthworm *Nicodrilus longus*, and reported that the output of $[^{14}C]CO_2$, in a 31 day laboratory incubation, was 3.2-fold greater in soils with worms than in those without worms (Fig. 8.2). Earthworms increase the mineralization of ^{14}C-labeled lignin, although the extent of the increase depends upon the species of earthworm, the length of the incubation period and the other types of organic matter present (Scheu, 1993a, b). During 253 day laboratory incubations, earthworms of the species *O. lacteum* increased the mineralization of labeled lignin for the first 10 weeks, but decreased mineralization later in the experiment, in four out of five limestone soils. In the same soils, earthworms increased the mineralization of ^{14}C-labeled holocellulose by factors of 1.5 and 1.4, in soils from 6- and 13-year-old fallows, but had only slight effects on such mineralization in soils of a wheat field and a

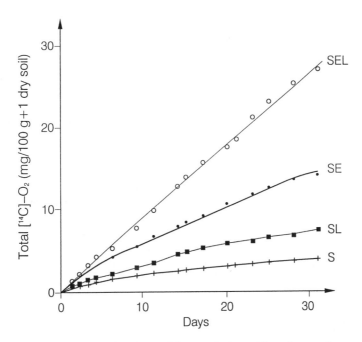

Figure 8.2 Cumulative CO_2 released from soil alone (s), soil + earthworms (SE), soil + plant litter (SL) and soil + earthworms + plant litter (SEL). (From Cortez, Hameed and Bouché, 1989.)

beechwood (Scheu, 1993a). Dietz and Bottner (1981) reported that earthworms increased the depth of incorporation of ^{14}C-labeled litter into soil.

A fundamental unanswered question regarding the effects of earthworms on cycling of soil carbon is whether the net effect of earthworms is to increase or decrease the overall storage of organic carbon (Blair *et al.*, 1994b). Earthworms can increase the amounts of carbon stored by increasing plant growth, and this subject is examined in detail in Chapter 10. On the other hand, most research suggests that earthworms increase the rates of loss of carbon from soil by stimulating the mineralization of organic matter, as described above. O'Brien and Stout (1978) estimated that the annual flux of carbon from a New Zealand pasture may have increased from 300 to 1000 kg/ha after earthworms were introduced. The mean residence time of organic carbon decreased from 180 to 67 years. However, more recent research suggests that the stabilization of organic matter in earthworm casts may lead to increased carbon storage and decreased mineralization of organic matter in the long term. Martin (1991) obtained fresh earthworm casts from *M. anomala* maintained in an homogenized soil. The fresh casts contained 2% less total carbon than the surrounding soil, demonstrating a short-term increase in the rates of mineralization of organic matter. However, in longer-term incubations of 1 year, carbon mineralization in the casts (3% per year) was much lower than in the non-ingested soil (11% per year). Lavelle and Martin (1992) claimed that the stabilization of organic matter in earthworm casts may be an important mechanism for the stabilization of organic matter in tropical soils (Chapter 10). This method of organic matter stabilization is probably important in temperate soils as well. Scheu (1993b) suggested that the stabilization of ^{14}C-labeled lignin in earthworm fecal pellets may have been responsible for the observed decrease in the mineralization of labeled lignin that occurred when *O. lacteum* was present, during the later stages of a laboratory incubation. Scheu and Wolters (1991a) examined the influence of *O. lacteum* on bioturbation and cumulative mineralization of ^{14}C-labeled beech litter in beechwood soils in Germany. They observed a significant reduction in cumulative carbon mineralization when the endogeic earthworm, *O. lacteum*, was present. They attributed the stabilization of organic matter at this site to the incorporation of litter fragments into the mineral soil by *O. lacteum*, and suggested that such bioturbation is a key process in the formation of carbon-rich mull soils of the beechwood that they investigated. Much more research needs to be done to determine the overall net influence of earthworms on the long-term storage or loss of carbon in soil.

8.3.2 NITROGEN

Significant amounts of nitrogen can pass directly through the earthworm biomass in terrestrial ecosystems. Satchell (1963) estimated that 60–70 kg nitrogen/ha/yr were returned to the soil in the dead tissues of *L. terrestris* in a woodland in England, and that an additional 30–40 kg nitrogen/ha/yr were returned in urine and mucus deposited by this species (Fig. 8.3). Keogh (1979) estimated that *A. caliginosa* contributed about 109–147 kg nitrogen/ha/yr to mineral nitrogen pools in a New Zealand pasture, or about 20% of the total amount of nitrogen mineralized in the pasture. Nowak (1975) estimated that the turnover of nitrogen through earthworm tissues in a pasture in Poland equalled 3–17% of the total nitrogen input from plant litter. Rosswall and Paustian (1984) calculated that 10 kg nitrogen/ha/yr flowed through an earthworm population that contained a mean annual standing stock of 3.0 kg nitrogen/ha. The direct flux of nitrogen through earthworm biomass in a no-till agroecosystem in Georgia, USA, was estimated to be 63 kg nitrogen/ha/yr, or nearly 38% of the total nitrogen uptake by the crop (Parmelee and Crossley, 1988). Christensen

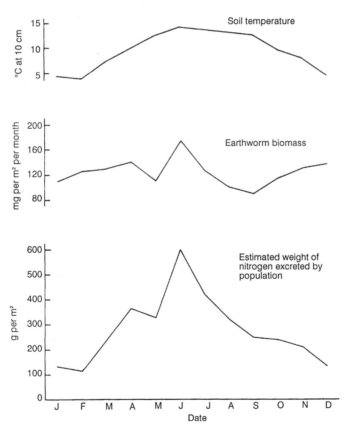

Figure 8.3   Estimated weight of nitrogen excreted by a population of *L. terrestris* at Merlewood Lodge Wood in 1960 (after Satchell, 1963).

(1988) reported that dead earthworm tissues contributed 20–42 kg nitrogen/ha to the soil during the autumn in three arable systems in Denmark.

Most nitrogen in earthworm tissues is associated with proteins. The protein content of earthworms ranges from 60 to 80% (Lawrence and Millar, 1945; Sabine, 1978; Bouché, 1981). Estimates for the nitrogen content of lumbricid earthworm tissues range from about 10% to 12% of their dry weight (Barley, 1961; Bouché, 1981). Parmelee and Crossley (1988) reported a nitrogen content of 8.45% of ash-free dry mass for *L. rubellus* and *A. turgida*.

Dead earthworms decompose rapidly and the nitrogen in earthworm tissues turns over rapidly and is mineralized readily. Satchell (1967) reported that nearly 70% of the nitrogen in dead earthworm tissue was mineralized in 10–20 days. Ferrière and Bouché (1985) labeled individuals of the anecic earthworm, *Nicodrilus longus*, by feeding them with algae that were double-labeled with ^{15}N and ^{14}C. They reported that the entire nitrogen (and carbon) content of the earthworms could turn over within 40 days. Barois *et al.* (1987) labeled individuals of *Pontoscolex corethrurus* with ^{15}N and reported that 14% of the incorporated label was lost within 5 days, and 30% was lost after 30 days. Hameed *et al.* (1994b) labeled *L. terrestris* with ^{15}N and a colored dye and then released them in an undisturbed grassland in France. The dye enabled the authors to recapture the labelled worms periodically, and they reported that the worms lost 80% of the ^{15}N label after 48 days in the field. From these results they calculated that the nitrogen flow through the earthworms was 16.1% of their total body nitrogen per day. This agreed with results from laboratory studies which showed that the daily nitrogen flux through *L. terrestris* varied from 13.6% to 14.3% of its total body nitrogen (Hameed *et al.*, 1994a).

Earthworms consume large amounts of plant organic matter that contain considerable quantities of nitrogen, and much of the nitrogen that they assimilate into their own tissues is returned to the soil in their excretions. Needham (1957) suggested that very little nitrogen is excreted in the feces of earthworms, but other workers have reported considerably more nitrogen in casts than in the surrounding soil (Lunt and Jacobson, 1944; Graff, 1971). When young worms of the species *A. caliginosa* were fed on soil containing finely ground plant litter, and their feces and urine collected, about 6% of the nonavailable nitrogen ingested by the worms was excreted in forms available to plants (Barley and Jennings, 1959) (Table 8.1). The presence of worms in cultures of well-aerated moist soil increased the rate of oxygen consumed and the rate of accumulation of ammonium and nitrate during the early stages of decay. These excretions, which include mucoproteins secreted by gland cells in the epidermis, and ammonia, urea, and possibly uric acid and allantoin, in a fluid urine excreted from the nephridiopores, contribute a significant amount of readily assimilable nitrogen to soil. Several estimates of the amounts of

Table 8.1   Influence of *A. caliginosa* on decomposition of organic matter (from Barley and Jennings, 1959)

	Oxygen consumed (30 days)		Nitrate and ammonium accumulated (50 days)	
	μl/g of medium	log value	p.p.m.	log value
With worms	2600	3.39	129	2.10
Without worms	2190	3.32	105	2.01
Sig. diff. (p=0.01)	–	0.05	–	0.09

nitrogen excreted by earthworms are based on the method of Needham (1957), who placed earthworms in a small volume of distilled water for 24 hours and then analyzed the water for its nitrogen content. Needham's figures were used to calculate that earthworms in a pasture containing 126 g fresh weight of earthworms/m^2 produced about 70 kg nitrogen/ha/yr in mineralizable excretions. Based on Needham's method, nitrogen excretion rates of 95 μg/g fresh weight/day (Needham, 1957), and 60–160 μg/g fresh weight/day (Tillinghast, 1967) were calculated for *L. terrestris*, and 10–60 μg/g fresh weight/day for *A. caliginosa* (Christensen, 1987). Lee (1983) used these values to estimate an annual nitrogen excretion rate of 18–50 kg nitrogen/ha for a typical population of lumbricid earthworms. More recently, Binet and Trehen (1992) used earthworms labeled with ^{15}N to calculate a nitrogen excretion rate of 21 μg/g fresh weight/day for *L. terrestris*. Less is known about the amounts of nitrogen produced in earthworm mucus that is deposited in casts or secreted from their body walls. However, Scheu (1991) estimated that *O. lacteum* produced, in 1 day, an amount of mucus equivalent to about 0.2% of its total body nitrogen. There are no reliable estimates of the nitrogen assimilation efficiencies of earthworms, and this represents a considerable gap in our understanding of basic earthworm biology (Blair *et al.*, 1994b).

The concentrations of inorganic nitrogen in fresh earthworm casts and around the lining of their burrows are usually much greater than in bulk soil, with ammonium usually being the dominant form of inorganic nitrogen in the casts (Lunt and Jacobson, 1944; Parle, 1963b; Graff, 1971; Lavelle and Martin, 1992). When young earthworms, of the species *A. caliginosa*, were fed on soil containing finely ground plant litter, and their feces and urine were collected, about 6% of the nonavailable nitrogen ingested by the worms was excreted in forms that are readily available to plants (Barley and Jennings, 1959). The increase in inorganic nitrogen in earthworm casts is due to excretory products and mucus from the earthworm, as well as through increased rates of mineralization of organic nitrogen by micro-organisms in the casts. The rates of nitrification in casts can be high, and several authors have noted a simultaneous increase in nitrate and decrease in ammonium as casts age (Parle, 1963b; Syers *et al.*, 1979; Lavelle *et al.*, 1992). Lavelle and Martin (1992) reported that the amount of total mineral nitrogen (NH_4^+ plus NO_3^-) in freshly deposited casts of *P. corethrurus* was nearly 5.5 times greater than in the surrounding soil. The initially high levels of mineral nitrogen fell substantially after only 12 hours but the amount of total mineral nitrogen in casts was still four times greater in the casts than in the bulk soil after 16 days.

A key question is whether the total amounts of available nitrogen deposited in earthworm casts can make a significant contribution to the total amounts of nitrogen available in soil for plant growth. Lee (1985), using data from Barley and Jennings (1959) and Barley (1959b), calculated that earthworm (*A. caliginosa*) casts would contribute only about 22–28 g nitrogen/ha/yr to soils in the Adelaide region of Australia. Aldag and Graff (1975) kept *L. terrestris* in pots and reported that the content of available nitrogen in their casts was 40% greater than in the surrounding soil. Lee (1985) applied these results to the known rates of casting of European lumbricids (5–7 kg/m²/yr) to calculate that the additional input of available nitrogen due to earthworm casts would be about 35–50 g/ha/yr. These estimates represent a small fraction of the annual nitrogen demand of plants. Syers *et al.* (1979) calculated that surface casts of *L. rubellus* in a pasture in New Zealand accumulated 3.0 kg ammonium-nitrogen/ha/yr and 0.9 kg nitrate-nitrogen/ha/yr, which represents a relatively small amount of mineral nitrogen from an agronomic point of view. However, they did not include subsurface casts in their calculations, which may have caused a gross underestimate of the overall amount of mineral nitrogen produced in the casts of all earthworm species in the pasture ecosystem. Other workers have reported significant turnover of nitrogen in earthworm casts. For example, James (1991) used earthworm population estimates, soil climate data and cast production–temperature relationships to estimate that the total amount of mineral nitrogen produced in earthworm casts (5–5.5 kg nitrogen/ha/yr) was 10–12% of the total nitrogen taken up by plants in the nitrogen-limited tallgrass prairie in Kansas, USA. Lavelle *et al.* (1992) reported that large amounts of nitrogen are deposited in earthworm casts in humid tropical pastures. They found that the casts of *P. corethrurus* had more than five times the amount of available nitrogen as the surrounding soil. These earthworms consume several tonnes of soil/ha each year in tropical pastures and, as a result, release a minimum amount of 50–100 kg of mineral nitrogen into the soil. It is clear that earthworms can make a substantial contribution to the overall turnover of available forms of mineral nitrogen, especially when the amounts produced in earthworm casts are considered as well as the amounts produced in mucus secretions and from the decaying tissues of dead earthworms.

Earthworms increase the rates of mineralization of nitrogen, but, sur-
prisingly, there are few estimates of the influence of earthworms on the
net mineralization of nitrogen in bulk soil. The enhanced mineralization
of nitrogen caused by earthworm activity is linked to the enhanced mineralization of carbon, suggesting that certain fractions of organic matter,
that are protected physically from mineralization, become mobilized
during passage through the earthworm gut (Scheu, 1994). Anderson *et al.*
(1983) measured nitrogen mineralization in forest soil incubated with oak
litter and with or without the earthworm, *L. rubellus*. The earthworms
increased the mobilization of nitrate-nitrogen by 10 times and that of
ammonium-nitrogen by 80 times, relative to soil without earthworms.
Ruz-Jerez *et al.* (1992) reported that mineral nitrogen concentrations were
about 50% greater in soils with earthworms than in soils without earthworms, in laboratory incubation of grassland soil with different plant
residues added. Earthworms of the species *L. terrestris* increased the
release of ^{15}N from ^{15}N-labeled ryegrass threefold, in a laboratory incubation of cultivated silty clay loam soils in France (Binet and Trehen, 1992).
Scheu (1987b) observed a direct relationship between the biomass of *A.
caliginosa* and the increased rate of nitrogen mineralization in a laboratory
incubation (Fig. 8.4). He used this relationship, combined with laboratory-
derived relationships between temperature and nitrogen mineralization,
to calculate that a field population of *A. caliginosa* could cause an additional mineralization of 4.23 kg nitrogen/ha/yr, in a beechwood site on
limestone soil. This increase was only 2.6% of the estimated amount of
nitrogen mineralized at the site, annually (160 kg nitrogen/ha). However,

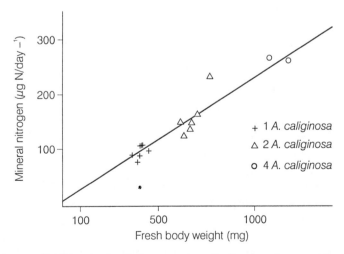

Figure 8.4 Additional rate of nitrogen mineralization in soil caused by the burrowing activity of *A. caliginosa* (one, two and four specimens/cage) in relation to the
total biomass of the animals (from Scheu, 1987b).

A. caliginosa accounted for only 17% of the total earthworm biomass, so the total amount of nitrogen mineralized by the entire earthworm community was probably much greater. Earthworms obviously can mobilize significant amounts of nitrogen, but much more research is needed, in a variety of ecosystems, if we are to increase our relatively sparse understanding of their net effects on nitrogen mineralization in the field.

Clearly, earthworms can increase the total uptake of nitrogen by plants, by increasing the amounts of available nitrogen in the soil and otherwise improving conditions for plant growth, although research data on this subject are scanty. Earthworms were shown to have no effect on the nitrogen content of black spruce needles (Marshall, 1971), ryegrass (McColl et al., 1982) or big bluestem (*Andropogon gerardii*) (James and Seastedt, 1986) in controlled laboratory experiments. Conversely, McColl et al. (1982) reported that activities of *A. caliginosa* significantly increased the total uptake of most major and trace elements (nitrogen, phosphorus, potassium, calcium, chlorine, magnesium and zinc) by ryegrass seedlings, although they did not increase the concentrations of these elements in the seedlings. In permanent pastures near Adelaide in Australia, Barley (1959b) reported that the period of the most intense surface-casting by earthworms corresponded with the most rapid growth period of the grasses in the pasture. Presumably, the nutrients in the earthworm casts were available for uptake by these grasses during their most intense growth period. Hameed et al. (1994b) introduced *L. terrestris* labeled with ^{15}N into a pasture in France. They recovered 24% of the original label in plant tissues after 48 days, demonstrating that the nitrogen excreted by the earthworms was available for uptake by plants. Spain et al. (1992) reported that the uptake of ^{15}N by *Panicum maximum*, from ^{15}N-labeled soils, increased significantly in soil to which they added earthworms, compared with soil without worms. The total nitrogen uptake by birch seedlings was twice as great in treatments with earthworms (*L. rubellus*) than in treatments with no earthworms, in microcosms containing a simulated coniferous forest floor (Haimi et al., 1992). Ruz-Jerez et al. (1992) showed that the uptake of nitrogen by ryegrass in the laboratory was significantly greater in soil that was incubated previously with earthworms than in soil incubated without worms. The discrepancies between the results of the various studies discussed above, some of which showed that earthworms increased the uptake of nitrogen and others which showed no such increase, may be due to the differences in the experimental techniques of the studies, but also suggest that other factors, such as soil type, organic matter and the species of plant and earthworm used, are important in determining the influence of earthworms on the uptake of nitrogen by plants. A more thorough coverage of the influence of earthworms on plant growth is presented in Chapter 10.

Earthworms can increase rates of loss of nitrogen by increasing the rates of denitrification and the leaching of nitrate and other mobile nitrogen compounds. Fresh earthworm casts usually have higher denitrification rates than the surrounding soil (Svensson et al., 1986; Elliot et al, 1990). Syers et al. (1979) reported a potential for increased losses of nitrogen from a New Zealand pasture, in the cooler winter months, due to denitrification in earthworm casts. Knight et al. (1992) estimated that earthworm casts on the soil surface in English pastures could account for 12% of the total denitrification losses from an unfertilized pasture and 26% of the losses from a fertilized pasture. They also reported that earthworms tripled the amounts of nitrate in leachates from these pastures. The degree to which earthworms increased the losses of nitrogen depended on the amounts and types of fertilizer added, losses being greater when large amounts of inorganic fertilizer were added to the soil. This is consistent with results from other workers which have shown that earthworms can increase the amount of mineral nitrogen in soil to a greater degree in soils fertilized with inorganic sources of nitrogen than in those fertilized with organic fertilizers (Blair et al., 1995; Bohlen and Edwards, 1995).

Several authors have demonstrated that earthworms can increase the loss of nitrogen in soil leachates in field lysimeter and microcosm studies (Anderson et al., 1983; Buse, 1990; Binet and Trehen, 1992; Robinson et al., 1992). Earthworms of the species A. caliginosa were shown to increase nitrogen in leachates from field lysimeters in limed peat soil in Cumbria, UK. A total of 60 kg nitrogen/ha/yr was collected in leachates from field lysimeters in limed peat soil containing earthworms, whereas only 36 kg nitrogen/ha/yr was collected from soils containing no earthworms (Robinson et al., 1992). Earthworms, particularly deep-burrowing species, create macropores which can act as preferential flow pathways for infiltrating water (Chapter 10). These pathways of preferential flow may be conduits for the loss of soluble forms of nitrogen. However, very few field studies have quantified the influence of earthworm macropores on the leaching of nitrogen and other nutrients. Shipitalo et al. (1994) reported that earthworm burrows collected more water, NO_3^- and NH_4^+ than an equivalent area of surrounding soil overlying a pan lysimeter. The total loss of inorganic nitrogen through earthworm burrows was estimated to be up to 14 kg/ha/yr.

8.3.3 NITROGEN FIXATION

Nitrogen fixation is mediated by micro-organisms, and early reports suggested that earthworms had no influence on nitrogen-fixing bacteria (Day, 1950; Khambata and Bhatt, 1957), and that, in contrast to some other groups of soil micro-organisms, the numbers of Azotobacter, an important nitrogen-fixing micro-organism, actually decreased when passing through earthworm guts (Ruschmann, 1953). However, Stöckli (1928) had reported that numbers of *Azotobacter* may be increased in earthworm casts. Most recent research suggests that earthworms can contribute to nitrogen fixation. Mba (1987) observed a great increase in nitrogen fixation during composting of ground Dallas grass (*Paspalum dilatatum*) by the earthworm *E. eugeniae*. The total nitrogen content of the substrate increased in the presence of these earthworms and the activity of the nitrogen-fixing enzyme, nitrogenase, increased tenfold in earthworm casts relative to the grass substrate. Šimek and Pižl (1989) also showed that activity of the nitrogenase enzyme in casts of *L. rubellus* was greater than in uningested soil, and that nitrogenase activity in soil was increased significantly by the feeding and burrowing activity of these worms. Shaw and Pawluk (1986a) reported that the nitrogen-fixing activity of the micro-organism *Spirillum* was significantly greater in the linings of earthworm burrows than in the surrounding soil. Bhatnagar (1975) investigated the microbial activity associated with earthworm burrows in a permanent grassland in France and estimated that 40% of the total aerobic nitrogen fixers and 13% of the total anaerobic nitrogen fixers in the soil occurred in the 2 mm zone surrounding earthworm burrows. The contribution of these interactions to total nitrogen fixation in soil may be small, although much more investigation is needed. The interactions between earthworms and symbiotic nitrogen-fixing micro-organisms are discussed further in Chapter 10.

8.3.4 EFFECTS ON THE C:N RATIO

The ratio of carbon to nitrogen (C:N ratio) in organic matter added to soil is of importance, because net mineralization of this organic matter does not occur unless the C:N ratio is of the order of 20:1 or lower. The C:N ratio of freshly fallen leaf litter is much higher than this, being 24.9:1 for elm, 27.6:1 for ash, 38.2:1 for lime, 42.0:1 for oak, 43.5:1 for birch, 54.0:1 for rowan and 90.6:1 for Scots pine (Wittich, 1953). Succulent leaf material often has much smaller C:N ratios, whereas tougher tree leaves with a high percentage of resistant constituents, such as cellulose and lignin, that are unpalatable to earthworms and other litter animals, often have high C:N ratios (Witkamp, 1966). During the process of leaf litter breakdown and decomposition, the C:N ratio of the litter decreases progressively, until the ratio falls to about 20:1, when net mineralization of nitrogen begins and the mineralized nitrogen can be taken up directly by plants (Edwards *et al.*, 1995b). Earthworms may lower the C:N ratio by combustion of carbon during respiration.

Earthworms can alter the C:N ratio of material that passes through their digestive tract although they have relatively low assimilation efficiencies for both carbon and nitrogen. Several authors have reported that earthworm casts often have a C:N ratio slightly greater than that of the surrounding soil (Wasawo and Visser, 1959; Graff, 1971; Czerwinski et al., 1974; Aldag and Graff, 1975). This could occur if earthworms ingest material that is enriched in carbon content selectively or have greater assimilation efficiencies for nitrogen than for carbon. However, other workers have found the C:N ratio of earthworm casts to be lower than that of the surrounding soil (Lavelle, 1978; Syers et al., 1979). Bouché et al. (1983) reported a C:N ratio in casts of 20:1 from earthworms that were fed litter with a C:N ratio of 45:1. However, such a comparison of the C:N ratio of earthworm casts to that of soil or leaf litter, does not always provide a valid assessment of the influence of earthworms on the C:N ratio of ingested material, because it is often very difficult to determine the exact nature of the heterogeneous material that is being ingested by the earthworms. The use of food materials labeled with stable isotopes helps to overcome this problem. Cortez et al. (1989) kept earthworms in microcosms containing wheat straw that was double-labeled with ^{14}C and ^{15}N, and had a $^{14}C:^{15}N$ ratio of 30.5. During the early period, the $^{14}C:^{15}N$ ratio of earthworm casts was 13.6, indicating that earthworms and the microorganisms associated with their casts enhanced the mineralization of litter-derived carbon. It seems likely that in most instances, earthworms probably decrease the C:N ratio in soil, because they increase combustion of carbon by enhancing total soil respiration.

Another way that earthworms influence the C:N ratio of soil organic matter is by feeding selectively on organic matter with a high nitrogen content (and thus a low C:N ratio). Bohlen et al. (1985b) studied the decomposition of crop residues in maize agroecosystems in which earthworm populations had been decreased, increased or left unmodified, in enclosed field plots. They showed that earthworms, in particular L. terrestris, actively selected crop residues that contained the greatest amount of nitrogen, thereby increasing the loss of nitrogen from the residues to a greater extent than they increased the loss of carbon. As a result, the crop residues remaining on the soil surface had higher C:N ratios, in plots with reduced earthworm populations, than in those with increased or unmodified populations. Ketterings et al. (1995) reported comparable results for particulate organic matter (particle size >2 mm) in the top 15 cm of soil in these same experimental plots. Particulate organic matter in plots with reduced earthworm populations contained nearly 30% more nitrogen and had a 20% lower C:N ratio than that in plots with unmodified or increased populations. Thus, earthworms can decrease the total amount of plant residues but appear to increase the C:N ratio of the remaining residues, which may have important consequences for patterns of nitrogen mineralization or immobilization by micro-organisms associated with the residues.

8.3.5 PHOSPHORUS

Most workers who have examined the available phosphorus (P) in earthworm casts, and in surrounding soil, have reported that casts usually have more available phosphorus than soils without earthworms. This was first demonstrated by Lunt and Jacobson (1944) (Table 8.2) and Graff (1971) (Table 8.3), and has been confirmed by many other workers throughout the world, in a variety of ecosystems (Powers and Bollen,

Table 8.2 Comparison of the available mineral elements in the casts of earthworms and in the upper layers of a ploughed soil in Connecticut, USA (from Lunt and Jacobson, 1944)

	Earthworm excreta	Depth of soil layer	
		(0–15 cm)	(20–40 cm)
Loss by ignition (%)	13.1	9.8	4.9
Carbon/nitrogen ratio	14.7	13.8	13.8
Nitrate nitrogen (p.p.m.)	21.9	4.7	1.7
Calcium:			
Total (%)	1.19	0.88	0.91
Exchangeable (p.p.m.)	2793	1993	481
Exchangeable calcium/total calcium (%)	25.6	24.4	6.1
Magnesium:			
Total (%)	0.545	0.511	0.548
Exchangeable (p.p.m.)	492	162	69
Exchangeable magnesium/total magnesium	9.19	3.24	1.29
Phosphorus available (p.p.m.)	150	20.8	8.3
Potassium available (p.p.m.)	358	32.0	27.0
pH	7.00	6.36	6.05

Table 8.3 Mineral elements in soil and worm casts (mg/100g dry wt) (from Graff, 1971)

	Casts	Soil
C	8550	3925
N	536	350
C : N ratio	16	11
P (aqua regia)	102	68
P (lactate-soluble)	13.7	2.2
K (soda extract)	1097	799
K (lactate-soluble)	44.6	7.0
pH (H_2O)	5.8	5.0
pH (KCl)	5.4	4.0

1935; Puh, 1941; Stöckli, 1949; Ponomareva, 1950; Finck, 1952; Nye, 1955; Atlavinyte and Vanagas, 1973; Czerwinski et al., 1974; Lal, 1974; Petal et al., 1977; Spiers et al., 1986; Tiwari et al., 1989; Krishnamoorthy, 1990). Increases in amounts of phosphorus in casts are commonly of the order of 5–10 times greater than surface soil (Lee, 1985). Sharpley and Syers (1976) collected fresh casts from a grassland in New Zealand and studied phosphorus availability in the casts relative to the surface (0–5 cm) soil. They reported that inorganic and organic phosphorus was concentrated particularly in the fine-size fractions (<4 μm and 4–20 μm) of the casts. The rates of release of inorganic phosphorus in the casts was about four times faster than that in the surface soil during 3 days of sequential extraction.

   Much of the increase in the availability of phosphorus in earthworm casts relative to that of surrounding soil is due to enhanced phosphatase activity in the casts (Satchell and Martin, 1984; Satchell et al. 1984), although it is not known whether the increase in activity is due to earthworm-derived enzymes or to increased microbial activity (Park et al., 1992). Krishnamoorthy (1990) examined the phosphorus-mineralizing capacity of the four species of earthworms, Lampito mauritii, Perionyx excavatus, Pontoscolex corethrurus and Pheretima elongata, in grasslands around Bangalore, India, and reported that, of the four species, P. excavatus showed the greatest capacity for increasing the availability of phosphorus. Phosphatase activity in the casts of P. excavatus was much greater than in soils without worms. This was true under both acid (pH 3–4) and basic (pH 9–9.5) conditions, indicating that the activities of both acid and alkaline phosphatases were greater in casts relative to uningested material.

   The contributions of earthworm casts to the total amount of available phosphorus in soils can be significant. An estimated total of 9.0 kg/ha/yr of inorganic phosphorus was produced in surface earthworm casts in a pasture in New Zealand (Sharpley and Syers, 1976). Most of the production of inorganic phosphorus in earthworm casts occurred in autumn (April–May) and decreased during winter (May–August), probably due to decreasing temperatures and a resulting decline in the activity of microbial phosphatase. Tiwari et al. (1989) also observed seasonal variations in the concentration of inorganic phosphorus in earthworm casts, with the highest levels relative to the surrounding soil occurring in the middle of the rainy season (July–August), when plant nutrient demand was great. Relatively large amounts of inorganic phosphorus in earthworm casts have been reported for tropical grasslands (55 kg/ha/yr) and woodlands (38 kg/ha/yr) in India (Krishnamoorthy, 1990). These are probably gross underestimates of the actual amount of inorganic phosphorus produced in earthworm casts, because they are based solely on the amount of surface casts produced by earthworms, and the majority of casting by a number of species probably occurs beneath the soil surface. James (1991) calculated that the total inorganic phosphorus produced in earthworm casts (surface and subsurface casts) ranged from 30 to 50% of the total plant uptake of phosphorus in a tallgrass prairie (about 0.3–1.5 kg/ha/yr).

The effect of earthworms on phosphorus availability is influenced by the phosphorus adsorption characteristics of the soil. Lopez-Hernandez *et al.* (1993) reported that the contents of water-soluble and exchangeable phosphorus were much greater in earthworm casts than the surrounding tropical soil, which had a low P-sorption capacity. The amounts of available phosphorus in casts continued to rise during 4 days of incubation. By contrast, the levels of available phosphorus in fresh earthworm casts increased only slightly, in a soil with a high P-sorption capacity, and decreased after 4 days to values that did not differ from uningested soil. Mouat and Keogh (1987) reported that subsurface casts, earthworm burrow linings and soil within 2 mm of the burrow wall had a much greater content of water-soluble phosphate compared with soil without earthworm activity. They reported that the addition of earthworm casts to soil increased the concentration of phosphate in the equilibrating soil solution, but that the casts had a progressively smaller effect with increasing soil depth, which they attributed to the increasing phosphate-adsorbing capacity of the soil with increasing depth. Nevertheless, they concluded that the continuing deposition of subsurface casts provided localized pulses of available phosphorus that could be of great importance to the nutrition of pasture and other plants that take up nutrients at greater soil depths.

Earthworms have been shown to increase the uptake by plants of phosphorus from plant litter. In a greenhouse experiment, MacKay *et al.* (1982) investigated the influence of earthworms on the uptake of phosphorus by perennial ryegrass over seven consecutive harvests of the grass. Earthworms increased the uptake of phosphorus from a superphosphate fertilizer by 20–40% at the first harvest, and by less than 10% by the seventh harvest, and increased the uptake of phosphorus from pelletized Chatham Rise phosphorite by 15–30% throughout the trial period. Mansell *et al.* (1981) exposed earthworms to ^{32}P-labeled ryegrass, and placed the resulting labeled casts from the earthworms or labeled ryegrass on the soil surface in pots in which ryegrass was growing. Earthworm casts increased the uptake of ^{32}P by ryegrass two- to threefold relative to that from uningested litter.

8.3.6 OTHER MACRONUTRIENTS AND MICRONUTRIENTS

The concentrations of exchangeable calcium, magnesium and potassium are usually significantly greater in earthworm casts than in uningested soil (Tables 8.2, 8.3) (Lunt and Jacobson, 1944; Stöckli, 1949; Ponomareva, 1950; Nye, 1955; Atlavinyte and Vanagas, 1973; Czerwinski *et al.*, 1974; Watanabe, 1975; Cook *et al.*, 1980; Tiwari *et al.* 1989). Nye (1955) reported that the casts of the tropical earthworm *Hippopera nigeriae* were much richer in exchangeable calcium and magnesium than the top 15 cm of soil in forests. In south-western Nigeria, Nijhawan and Kanwar (1952) reported that there was more calcium in casts than in soil, but that there was more available potassium, manganese and exchangeable calcium, magnesium, potassium and sodium in large earthworm casts than in either soil or in small casts. Earthworms have also been shown to have an important influence on the availability of molybdenum. In New Zealand pastures that are rich in total molybdenum, but poor in available molybdenum, the introduction of European species of earthworms increased the amounts of molybdenum available for plants greatly. This is particularly important, since molybdenum is an essential element in the nitrogenase enzyme complex of nitrogen-fixing plants. Thus, most of the available evidence demonstrates that earthworms make more mineral nutrients available for plant growth, and therefore they are important in improving soil fertility.

Lee (1985) postulated that the increases in available cations in earthworm casts are related to the greater amounts of plant tissues in earthworm casts than in soil. Accordingly, litter-feeding earthworm species have greater amounts of cations in their feces than geophagous species, whose diet consists primarily of soil. Piearce (1972) reported that the calcium contents were much greater in the casts of *L. rubellus*, a litter-feeding species, than they were in the casts of *A. caliginosa*, a soil-feeding species. Basker *et al.* (1993) reported that *L. rubellus* was more effective than was *A. caliginosa* in increasing the amount of available potassium in four different types of silt loam soil in New Zealand, even though the earthworms were maintained in homogenized soil in the laboratory, with no plant litter added as food. Even greater differences in chemical composition between the casts of the two species would be expected under field conditions. In another incubation experiment, Basker *et al.* (1994) compared the influence of *L. rubellus* and *A. caliginosa* on potassium levels in two different soil types and reported that the earthworms actually reduced the concentration of potassium in one of the soils. The effects of the worms on all of the cations (K^+, Ca^{2+}, Mg^{2+} and Na^+) showed no consistent trends and were of lower magnitude than those reported from field soils, suggesting that the elevated levels of cations often observed in earthworm casts in the field are probably due to selective feeding by earthworms on materials enriched in those cations.

Many species of earthworm possess calciferous glands that are involved in the production of calcium carbonate spherules, which may have an important influence on calcium availability in some soils (Piearce, 1972). Wiecek and Messenger (1972) calculated that calcite spheroids produced by earthworm calciferous glands contributed up to 56 kg calcium /ha to the A_1 horizons of some deeply leached, forested soils in the north central region of the US. Spiers *et al.* (1986) reported that earthworms (*Arctiostrotus* spp.) in the acid forest soils of Vancouver Island can convert calcium oxalate crystals of ingested fungi to calcium bicarbonate. The calcium bicarbonate is egested in earthworm casts, thereby temporarily increasing the availability of calcium for uptake by the roots of the forest vegetation.

Earthworms and micro-organisms

9

Earthworms have many complex interrelationships with micro-organisms. They depend upon micro-organisms as their major source of nutrients, they promote microbial activity in decaying organic matter by fragmenting it and inoculating it with micro-organisms, and they disperse micro-organisms widely through soils.

9.1 EFFECTS OF EARTHWORMS ON THE NUMBER, BIOMASS AND ACTIVITY OF MICRO-ORGANISMS

9.1.1 MICRO-ORGANISMS IN THE INTESTINES OF EARTHWORMS

Most of the species of micro-organisms that occur in the alimentary canal of earthworms are the same as those in the soils in which the worms live. For instance, Bassalik (1913) isolated more than 50 species of bacteria from the alimentary canal of *L. terrestris*, and found none that differed from those in the soil from which the worms had been taken. This was confirmed for three species of earthworms by Parle (1963a), who reported that most of the cellulose and chitinase enzymes that occur in the alimentary canals of earthworms are secreted by the earthworms and not by symbiotic micro-organisms, as they are in some arthropods. From such observations, Satchell (1967) concluded that it was unlikely that earthworms have an indigenous gut microflora. Stöckli (1928) reported that there was a great increase in the total numbers of bacteria and actinomycetes occurring in the earthworm gut compared with those in the surrounding soil, and other workers showed that numbers increase exponentially from the anterior to the posterior portions of the earthworm gut (Table 9.1) (Parle, 1959, 1963a).

Teotia *et al.* (1950), Kollmannsperger, (1952, 1956), Ruschmann (1953) and Schultz and Felber (1956) all reported greatly increased microbial populations in earthworm casts compared with the surrounding soil. The

Table 9.1 Numbers of micro-organisms in different parts of the intestine of *L. terrestris* (from Parle, 1959)

	(×10⁶)		
	Foregut	Midgut	Hindgut
Actinomycetes	26	358	15 000
Bacteria	475	32 900	440 700

large increases in the numbers of micro-organisms in the earthworm gut may be partially due to the large amounts of water and mucus that earthworms secrete into their guts. Barois and Lavelle (1986) showed that the intestinal mucus produced by the earthworm, *Pontoscolex corethrurus*, contained large amounts of water-soluble, low molecular weight organic compounds that could be assimilated easily by the rapidly multiplying microbial community in the gut.

Several researchers have shown that particular groups of micro-organisms are stimulated selectively during passage through the earthworm gut. These include the actinomycetes *Nocardia*, *Oerskovia* and *Streptomyces* spp. and the bacteria, *Vibrio* spp. (Máriaglieti, 1979; Contreras, 1980; Szabó *et al.*, 1990; Hossein *et al.*, 1991; Krištůfek *et al.*, 1993). Of the microbes isolated from the gut of *Eisenia lucens*, *Vibrio* spp. accounted for 73% of the total bacteria and *Streptomyces lipmanii* accounted for 90% of the actinomycetes, and these species were of relatively low abundance in the wood substrate where the earthworms were living (Contreras, 1980). Gest and Favinger (1992) reported that several species of purple photosynthetic bacteria (Rhodospirillaceae) became enriched in the alimentary canal of earthworms, and suggested that the intestinal contents of earthworms could be used to make highly selective enrichments for isolating *Rhodomicrobium vannielii* and *Rhodopseudomonas palustris*, and possibly other purple bacteria. The species composition and relative abundance of actinomycetes in the hindgut differs between different species of worm, at different times of year and with respect to different types of food ingested (Krištůfek *et al.*, 1990, 1993; Ravasz *et al.*, 1987a), and this is probably true for bacteria and fungi also. Striganova *et al.* (1989) reported that certain species of fungi, namely *Aspergillus fumigatus* and *Penicillium roqueforti*, were abundant in the digestive tracts of *Nicodrilus caliginosa*, but were absent from the surrounding soil in turf-podzolic soils in Russia. These fungal species may be suppressed in the soil and be obligate inhabitants of the earthworm intestine.

There is still controversy over whether the earthworm gut contains a truly indigenous microflora. This stems from the general difficulty in culturing all of the micro-organisms that live either in soil or in the intestines of earthworms. Certain groups of micro-organisms are clearly more abundant in the earthworm gut than in the surrounding soil, but the extent to which this selective stimulation of particular microbial species constitutes a true mutualistic association, remains to be demonstrated. Recent work by Jolly et al. (1993), who used scanning and transmission electron microscopy to examine the hindgut epithelium of O. lacteum and L. terrestris, provides evidence of a physical link between bacterial cells and the epithelium of the hindgut. The electron micrographs revealed segmented filamentous bacteria that were connected to the hindgut via a socket-like structure, as well as cocci and bacilli attached to the gut wall via a mucopolysaccharide-like material. These physical links do not necessarily prove the existence of truly indigenous microbial strains, but they do indicate that some microbial strains are highly adapted to living in the alimentary tract of earthworms.

Although many micro-organisms can survive passage through the earthworm gut, not all emerge in a viable form. Aichberger (1914a) reported that the crops, gizzards and intestines of earthworms contained few live organisms that did not possess a firm outer coat, and found no diatoms, desmids, blue-green algae, rhizopods or live yeasts. Dawson (1947) reported that the number of species of bacteria in soil that had passed through the gut of an earthworm were decreased, whereas those of fungi seemed unaffected. Day (1950) stated that when soil with a large inoculation of *Bacillus cereus* var. *mycoides* passed through the gut of *L. terrestris*, the numbers of these bacilli decreased greatly, although a few survived passage through the gut, indicating that vegetative cells, rather than spores, were destroyed. Two other bacteria, *Serratia marcescens* (Day, 1950) and *Escherichia coli* (Brüsewitz, 1959), which had been introduced into soil by inoculation, were killed after the soil had been ingested by *L. terrestris*. Khambata and Bhatt (1957) reported that the bacillus *E. coli* was usually absent from the intestines of *Pheretima*, although these worms often live in soil that is regularly manured with human excreta, and they suggested that secretions in the intestine of the earthworms possibly prevented the growth of this and other human pathogens. Spores of *Pithomyces chartarum*, the causal fungus in facial eczema of livestock, were nonviable after passage through the gut of mature *L. rubellus* (Keogh and Christensen, 1976). Dash et al. (1979) examined the number of species of microfungi in the soil of a tropical pasture in India and in the gut and casts of an earthworm (*Drawida calebi*) living in the pasture. These authors isolated 19 species of microfungi from the soil, 16 species from the anterior portion of the earthworm's gut, and 8 species from the posterior portion of the gut. All of the microfungal species encountered in the earthworm gut were also found in the surrounding soil. The reduction in the number of species that occurred during passage from the anterior to the posterior portion of the gut indicates that half of the ingested microfungal species were killed during gut passage, probably due to selective digestion of fungal mycelia and spores. However, Krištůfek *et al.* (1992) reported that the numbers of some fungi increased during passage through the gut of *L. rubellus*, indicating that the viability or culturability of some fungal species is enhanced during passage through the gut. Several workers have shown that spores of some species of fungi can survive passage through the alimentary canal of earthworms (Hutchinson and Kamel, 1956; Harinikumar *et al.*, 1991; Reddell and Spain, 1991a; Harinikumar and Bagyaraj, 1994). Various fungal spores that have thick walled or wrinkled coats (Dash *et al.*,1979), and the spores and mycelia of certain dark-colored fungi are resistant to breakdown by the intestinal enzymes of earthworms (Striganova *et al.*, 1989).

Antibiotic substances produced by actinomycetes in the intestines of earthworms can inhibit the growth of fungi and Gram-positive bacteria, and may explain why some actinomycetes and antibiotic-resistant, Gram-negative bacteria predominate in the gut (Ravasz *et al.*, 1986; Krištůfek *et al.*, 1993). The micro-organisms *Nocardia polychromogenes*, *Actinomyces* spp. and *Streptomyces coelicolor,* which have been isolated from the gut contents and casts of earthworms, were reported by Ruschmann (1953) to be particularly antagonistic towards the anaerobic spore-forming bacteria. Other workers have noted that earthworms may produce antibiotic substances. For instance, it has been shown that the growth of certain fungi on soil in a Petri dish ceased whenever an earthworm was introduced (van der Bruel, 1964). Ghabbour (1966) reported that when earthworms were placed in dilute glucose or glycine solutions, fungi did not grow until the earthworms died. Kobatake (1954) reported that earthworm extracts were anti-bacterial against several strains of non-acidfast pathogenic micro-organisms and saprophytic mycobacteria (a petroleum ether extract was bacteriocidal at a dilution of 1 in 1000 and bacteriostatic at 1 in 3200).

9.1.2 MICRO-ORGANISMS IN EARTHWORM CASTS AND BURROWS

Many workers have observed that there are greater numbers of micro-organisms in earthworm casts than in bulk soil (Stöckli, 1928; Ghilarov, 1963) (Table 9.2, Fig. 9.1). Ponomareva (1953) stated that there is an increase in numbers of actinomycetes, pigmented bacteria and other bacteria of the *Bacillus cereus* group after passage through the earthworm intestine, and the same worker found that the number of bacteria in earthworm feces was 13 times as high as in the surrounding soil (Ponomareva, 1962). It has been suggested that the large numbers of bacteria (5.4×10^6 per g) that occur in the B layer of chernozem soil (depth 50–60 cm), can be explained by the burrowing activities of earthworms that penetrate this horizon and leave their casts. Teotia *et al.* (1950) reported that worm casts had a bacterial count of 32.0 million/g,

Table 9.2   Numbers of micro-organisms in earthworm casts and soil (from
Ghilarov, 1963)

	(thousands per g)		
	Oak forest	Rye field	Grass field
Earthworm casts	740	3430	3940
Soil	450	2530	2000

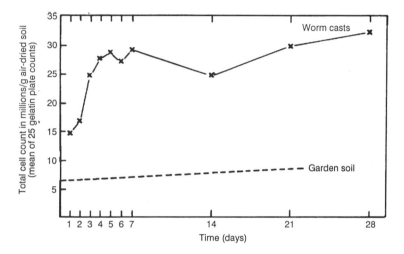

Figure 9.1   Total cell counts from earthworm casts of different ages, and garden
soils (Stöckli, 1928).

compared with 6.0–9.0 million/g in the surrounding soil. There were few
actinomycetes and fungi in the casts, but more *Azotobacter* and other bacteria. Many other workers have reported that the microbial population of
cast soil is much larger than that of the surrounding soil (Bassalik, 1913;
Zrazhevskii, 1957; Went, 1963; Shaw and Pawluk, 1986a; Tiwari *et al.*,
1989). Daniel and Anderson (1992) kept *L. rubellus* in four different soils,
containing different amounts of particulate organic matter, and observed
higher bacterial plate counts in the earthworm casts than in the surrounding soils. Differences in the size of microbial populations of casts, soil and
the surrounding field soil may be the result of changes in numbers of
micro-organisms occurring in the earthworm's intestine, or because the
selected food material ingested by the worm forms a richer substrate for
microbial activity, and it is not usually easy to determine which of these is
the major factor involved.

Comparisons of field-collected earthworm casts with the surrounding
soil has revealed that the casts usually have greater populations of fungi,
actinomycetes and bacteria and higher enzyme activity than the sur-

rounding soil (Dkhar and Mishra, 1986; Tiwari *et al.*, 1989; Tiwari and Mishra, 1993). Tiwari and Mishra (1993) sampled earthworm casts and adjacent soil at 30 different sites in India and reported that the casts usually contained larger fungal populations and a greater number of fungal species than did the soil. Kozlovskaya and Zhdannikova (1961) reported that the ratios between different groups of micro-organisms in soil differed from those in casts; the spore-forming bacteria and actinomycetes predominated in casts, whereas the numbers of *Bacillus idosus* and *B. cereus* were greater in casts than in soil, but those of *B. agglutinatus* were less. Other workers have shown that earthworm casts contain greater numbers of cellulolytic, hemicellulolytic, nitrifying and denitrifying bacteria than does the surrounding soil (Bhatnagar, 1975; Loquet *et al.*, 1977). Shaw and Pawluk (1986a) examined the micro-organisms in the casts of the anecic earthworm species *L. terrestris*, and two endogeic species, *O. tyraeum* and *A. turgida*, after keeping the worms for 1 year under controlled conditions in three different types of calcareous soil (sandy loam, clay loam and silty clay loam). The casts of both the anecic and endogeic earthworms had densities of bacteria and actinomycetes between one and three orders of magnitude greater than in soils kept without worms. The densities of anaerobic micro-organisms, filamentous fungi and *Cytophaga* were also higher in the casts, although this varied with soil type. Yeasts were more numerous in casts of all the species of earthworms in the clay loam soil, but in the other two soil types they were more numerous in the casts of *L. terrestris*, but not in those of the endogeic species. Such results demonstrate that the influence of earthworms on the density of micro-organisms can vary with different types of soil and with different species of earthworm.

Zrazhevskii (1957) observed that the density of the bacterial population, in turf soil without earthworms was 2.8×10^6 per g, whereas in the casts of worms added to this soil it was 9.88×10^6 per g, and, in the same soil after long colonization by worms, 9.8×10^6, or three times greater than initially. Atlavinyte and Lugauskas (1971) reported that in pot tests, earthworms increased the number of micro-organisms in soil as much as five times. Murray *et al.* (1985) reported that the number of plasmodium-forming units of myxomycetes was five times greater in earthworm casts than in adjacent, noncast soil. The gut contents of field-collected worms contained numbers of myxomycetes of the same order as those in surrounding soil, indicating that the increased numbers in the casts were due to favorable conditions for microbial growth in the casts, and not to enrichment of myxomycetes during passage through the gut. Hutchinson and Kamel (1956) isolated 17 species of viable fungi from the alimentary canal and rectum of 10 individuals of *L. terrestris*. Despite the limited number of species of fungi represented, some occurred consistently in worms from different areas, and at different times of the year. Fewer micro-organisms were isolated from worms examined in mid-winter than in late autumn, which corresponded to the seasonal fluctuations in microbial populations in soil, as reported by other workers. Brüsewitz (1959), Jeanson-Luusinang (1963) and Went (1963), for example, reported little or no difference between microfloral populations of earthworm casts and soil, but they used soil to which was added a readily available organic energy source that would minimize potential differences in the microflora of the earthworm casts and soil.

It has been suggested that microbial activity in earthworm casts may have an important effect on soil crumb structure, by increasing the stability of the worm-cast soil relative to that of surrounding soil. Many workers have shown that earthworm casts contain more water-stable aggregates than noncast soil, and part of this may be due to polysaccharide gums, produced by the bacteria of the intestine (Swaby, 1949; Satchell, 1958) and by the proliferation of fungal hyphae on the surface of the casts (Marinissen and Dexter, 1990).

The walls of earthworm burrows may also be enriched in micro-organisms compared to the surrounding soil. Bhatnagar (1975) analyzed the micro-organisms associated with earthworm burrows in a grassland in France, and reported that 42% of the total soil aerobic nitrogen-fixing bacteria, 13% of the anaerobic nitrogen-fixing bacteria and 16% of denitrifying bacteria were associated with the burrows. The numbers of ammonifying, denitrifying, nitrogen-fixing and proteolytic bacteria were greater in the burrow walls than in the surrounding soil.

Kozlovskaya and Zhdannikova (1961) reported that not all species of earthworms had similar relationships with soil micro-organisms. They compared the microbial populations of the gut contents of *O. lacteum* and *L. rubellus. Octolasion lacteum* lives at a depth of 10–40 cm in soil, and its gut contents had about the same density of bacteria as that in the soil in which it lives, although there were rather more spore-forming bacteria and actinomycetes, and fewer fluorescing bacteria in the gut contents than in the soil. By contrast, *L. rubellus* lives in the top 5 cm of soil, and the total density of bacteria in its gut was more than 10 times larger than that in soil, depending very much on the food of the worm. The excreta of both species contained more fungi, actinomycetes, butyric acid-forming bacteria of the *Clostridium* type, and cellulose-decomposing bacteria, than did the surrounding soil.

Casts are usually rich in ammonia and partially digested organic matter, and thus provide a good substrate for growth of micro-organisms. Some of the intestinal mucus secreted during passage through the earthworm gut is egested with the casts, where it continues to stimulate microbial activity and growth (Barois and Lavelle, 1986; Scheu, 1991). Increases in microbial populations in the earthworm casts were first reported by Stöckli (1928), who estimated that the total microbial cell count in the earthworm casts doubled in the first week after they were deposited. Thereafter, for a further 3 weeks, numbers did not increase overall, although they fluctuated considerably during this period. Parle (1963b) reported that yeasts and fungi, which occurred in the soil as spores, germinated as soon as they were in worm casts, and most hyphae were formed in 15-day-old casts. No consistent changes in numbers of actinomycetes or bacteria were found in old casts. Microbial activity, as shown by oxygen consumption, was stated to decline from the moment casts were produced (Parle, 1963b), and the reason suggested for the simultaneous decline in oxygen consumption and increase in microbial population, is that, as the casts age, an increasing proportion of the micro-organisms form resting stages.

There is more recent evidence that the microbial communities and activity associated with contents of fresh earthworm casts begin to change quite soon after the casts are deposited by the worms. Scheu (1987) observed an ephemeral increase in microbial biomass in the freshly deposited casts of A. caliginosa, followed by a decline in microbial biomass in older casts. Within 4 hours of being deposited, fresh casts contained 1.28-fold more microbial biomass than soil, and then decreased to approximately 90% of that in soil after 2 weeks. However, the microbial respiration in casts was higher than in soils throughout the 30 day incubation period; this suggests that the microbial biomass of casts is more metabolically active than that of soil. Lavelle et al. (1992) observed a six- to sevenfold increase in the amount of microbial biomass nitrogen in fresh casts relative to uningested soil, but within 12 hours, the amount of microbial biomass nitrogen in the casts decreased to slightly more than twice the amount in surrounding soil, and declined only slightly during the remainder of the 16 day incubation. The greater relative stimulation of microbial biomass observed by Lavelle et al. (1992), compared with that observed by Scheu (1987a), may be because the soil used by Lavelle had much lower amounts of organic matter and microbial biomass than the soil used by Scheu; the stimulatory effect of earthworms on microbial biomass and activity in their casts might be expected to be greater in soil with low amounts of organic matter (Blair et al., 1994b). This hypothesis is supported by some of the results of Lavelle et al. (1980), who reported that casts of earthworms in soil containing low concentrations of soluble organic matter had greater microbial activity than uningested soil, whereas the casts of earthworms provided with soil containing large concentrations of soluble organic matter had decreased microbial activity relative to noningested soil.

There is some evidence that earthworms can decrease the total biomass of micro-organisms in soil, by causing a change to a smaller, more metabolically active microbial community. Wolters and Joergensen (1992) kept earthworms in six different soil types for 21 days and observed that the microbial biomass was lower, relative to control soils without earthworms, in five of the six soils. Earthworms increased the metabolic activity per unit of microbial biomass in all six soils. In the same experiment, Wolters and Joergensen (1992) removed earthworms from soil after 21 days and incubated the soil for a further 21 days to determine the longer-term effects of earthworms on microbial biomass, following cessation of earthworm activity. After this additional period of incubation, the response of the microbial biomass varied among soil types, being significantly greater in those soils that had supported earthworms than in control soils in three of the soils, but significantly less than in control soils in another soil. Bohlen and Edwards (1994) observed that earthworms caused microbial biomass to decrease in a laboratory incubation of silty loam soils with organic or inorganic nutrient inputs. After 112 days of incubation, soil with earthworms had lower microbial biomass nitrogen than soil without earthworms, particularly in soil that had been treated with inorganic fertilizer. Bohlen and Edwards (1995) speculated that earthworms fed upon microbial biomass, releasing the nutrients bound in microbial tissues. This process was overshadowed in soils that received organic inputs, because of the stimulatory effects of these inputs on the microbial community. The results from the various studies cited above underscore the complexity of earthworm–microbial interactions and emphasize that the interactions between earthworms and the microbial community are often specific to soil type, organic matter resources and the time scale being considered. Differences among these various factors may explain some of the conflicting results reported in the literature.

9.2 IMPORTANCE OF MICRO-ORGANISMS AS FOOD FOR EARTHWORMS

Micro-organisms constitute an important nutritional component of the earthworm diet. Edwards and Fletcher (1988) summarized the experimental evidence that micro-organisms provide a source of nutrients for earthworms, and concluded that bacteria are of minor importance, algae are of moderate importance and fungi, and to a lesser extent protozoa, are major sources of nutrients. Moreover, they emphasized that earthworms cannot grow on pure cultures of micro-organisms and need mixed groups or species of micro-organisms to develop satisfactorily. Feeding-preference studies have shown that earthworms prefer to feed on materials inoculated with particular groups of micro-organisms. For instance, Cooke and Luxton (1980) and Cooke (1983) showed that L. terrestris preferred to feed on paper disks that were inoculated with particular species of fungi such as Fusarium oxysporum, Alternaria solani, and Trichoderma viride, and rejected, or were not stimulated by, disks inoculated with other species such as Cladosporium cladosporoides, Poronia piliformis, and

Chaetonia globosum. Large numbers of fungal hyphae can be observed in the intestines of earthworms, and many of these hyphae are digested as they pass through the gut (Dash *et al.*, 1979, 1984; Spiers *et al.*, 1986). Domsche and Banse (1972) reported that fungal hyphae were digested completely during passage through the earthworm gut, although most other workers have reported that some of the fungi in the gut remain undigested. Dash *et al.* (1984) calculated that earthworms digested and assimilated 54% of the fungal material that they ingested. Most of the digestion of fungal hyphae occurs in the anterior portion of the gut (Tiwari *et al.*, 1990). By examining the gut contents of field-collected earthworms, Piearce (1978) concluded that fungi and algae composed a significant component of the food of six different lumbricid species. Atlavinyte and Pociene (1973) reported that earthworms grew best in soil with green and blue-green algae, indicating the importance of algae to the earthworm diet.

The best experimental evidence for the importance of micro-organisms to the diet of earthworms comes from studies on *E. fetida*. Miles (1963a) introduced this species into soils inoculated with fungi and bacteria and showed that the worms were unable to reach sexual maturity unless protozoans were also added to the cultures. Protozoa are normally abundant in the habitat of this worm, which lives in compost and manure heaps, and it has also been suggested that protozoa were essential components of the diet of *E. fetida*. Other workers have shown that ciliates and amoebae exposed to the digestive juices of *E. fetida* and *L. terrestris* are killed and digested (Piearce and Phillips, 1980; Rouelle, 1983). Neuhauser *et al.* (1980) reported that *E. fetida* increased in weight in the presence, but not in the absence, of seven species of micro-organism (two bacteria, two protozoa and three fungi). The growth of earthworms on certain specific micro-organisms was not significantly different whether dead or live organisms were offered as food. Flack and Hartenstein (1984) showed that earthworms grew well when provided with many species of protozoa and bacteria, although growth rates were 20% greater in the presence of protozoa than in the presence of bacteria alone. Hand and Hayes (1988) provided 18 individual species of bacteria and 22 different species of fungi to *E. fetida* and showed that earthworm growth was improved in the presence of some, but unaffected by others. Certain species of bacteria, such as *Flavobacterium lutescens*, *Pseudomonas fluorescens*, *Ps. putida* and *Streptomyces* spp. actually had a toxic effect on earthworms, leading to mortality. In general, fungi had a greater nutritional value than bacteria, although one bacterial species, *Acinetobacter lwoffi*, produced significant weight gains in the worms. The importance of particular groups of micro-organisms as food for earthworms probably differs between different earthworm species, particularly those that have markedly different feeding habits. However, the data on this subject are limited to studies on only a few earthworm species, and the importance of different micro-organisms in the diet of most earthworm species has yet to be established.

9.3 DISPERSAL OF MICRO-ORGANISMS BY EARTHWORMS

Earthworms can enhance the dispersal of micro-organisms by ingesting them at one location from a particular food source and egesting them elsewhere, or by transporting microbes that adhere to their body surface. Many of the micro-organisms transported by earthworms are those involved in the decomposition of organic materials, but earthworms also consume and transport other beneficial microbial groups, such as the plant-associated mycorrhizae and other root symbionts, biocontrol agents and microbial antagonists of plant pathogens. Thick- and thin-walled spores tend to lose little of their viability during passage through the intestines of earthworms (Hoffman and Purdy, 1964). For example, the spores of dwarf bunt (*Tilletia controversa*) lose none of their viability during passage through the earthworm gut. It has also been suggested that earthworms disperse spores of harmful fungi such as *Pythium* (Baweja, 1939) and *Fusarium* (Khambata and Bhatt, 1957). The influence of earthworms on the dispersal of pathogenic micro-organisms is discussed further in Chapter 11.

Hutchinson and Kamel (1956) inoculated sterilized soil with several different species of fungi, and reported that the rate of spread of the fungi through the soil was much greater when there were worms present than when they were absent. Huss (1989) isolated 11 species of slime mold from the guts of *A. caliginosa* and *O. tyrteum* collected in north-eastern Kansas, USA. He force-fed adult *L. terrestris* with separate suspensions of spores and myxamoebae of *Dictyostelium mucoroides* and found that spores were the life stage that survived passage through the gut best, suggesting that earthworms may play an important role in the short-range dispersal of slime mold propagules.

Earthworms may also have adverse effects upon the spread of fungi. For example, the ascospores of *Ventura inaequalis* (Coole) Wint, which causes apple scab, are released from perithecia on overwintering dead leaves lying on the soil surface in the spring, and these infest the new growth (Hirst *et al.*, 1955). However, a large population of *L. terrestris* can remove most of these leaves from the soil surface during the winter, thus preventing at least a proportion of the ascospores from being able to infect trees. There are probably many other important relationships between earthworms and plant pathogens, but much more work is required to assess these relationships.

Earthworms have been shown to have a significant influence on the dispersal of vesicular-arbuscular mycorrhizae (VAM) fungi, which form an important mutualistic association with plant roots. Rabatin and Stinner

(1989) reported that 25% of earthworms in conventional no-tillage corn, 83.3% from no-tillage corn, and 50% from pasture contained VAM fungi in their guts. Reddell and Spain (1991a) surveyed the casts of 13 earthworm species from 60 sites in Australia and found intact spores of VAM fungi in all but one collection. They also found VAM root fragments in the casts. The diversity of casts was similar to that of surrounding soil, but the numbers of spores were highest in casts of *P. corethrurus* and *Diplotrema heteropora*. Greenhouse experiments verified that VAM spores and some root fragments recovered from casts initiated mycorrhizal infection in *Sorghum bicolor*. Earthworms have also been shown to enhance the spread of VAM on roots of soybean (McIlveen and Cole, 1976) and seedlings of tropical fruit trees (Ydrogo, 1994), in greenhouse experiments. Propagules of VAM can survive for several months in the air-dried casts of *E. eugeniae* (Harinikumar *et al.*, 1991), and for at least 12 months in those of *L. terrestris* (Harinikumar and Bagyaraj, 1994). Gange (1993) found that earthworms had a significant impact on the distribution of VAM propagules in early (1 and 3 year) and later (5, 8 and 11 year) successional plant communities. Earthworm casts contained nearly twice as many spores as surrounding soil in most communities. The influence of earthworms on infective VAM propagules was even greater, particularly at the later successional sites, in which casts contained up to 10 times as many infective VAM propagules as did surrounding soil. The high casting rates of earthworms in the early successional sites and the abundance of VAM spores in later successional sites, indicate the very considerable potential of earthworms to affect the establishment and competitive ability of mycorrhizal plants in these communities.

Dispersal of nitrogen-fixing bacteria that form mutualistic associations with plant roots can also be enhanced by earthworm activity. Reddell and Spain (1991b) investigated the ability of the earthworm, *P. corethrurus*, to transfer infective propagules of *Frankia*, an endophytic actinomycete that fixes nitrogen in association with roots of certain nonleguminous plants. They inoculated seedlings of *Casuarina equisetifolia* with either a crushed nodule suspension of *Frankia* or casts of *P. corethrurus*, raised in sterilized soil in which crushed *Frankia* nodules had been thoroughly mixed. The shoot and nodule dry weight of seedlings treated with casts from 11 species of earthworms were similar to those of seedlings inoculated with crushed nodules. Another very important group of beneficial nitrogen-fixing micro-organisms influenced by earthworms are the rhizobium bacteria, which fix nitrogen in nodules formed on the roots of leguminous plants. For instance, *L. rubellus* has been shown to enhance the translocation of *Bradyrhizobium japonicum* to greater soil depths (Madsen and Alexander, 1982). Rouelle (1983) reported that *L. terrestris* can increase the spread of *Rhizobium japonicum* and the formation of nodules on soybean roots. More recently, Stephens *et al.* (1994a) reported that *A. trapezoides* increased the rates of dispersal of *R. meliloti* as well as the levels of root nodulation in infected alfalfa plants. Doube *et al.* (1994a) showed that *A. trapezoides* greatly increased the number of nodules on roots of *Trifolium subterraneum*, in pot experiments in which sheep dung, inoculated with *R. trifolii*, was applied to the soil surface in pots with or without earthworms. Thompson *et al.* (1993) observed that *L. terrestris* and *Aporrectodea* spp. increased root nodulation on *Trifolium dubium* by up to 100 times and increased the proportion of *Trifolium* threefold, in simple plant communities grown in controlled environmental chambers.

Thus, there is good evidence that earthworms are very important in inoculating soils with micro-organisms and that their casts are foci for dissemination of many species of soil micro-organisms (Ghilarov, 1963). Earthworms may enhance the dispersal of several important groups of beneficial micro-organisms, sometimes by several orders of magnitude. The research on this subject is relatively sparse, which suggests that there is an excellent opportunity for important developments to be made through future research. Knowledge gained in this area may provide the basis for a new technology to introduce and disperse beneficial micro-organisms in soil (Doube *et al.*, 1994b,c).

9.4 STIMULATION OF MICROBIAL DECOMPOSITION BY EARTHWORMS

The decomposition of organic material in the soil is accelerated when simple nitrogenous compounds are added to soil (Tenney and Waksman, 1929; Harmsen and van Schreven, 1955), particularly if the organic material is poor in nitrogen. Therefore, because the excreted cast material from earthworms is usually rich in nitrogenous compounds, large numbers of earthworms in an organic soil can not only help to decompose organic material in the soil by ingestion, disintegration and transport, but their waste products may also stimulate other microbial decomposition processes as well.

Many micro-organisms in the soil are in a dormant stage, with low metabolic activity, awaiting suitable conditions to become active (Lavelle *et al.*, 1992). The earthworm gut provides suitable conditions for the vigorous multiplication of particular micro-organisms, which are stimulated to decompose ingested organic matter. Earthworms secrete large amounts of water-soluble organic compounds, which can be assimilated readily by micro-organisms in the gut (Barois and Lavelle, 1986). The addition of these compounds may help to prime the micro-organisms in the gut to break down more complex organic compounds in the ingested soil (Barois and Lavelle, 1986; Lavelle *et al.*, 1993). This process, in which the microbes benefit from the mucus secretions of the earthworm, and the earthworm benefits from the enhanced microbial decomposition of ingested organic matter, has been described as a mutualistic digestive system (Barois, 1992; Trigo and Lavelle, 1993).

Since microbial activity remains higher in earthworm casts than in surrounding soil, the enhanced decomposition of organic matter, which begins in the earthworm gut, continues for some time after the gut contents are egested. Kozlovskaya and Zhdannikova (1961) reported that the decomposition of organic matter was much faster and more intensive in earthworm casts than in the soil. Parle (1959) showed that oxygen consumption, which is an indicator of microbial activity, was still considerably higher in cast soil than in the surrounding soil, even 50 days after being excreted. Scheu (1987a) observed that microbial respiration was, on average, 1.86 times higher in earthworm casts than in surrounding soil during 30 day incubations. Daniel and Anderson (1992) also showed that rates of CO_2 production were greater in earthworm casts than in the soil ingested by the worms. The increased respiration rates in earthworm casts were accompanied by an increase in the numbers of bacteria and soluble organic carbon in the casts. After this preliminary stage of high respiratory activity in fresh casts, microbial activity tends to decrease, gradually returning to the same level as the surrounding soil.

As earthworm casts age, there is a reorganization of mineral and organic components of the casts, which results in a lower rate of decomposition in casts relative to the surrounding soil, due to the physical protection of organic matter in the compact structure of the casts (Martin, 1991; Lavelle and Martin, 1992). This process may be more important in poorly aggregated soils where climate and soil texture favor rapid mineralization of soil carbon. The physical changes that occur in aging casts are accompanied by biological changes, in which slower-growing soil fungi and dormant microbial stages begin to predominate in older casts. Thus, whereas the short-term effect of earthworms is to stimulate microbial decomposition of organic matter, the long-term effect may be to decrease microbial decomposition by increasing the physical protection of organic matter. The consequences of this for the long-term net storage or loss of organic matter in soil still remain unknown (Chapter 8).

Another way in which earthworms may affect the microbial decomposition of soil organic matter is by influencing the ratio of fungi to bacteria in the soil (Blair et al., 1994b; Brown, 1995). Earthworms may change the fungal to bacterial ratio in soil by increasing the amount of soluble organic carbon that can be rapidly mineralized by bacteria and other microbial groups, and also by feeding preferentially on fungi. Changes in the ratio of fungi to bacteria are important because of the differences between fungi and bacteria in the efficiency with which they assimilate carbon. Bacteria tend to be less efficient than fungi at assimilating carbon and thus respire more carbon as CO_2 for each unit of carbon consumed (Adu and Oades, 1978). Furthermore, fungal hyphae contain carbon compounds that are resistant to degradation. Evidence from short-term incubations and investigations of the microflora in earthworm casts indicates the potential for earthworms to stimulate soil bacteria preferentially. There is a need to link earthworm-induced changes in the ratio of fungi to bacteria to carbon cycling processes under natural field conditions.

Earthworms have been shown to increase overall microbial respiration in soil, thereby enhancing microbial degradation of organic matter. Barley and Jennings (1959) added grass and clover litter and dung pellets to soil culture pots with and without earthworms of the species *A. caliginosa*, in numbers equivalent to a normal field population, and maintained them at 15 °C for a period of 45 days. The cultures were kept moist and aerated during this time, and their oxygen consumption recorded. The rate of nitrate and ammonium accumulation was measured for a period of 50 days. The cultures with earthworms consumed 410 litres of oxygen/g of soil more than those without earthworms, and, of this, 200 litres/g was estimated to have been consumed by the earthworms, the remainder being due to increased microbial activity. Thus, the rate of litter decomposition estimated from the accumulation of nitrate and ammonia was 17–20% greater in the earthworm cultures than in soil with no earthworms; half of this was thought to be due directly to the earthworms and the other half to the stimulation given to other decomposer organisms. Anstett (1951) showed that *E. fetida* could stimulate microbial decomposition in soil. After 5 months, the microbial population in soil with earthworms, and to which grape husks had been added, was 4–5 times greater than in similar soil without earthworms. Haimi and Huhta (1990) showed that earthworms increased microbial decomposition in the humus layer of coniferous forest floor, increasing microbial respiration by 10–15% in laboratory incubations. Shaw and Pawluk (1986a) also showed that earthworms increased decomposition and soil respiration rates, but that the amount was dependent on soil type. They reported that *L. terrestris* actually lowered soil respiration rates and decreased the decomposition of grass residues added to a sandy loam soil. This was accompanied by an increase in the proportion of carbon associated with clay particles, resulting in a net storage of organic matter. The opposite effect was observed in a clay loam and silty clay loam soil, in which *L. terrestris* accelerated the decomposition of the added grass residues. These results reinforce the conclusion that soil type can have an important influence on the interaction between earthworms and micro-organisms. Other factors that influence the stimulatory effects of earthworms on microbial decomposition include organic matter availability, climate and physico-chemical features, all of which need to be considered carefully when interpreting any data on earthworm–microbial interactions as they relate to the decomposition of soil organic matter.

Role of earthworms in soil structure, fertility and productivity

10

Earthworms have a critical influence on soil structure, forming aggregates and improving the physical conditions for plant growth and nutrient uptake. Rarely has a direct link been established between the improvement of soil structural attributes by earthworms and enhanced plant productivity. Nevertheless, it is generally agreed that earthworms improve soil structure to the benefit of soil productivity (Barley, 1959b; Lee and Foster, 1991; Edwards et al., 1994). Earthworms also improve soil fertility by accelerating decomposition of plant litter and soil organic matter and, consequently, releasing nutrients in forms that are available for uptake by plants (Curry, 1987). It is difficult to separate the biological effects of earthworms from their physical effects, but most research has analyzed them separately. The great influence of earthworms on nutrient and organic matter cycles and their interaction with soil microbes in the decomposition of organic matter are discussed in Chapters 8 and 9.

10.1 EARTHWORM BURROWS AND CASTS

10.1.1 EARTHWORM BURROWS

Earthworms form their burrows by literally eating their way through the soil and pushing through cracks and crevices. However, different species of earthworms differ greatly in their burrowing behavior. Some burrows form the permanent home of an individual earthworm as it moves from the surface to the subsoil gathering food, depositing feces and estivating. Other burrows are temporary, and are produced by the earthworm as it moves from one place to another in its preferred soil stratum. All earthworms burrow in much the same way, by first anchoring their posterior setae, increasing the hydrostatic pressure in their coelomic cavity, and stiffening the longitudinal muscles of the body wall so as to project the anterior segments forward. The posterior segments are then drawn up towards the anterior part of the body by contraction of the longitudinal muscles, and the sequence begins again. In this way, earthworms can exert considerable pressures to enable them to penetrate relatively compacted soils. The pressure that can be exerted depends upon the hydrostatic pressure that can be generated in the fluid-filled coelomic cavity. As the earthworm exerts this pressure, it ingests soil, sometimes by everting the pharynx, filling it with soil, and then retracting it. The whole mechanism enables the earthworm to penetrate soils readily, even when they are relatively compacted.

The semi-permanent burrows of some species are continuous, often branched, and can extend from the soil surface deep into the lower layers of soil. Burrow sizes differ with the species and stage of development of the earthworm, but can range from 1 to 12 mm in diameter. The numbers of earthworm burrows that have been counted on horizontal surfaces exposed at various depths in soil range from a few (van der Westeringh, 1972) to as many as 800 per m² (Bouché, 1971), with many soils having between 50 and 200 burrows/m² (Edwards *et al.*, 1990, 1992). Not all such burrows may be occupied by live earthworms since they can persist long after they have been abandoned by the earthworms that constructed them.

Usually only those species that penetrate deep into the soil, such as *L. terrestris*, *A. longa*, and *A. nocturna* have permanent burrows, with smooth walls, cemented together with mucous secretions and ejected soil, pressed into the soil interspaces. Often, the mucous secretions serve as a substratum for growth of fungi and bacteria on the walls of the burrow (or drilosphere). Some of these earthworm species excrete their feces around the walls of the burrows and some deposit them at the mouth of their burrows.

To study burrowing activity, earthworms can be induced to form burrows in soil between two sheets of glass (Evans, 1947b). In a series of such experiments, *A. caliginosa* formed an extensive burrow network through the top 20 cm of soil in a few days, whereas *L. terrestris* took 4–6 weeks to do this. The speed of burrowing depended very much on the texture of the soil, deep-burrowing species taking 4–5 times as long to burrow in clay as the same species in light loam (Edwards and Lofty, 1977). Gardner (1953) described a method of following the burrow network by making latex casts of the burrow of earthworms by pouring liquid latex (thinned with ammonia and diluted 1:8 with distilled water) into the burrows. In later studies, latex paint of different colors has been used to trace burrows by studying the distribution of colored openings after the paint was applied to successive horizontal sections. Other workers have used minirhizotrons (Ligthart *et al.*, 1993) and X-ray computed tomography (Joschko *et al.*, 1991) to study earthworm burrow systems.

It has been suggested that earthworms can leave trails that contain pheromones (Rosenkoetter and Boice, 1973). Such chemicals might well account for the different burrowing activities of different species.

Diplocardia riparia excretes a pheromone in its trail that is a repellent, whereas *L. terrestris* produces an attractant. These may be important in the controlling habits of the animals; *L. terrestris* tends to inhabit a permanent burrow system so its attractant trail would provide a means of returning safely to its burrow, whereas *D. riparia*, which is more of a scavenger, would obtain no advantage from following its previous trail.

Earthworms of the genus *Lumbricus* often do not burrow extensively, so long as an adequate food supply is present on, or close to, the surface, but when food is scarce their burrowing activity is stimulated greatly (Evans, 1947b). The burrows of some species, such as *L. terrestris* and *A. nocturna*, penetrate down to a depth of 150–240 cm, tending to be vertical for most of their depth, but often branching extensively near the soil surface. Burrows range from about 1–12 mm in diameter, but it is not certain whether worms increase the size of their burrows as they grow, or whether they make new ones. Some earthworms can burrow very deeply into the soil; *Drawida grandis* has been reported to make burrows to a depth of 2.7–3.0 m (Bahl, 1950). Most shallow-working species, which are usually smaller, do not have well-defined burrows, although some of the surface species, e.g. *P. hupeiensis*, have complex burrow systems in the top 7.5–15.0 cm of soil (Grant, 1956). Earthworm burrows have major influences on soil macroporosity and movement of water and nutrients. These are discussed in section 10.2.5.

10.1.2 EARTHWORM CASTS

It is usually the deep-burrowing species of earthworms that produce casts on the soil surface near the exits to their burrows. Of the 8–10 common field species of lumbricids in England, only three, *L. terrestris*, *A. longa* and *A. nocturna*, particularly the latter two species, produce significant amounts of casts on the surface of the soil. Usually, they cast more on heavy soils than on light, open ones, because in the latter much of the feces are passed into spaces and crevices below the soil surface. It was shown in India and England that the number of casts produced varies seasonally (Fig. 10.1) and is a good index of earthworm activity (Roy, 1957). Workers in Canada have reported that another lumbricid species, *E. rosea*, also produces surface casts in soils with high bulk density and silt content and much organic matter (Thomson and Davies, 1974) and so does *A. trapezoides*. Most surface-dwelling earthworms probably deposit some surface casts, although the proportions of above- and below-ground casts produced by various species have not been adequately studied.

There are very many different forms of casts which are often typical of the species that produced them. They range from the small, heterogeneous masses excreted by *A. longa*, and individual pellets produced by *P. posthuma*, to short threads from *Perionyx millardi*. Basically, there are three

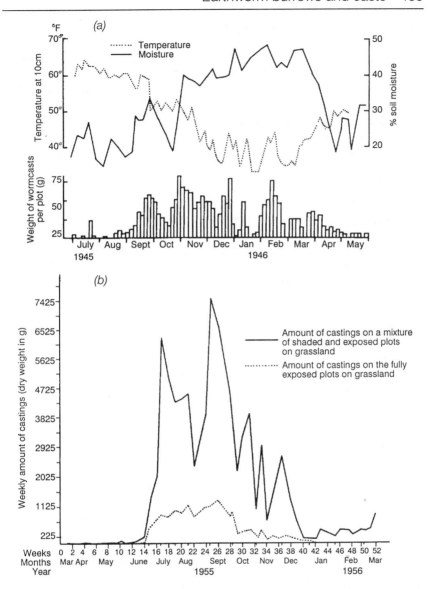

Figure 10.1   The seasonal changes in earthworm cast production: (a) in England (after Evans and Guild, 1947c); and (b) in India (after Roy, 1957).

types of casts: spheroidal or sub-spherical pellets ranging from 1 to 12 mm in diameter; paste-like slurries which form composite irregular shapes; tall heaps or columns with a range of shapes. Ljungström and Reinecke (1969) described round hollows (1 m diameter, 30–100 cm deep) called *'kommetjies'*, in South Africa. They believed these to be formed by a very large species of *Microchaetus* casting around the margins of their burrows. *Eutyphoeus waltoni*, an Indian species, has casts that look like a twisted, coiled tube, and an African species *E. eugeniae* produces casts that take the form of pyramids of very finely divided soil.

Some earthworms excrete a long, thick column of feces which produces a hollow column about 5–10 cm high and 2.5–3.5 cm in diameter (Plate 2a). Charles Darwin (1881) described an irregular tower-like casting about 9 cm high by 4 cm in diameter produced by a species of *Perichaeta*.

Earthworm casts can be very large; those of European worms seldom exceed 10.0 g, but some African worms, such as *Dichogaster jaculatrix*, have casts in the form of red clay chimneys about 10–12 cm high and 4 cm in diameter (Baylis, 1915). The giant *Notoscolex* earthworms in Burma produce large tower-shaped casts 20–24 cm high and 4 cm in diameter; one such cast weighed 1.6 kg (Gates, 1961). *Hyperiodrilus africanus* also produces large tower-like casts from 2.5 to 8.0 cm high, and 1.0 to 2.0 cm in diameter (Madge, 1969) (Plate 2a). Both Madge and Gates (1961) agreed that casting by tropical species of earthworms is limited to the wet season. Most species of earthworms cast during the night, but *P. hupeiensis* usually deposits its casts during daylight hours (Schread, 1952).

The amounts of cast produced in different regions and habitats vary greatly and are summarized in Table 10.1. Darwin (1881) estimated that

Table 10.1   Amounts of casting in different habitats and areas

Habitat	Locality	Casts produced		Reference
		(Tons/acre)	(Tonnes/ha)	
Arable soil	Germany	36.5	91.6	Dreidax (1931)
Arable (sandy loam)	India	0.6–2.0	1.4–5.0	Roy (1957)
Garden soil (clay loam)	Switzerland	7.1–32.4	17.8–81.2	Stöckli (1928)
Soil near White Nile	Egypt	170.0	268.3	Beauge (1912)
Old grassland	England	7.5–16.1	18.8–40.4	Darwin (1881)
Old grassland	England	11.0	27.7	Evans and Guild (1947c)
Grassland	Switzerland	7.1–32.4	17.8–81.0	Stöckli (1928)
Grassland	Germany	36.5	91.4	Dreidax (1931)
Grassland	India	1.6–31.0	3.9–77.8	Roy (1957)
Heathland (mainly loam)	Germany	2.1	5.2	Kollmannsperger (1934)
Oakwood (sandy soil)	Germany	2.3	5.8	Kollmannsperger (1934)
Beechwood	Germany	2.7	6.8	Kollmannsperger (1934)

(a)

(b)

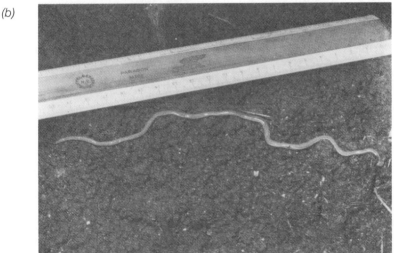

Plate 2a   Casts of *Hyperiodrilus africanus*
Plate 2b   *Bipalium kewensis* - a predator of earthworms

the annual production of worm casts in English pastures was 18.7–40.3 tonnes/ha (7.5–16.1 tons/acre), which is equivalent to a soil layer 5 mm deep being deposited annually. Guild (1955) calculated that 27 tonnes/ha were produced in another English pasture. Similar estimates for the amount of casts deposited per annum in other parts of Europe range from

5.8–6.8 tonnes/ha (Kollmannsperger, 1934) to 25.7 tonnes/ha for Germany (Graff, 1971), to 75–100 tonnes/ha for Zurich (Stöckli, 1928).

In the tropics, even larger amounts have been reported, ranging from: 50 tonnes/ha in Ghana (Nye, 1955); 17.6–29 tonnes/ha in Nigeria (Madge, 1969); 507 tonnes/ha in the Ivory Coast (Lavelle, 1975) and 0.47–12.76 tonnes/ha in India (Roy, 1957); to as much as 2100 tonnes/ha in the Cameroons (Kollmannsperger, 1956); 3.0–23.6 tonnes/ha in the Ivory Coast and 2600 tonnes/ha in the Nile Valley (Beauge, 1912); and 5.0–27.0 tonnes/ha in South Africa (Reinecke and Ryke, 1970). Although North American species are known to cast on the surface, we still lack data on the quantities of soil moved in this way in different parts of the US. However, James (1991) estimated that total earthworm cast production in a Kansas tallgrass prairie, by populations of native *Diplocardia* spp., ranged from 1276 to 7571 g/m^2, depending on the year and soil type.

The amounts of soil actually turned over by worms may be even greater, because some species void most of their casts underground (Evans, 1948a). Nevertheless, the presence of casts is a good indication of the amount of earthworm activity; usually many more casts are found on the surface of pasture than on arable soils (Table 10.1). In one field at Rothamsted (Highfield) which had bare fallow and pasture, 12.4 casts/m² were recorded from the fallow and 64.0 per m² from pasture. In terms of weight, this was 29.0 g/m^2 on the fallow, and 178.8 g/m^2 on the pasture.

10.2 EFFECTS OF EARTHWORMS ON SOIL STRUCTURE

The activities of earthworms that have the greatest influence on soil structure are:

1. ingestion of soil, partial breakdown of organic matter, intimate mixing of these fractions and ejection of this material as surface or sub-surface casts; and
2. burrowing through the soil and bringing subsoil to the surface.

During these processes they thoroughly mix the soil, form water-stable aggregates, increase soil macroporosity, aerate the soil and improve its water-holding capacity.

10.2.1 TURNOVER OF SOIL

Large amounts of soil from deeper layers are brought to the surface and deposited as casts. The amounts turned over in this way differ greatly with habitat and geographical region, ranging from 2 to 268 tonnes/ha (Table 10.1). This is equivalent to bringing up layers of soil between 1 mm and 5 cm thick, to the surface every year, with typical amounts probably ranging from 2 to 5 cm for many temperate ecosystems. In addition, large amounts of soil are deposited either as subsurface casts, or within burrows, so the total soil turnover and mixing are even greater. The long-term effect of such soil turnover is to modify the overall structural properties of the soil. In some instances, surface casting leads to the burial of a layer of stones on the soil surface, particularly in old pasture, which may have such layers 10–15 cm deep. The importance of this turnover, which was discussed first by Charles Darwin (1881), can be seen by comparing the profile of a stratified mor soil (with few earthworms) with that of a well-mixed mull soil.

10.2.2 BREAKDOWN OF SOIL PARTICLES

The presence of smaller amounts of sand and other larger soil fractions in worm casts than in the surrounding soil, has been taken as evidence that worms can comminute or break down mineral particles to smaller units (Teotia et al., 1950; Shrikhande and Pathak, 1951; Joshi and Kelkar, 1952). Evans (1948a) reported that the proportions of coarse sand relative to silt and clay, in two pastures with large numbers of earthworms, increased with depth, and they suggested that earthworms might be breaking down the coarse sand in the surface soil. However, others have suggested that the reduction of the coarse sand fraction in earthworm casts is due to selective ingestion by earthworms of finer particles (Zhang and Schrader, 1993). Therefore, the enrichment of silt and clay in the surface soil observed by Evans (1948a) may have been due to the long-term deposition of surface casts that were depleted in numbers of coarse particles. Such an effect was demonstrated in a long-term investigation of field lysimeters filled with homogenized sandy loam soil, in southern California (Graham and Wood, 1991). After 41 years, earthworm activity in the lysimeters had produced a surface soil horizon, from 2 to 7 cm thick, that consisted primarily of worm casts and was enriched in clay particles compared with the underlying fill material. Lysimeters without worms had no such clay-enriched surface horizons. Nijhawan and Kanwar (1952) reported that the castings of some earthworm species actually contained more coarse fractions than the parent soil, indicating that different species of earthworms may feed preferentially on different sized particles.

In pot experiments with earthworms, it was shown that the size of particles of granite in the soil was less when earthworms were present than when there were none (Bassalik, 1913). Similar experiments with soil (Blancke and Giesecke, 1924) and basalt (Meyer, 1943) confirmed these results, although the number of particles broken down was small. It seems unlikely that the amount of comminution that occurs in this way is either sufficient or rapid enough to be of any importance compared with normal mineralogical weathering processes. It is by no means certain that earthworms do in fact comminute the mineral fraction of soils to any significant extent. Their selective feeding on particular size fractions and redistribution of particles is probably of much greater importance.

10.2.3 FORMATION OF AGGREGATES

Soil aggregates are formed by the adhesion of mineral granules and soil organic matter into composites of various sizes that resist breakdown when exposed to internal or external stress such as wetting, drying, compaction or other physical disturbance. Aggregation influences structural characteristics that have a major influence on soil fertility, such as water infiltration, water-holding capacity, porosity and aeration. Soil type has an overriding influence on the degree of aggregation, with sandy soils exhibiting the least potential for formation of aggregates and clay soils the greatest aggregation. The influence of earthworms and other biological factors on aggregate stability may be greatest in poorly and moderately aggregated sandy soils and loams, and relatively less important in heavy clay soils, which are already well aggregated (Oades, 1993). Earthworms are not essential for soils to be well aggregated, and many soils attain good structural properties without earthworms. However, where earthworms are abundant, they can have a predominant influence on aggregate formation and stabilization.

Earthworms contribute to soil aggregation mainly through the production of casts, although earthworm burrows can also contribute to aggregate stability, since they are often lined with oriented clays and humic materials (Jeanson, 1964) which can form a stable structure. Most workers have agreed that earthworm casts contain more water-stable aggregates than the surrounding soil (Table 10.2) (Bassalik, 1913; Gurianova, 1940; Hopp and Hopkins, 1946a; Dawson, 1947; Chadwick and Bradley, 1948; Dutt, 1948; Joachim and Panditesekera, 1948; Swaby, 1949; Bakhtin and Polsky, 1950; Ponomareva, 1950; Teotia et al., 1950; Finck, 1952; Nijhawan and Kanwar, 1952; Mamytov, 1953; Guild, 1955; Low, 1955; Parle, 1963b; Shipitalo and Protz, 1988; Marinissen and Dexter, 1990). The casts of earthworms often make up the majority of structural aggregates in the upper 10–20 cm of soil,

Table 10.2 The stability of earthworm casts and the surface soil (from Dutt, 1948)

Source of sample	Percentage aggregates					
	Surface soil (0–7.5 cm)			Earthworm casts		
	Sample			Sample		
	a	b	Mean	a	b	Mean
Cultivated (silt loam)	5.8	7.4	6.6	19.0	19.6	19.3
Pasture (3 years)	43.4	48.0	45.7	56.6	59.6	58.1
Forest	58.2	59.0	58.6	65.6	70.8	68.2

although this depends on a range of factors (Lee and Foster, 1991). A variety of methods have been used to assess the stability of earthworm casts, and these include wet sieving, dry sieving, falling water droplets, crushing, permeability and clay dispersion techniques (Tomlin *et al.*, 1994)

In one typical experiment, the percentage of aggregates in soil to which earthworms were added, was compared with those in soil without earthworms (Hopp, 1946). After 3 days, the soil with worms had 12% of large water-stable aggregates, whereas the soil without worms contained only 5.9%. In a different experiment, Blanchart (1992) took monoliths of sieved, destructured soil from a savanna in the Ivory Coast, and placed them in the field, under natural conditions with or without earthworms. In soil with no earthworms, large aggregates (>2 mm) comprised only 12.9% of soil, whereas in soil with worms, large aggregates comprised 60.6% of soil, after 30 months in the field. These results were very similar to those obtained in laboratory studies (Blanchart *et al.*, 1990), and confirmed that *M. anomala* has an overriding influence in aggregate formation and stabilization of soil structure in highly weathered tropical soils.

Tomlin *et al.* (1994) reviewed possible mechanisms of stabilization as a result of earthworm activity (Table 10.3). For instance, the addition of succulent material such as alfalfa hay to soil greatly increases the amount of aggregation that occurs in response to earthworm activity (Dawson, 1948; Dutt, 1948) and the stability of casts may depend very much on the availability of certain types of plant organic matter (Hoeksema *et al.*, 1956). Teotia *et al.* (1950) reported that worm casts collected from pots were most

Table 10.3 Proposed mechanisms of stabilization in earthworm casts (adapted from Tomlin *et al.*, 1994)

Mechanisms of stabilization in casts	Reference
Internal secretions of earthworms	Dawson (1947)
Mechanical stabilization by plant fibres in casts	Dawson (1947), Lee and Foster (1991)
Mechanical stabilization by fungal hyphae	Parle (1963a), Marinissen and Dexter (1990), Lee and Foster (1991)
Bacterial gums in casts	Swaby (1949)
Formation of organo-mineral bonds	Dutt (1948), Shaw and Pawluk (1986a)
Wetting–drying cycles	Shipitalo and Protz (1988), Marinissen and Dexter (1990)
Age-hardening combined with organic bonding	Shipitalo and Protz (1988, 1989)

stable in soil with alfalfa residues, less stable in soil with wheat straw and least stable in soil with no added residues. Shipitalo and Protz (1988) found that the type of organic residues offered to earthworms in relation to the amounts remaining in their casts influenced cast stability. Casts were slightly more stable than undigested soil when no additional food was provided, but were much more stable when food was added. This suggests that plant remains that have passed through earthworm guts reinforce and hold the aggregates together. When *E. fetida* was kept in artificial soil mixed with beech litter, about two-thirds of the total organic and mineral matter occurred as 200–2000 μm water-stable aggregates, after 446 days (Ziegler and Zech, 1992). Earthworms may have different effects on aggregation in different types of habitats. For example, it has been well established that the stability of earthworm casts from land under grass or forest is greater than those under arable crops such as lucerne or cereals (Table 10.2) (Dutt, 1948; Teotia *et al.*, 1950; Mamytov, 1953). Probably, aggregates are formed readily in grasslands without the intervention of earthworms, but in forest soils earthworms have a much more important role to play in the formation of aggregates (Jacks, 1963).

Not all species of earthworms are equally efficient in producing aggregates; the degree of stability of their casts depending very much on the food and behavior of the earthworms. For instance, Guild (1955) showed that the surface-feeding *A. longa* and *L. terrestris* formed the largest aggregates in soil, whereas the nonsurface-casting species *L. rubellus* and *D. subrubicunda* formed only a few small aggregates. Shipitalo and Protz (1988) reported that casts produced by *L. rubellus* tended to be more stable than those formed by *L. terrestris*, possibly because they contained more organic matter, even though both species were offered the same amounts and type of food.

Another mechanism to explain the formation of water-stable soil aggregates from earthworm casts is that the aggregates are formed by internal secretions, which cement soil particles together as they pass through the intestines of earthworms (Bakhtin and Polsky, 1950). An alternative explanation is that the stable aggregates in earthworm casts are produced from soil particles, cemented together by calcium humate that is synthesized in the earthworm intestine from decaying organic matter, and calcium from the calciferous glands. Certainly, calcium humate can stabilize soils (Swaby, 1949) and this could account for the correlation between the stability of worm casts and their humus content (Ponomareva, 1950). The direct examination of the micromorphology of casts has shown that humified portions of ingested organic matter serve as foci for aggregate formation (Shipitalo and Protz, 1989; West *et al.*, 1987; Barois *et al.*, 1993).

Some workers have suggested that the stability of aggregates in worm casts and worm-worked soil is due to bacteria that produce stabilizing materials in the casts. Although organic matter can cause aggregation, it does so mainly in soils where micro-organisms are present (Waksman and Martin, 1939). Some bacteria are known to produce secretions such as polysaccharide gums which can contribute to aggregate stabilization (Geoghegan and Brain, 1948). However, Dawson (1947) could find no direct relationship between the numbers of bacteria in soils and casts, and the proportion of water-stable aggregates in them. Direct microscopic examination of casts has shown that bacteria and bacterially produced gums or polysacharrides are abundant in casts, and may serve to bond mineral particles together in the casts (Shaw and Pawluk, 1986b; Altemuller and Joschko, 1992).

Fungi also probably contribute to the stabilization of earthworm casts. When pasture soil was incubated together with fresh earthworm casts, there was a rapid growth of fungi and an increase in the degree of aggregation (Swaby, 1949). Parle (1963b) showed that fungi are usually fewer in newly deposited earthworm casts, but he noted that the stability of casts of *A. longa* increased gradually for 15 days, then began to decrease, and these changes were correlated with the growth of fungal hyphae. Marinissen and Dexter (1990) also reported that fungal hyphae growing on the surface of aging casts contributed to cast stability.

The stabilization of earthworm casts also increases with aging and drying of the casts (Shipitalo and Protz, 1988, 1989; Marinissen and Dexter, 1990; Barois *et al.*, 1993). For example, Marinissen and Dexter (1990) reported that the stability of earthworm casts increased with age for up to 42 days after they were deposited. Shipitalo and Protz (1988) showed that the effects of aging depended on the type and amounts of organic matter ingested by the worm. When no organic matter was provided to the worms, aging for up to 32 days did not increase cast stability, but cast stability increased gradually with time when organic residues were added to the soil. The stability of casts also increased when the casts were dried and rewetted. Drying of casts can override the influence of aging on stability (Shipitalo and Protz, 1988; Marinissen and Dexter, 1990). In fact, fresh casts are often less stable than surrounding soil, which may explain why earthworm casting activity can increase surface erosion of soil particles in some situations (Madge, 1969; Sharpley *et al.*, 1979).

It is by no means certain how long the stability of aggregates in earthworm casts lasts, and how important this is in determining the long-term effects of earthworms on soil aggregation. The available evidence suggests that the stabilization of earthworm casts occurs by one or more of several different mechanisms (Table 10.3). The relative importance of any one of these mechanisms can vary under different conditions and with different earthworm species.

10.2.4 AERATION, POROSITY AND DRAINAGE

Earthworms improve soil aeration by their burrowing activity, but they also influence the porosity of soils. Wollny (1890) calculated that earthworm burrows increased the soil-air volume from 8% to 30% of the total soil volume, but this was under rather unnatural culture conditions. Stöckli (1928) estimated that in a garden soil with 2.4 million earthworms/ha, earthworm burrows occupied 9–67% of the total soil-air space, which can be compared with 38% and 66% for a ley and a pasture, respectively (Evans, 1948a). Probably, a more realistic estimate is that earthworm burrows constitute only about 5% of the total soil volume (Stöckli, 1949). Teotia *et al.* (1950) claimed that earthworm activity increased the porosity of two soils from 27.5% to 31.6% and 58.8% to 61.8%, respectively. Hoeksema and Jongerius (1959) compared the porosity of orchard soils with no earthworms to that of similar soils in which earthworms were abundant. Their results indicate that earthworms increased the total pore space by 75–100%. Satchell (1967) estimated that earthworm burrows could account for up to two-thirds of the total air-filled pore space in soil. However, other workers have reported that the introduction of earthworms into soils with no earthworms results in only small increases in total porosity (Springett *et al.*, 1992).

The main effect of earthworms on soil porosity is to increase the proportion of large pores in the soil, which are commonly referred to as macropores. Several authors have reported that the number of macropores in the soil is correlated positively with the number of earthworms (Hopp, 1973; Lal, 1974; Ehlers, 1975; Edwards and Lofty, 1982). For instance, Knight *et al.* (1992) introduced earthworms into pastures in Devon, England, and reported that the earthworms increased the soil's macroporosity considerably. Estimates of the number of earthworm burrows in soil are as high 624 per m^2 for cropland in the midwestern US (Hopp, 1973), 800 per m^2 in pastures in France (Bouché, 1971), and 1204 per m^2 in heavily manured sugar-beet field in Bavaria (Becher and Kainz, 1983). Kretzschmar (1987) did a detailed three-dimensional analysis of the burrows of *Allolobophora* spp. in grassland soil, and calculated that there were 10 000 burrows/m^3 of soil. In an earlier investigation, Kretzschmar (1978) calculated that in one soil, earthworm burrows comprised a total volume of 5 liters/m^3 of soil, making a small, but significant contribution to soil aeration. Edwards *et al.* (1988) estimated that there were 1.6 million *L. terrestris* burrows/ha in a no-tillage maize field. These burrows are particularly important from the standpoint of water infiltration, because they open to the surface, are nearly vertical, and can be up to 2 m deep.

Macropores formed from earthworm burrows, such as those of *L. terrestris*, which have a large diameter (>5 mm) and are open to the soil surface, can act as pathways for the preferential flow of water. Water that flows down earthworm burrows has been referred to as bypass flow, because it bypasses the soil matrix (Bouma *et al.*, 1982; Bouma, 1991). Burrows do not transport water to the same degree during all rainfall events. The moisture status at the time of rainfall, rainfall intensity, soil type and residue cover, and microrelief all influence the degree to which water will flow down earthworm burrows (Zachmann et al., 1987; Shipitalo et al., 1990; Edwards et al., 1992; Trojan and Linden, 1992). Moreover, the preferential flow of water down earthworm burrows is not limited to burrows of anecic species, such as L. terrestris. Joschko et al. (1992) demonstrated that burrows made by endogeic earthworm species (A. caliginosa and A. rosea) can also conduct water very effectively, even though their burrows are subvertical and only infrequently open to the soil surface. Similarly, Trojan and Linden (1992) showed that burrows of the endogeic earthworm, A. tuberculata, were effective in conducting water. The majority of the water flux down earthworm burrows may be due to only a few, very effective burrows per square metre. Burrows must be continuous to be effective at transporting water to deeper soil depths (Ela et al., 1992), and tillage of the soil can significantly reduce the number of burrows and macropores that transmit water to the groundwater (Chan and Heenan, 1993).

Water infiltration is from 4 to 10 times faster in soils with earthworms than in soils without earthworms (Hopp and Slater, 1948; Teotia et al., 1950; Guild, 1952a; Stockdill, 1966; Carter et al., 1982). In Wisconsin, USA, L. terrestris activity increased cumulative rainfall intake by one-half (Peterson and Dixon, 1971). Earthworms have increased the field capacity of some New Zealand soils by as much as 17%, compared to soils without earthworms, (Stockdill and Cossens, 1966). Similar results have been recorded from savanna in Nigeria, where cropping was shown to influence earthworm populations which in turn affected infiltration rates (Wilkinson, 1975). Ehlers (1975) measured water infiltration in no-tilled and conventionally tilled corn fields in Germany and reported that the infiltration rate due to earthworm burrows in zero-tilled soils was 0.12 mm water/minute, whereas in tilled soil with low earthworm populations, the infiltration rate was only 0.02 mm water/minute. In a similar comparison of differently managed wheat fields in New South Wales, Australia, Chan and Heenan (1993) observed much greater hydraulic conductivity in no-tillage than in conventionally tilled fields. They reported a significant correlation between the number of pores (>1 mm) that transmitted water and the number of earthworms in the field. Edwards et al. (1990) attributed the large reductions in runoff and increased water infiltration in long-term no-tillage maize fields to increased earthworm activity in those fields. Becher and Kainz (1983) reported that sugar-beet plots that received manure applications for 15 years exhibited over a 4.2-fold increase in saturated hydraulic conductivity, which they attributed mainly to a 2.7-fold increase in earthworm channels. The elimination of earthworms from English grassland, due to the repeated use of the pesticide phorate, decreased the water infiltration rate by 93% (Clements, 1982). Hoogerkamp et al. (1983) measured huge increases in water infiltration in Dutch polders, 8–10 years after adding earthworms. Water infiltration over a 24 hour period was 118–136 times greater in plots with earthworms than in plots without earthworms. Such large increases represent an extreme case of the effects of earthworms in unstructured soils, recently reclaimed from the sea, but water infiltration in other extreme situations, such as in reclaimed mine spoils, might provide similar benefits after the introduction of earthworms.

Increases in water infiltration due to earthworm activity can translate into decreases in surface runoff. In soil columns subjected to artificial rain, the presence of earthworm burrows open to the soil surface decreased runoff considerably (Roth and Joshko, 1991). The dramatic influence of earthworms on surface runoff in the field was demonstrated by Sharpley *et al.* (1979) who used a broad-spectrum insecticide (carbaryl) to eliminate earthworms from a permanent pasture in New Zealand. There was a twofold increase in the volume of runoff and a threefold decrease in water infiltration rate in response to the elimination of the earthworm populations.

Changes in soil porosity and water infiltration due to earthworm activity can also increase the water-holding capacity of soil. Van Rhee (1969a) measured the soil water content at the wilting point and at field capacity in three plots with earthworms and three plots without earthworms in Dutch orchard soils. The amount of soil water available to plants (the difference between moisture content at the wilting point and at field capacity) was greater in plots with worms than in plots without worms, apparently due to the higher percentage of large pores in the soils with worms, as compared to the soils with no worms. The introduction of lumbricid earthworms into pastures in New Zealand led to a 17% increase in the moisture content at field capacity. The availability of soil water to plants was consistently greater in soils with introduced earthworms than in soils with no worms. Clearly, earthworms influence both the drainage of water from soil and the moisture-holding capacity of soil, both of which are important factors for plant production.

10.2.5 EFFECTS ON SOIL EROSION

Darwin (1881) was the first to note that earthworm casts deposited on the soil surface were susceptible to erosion. He observed that much of the soil associated with surface casts moved downhill, on sloping permanent grasslands in England following rains. The impact of the rain caused the casts to spread out, down the slope from the mouths of the earthworm burrows, and Darwin estimated that considerable amounts of soil could move downhill in this way, over long periods of time. He also reported observations from localities in India, where large amounts of earthworm casts were removed from slopes during monsoonal rains.

The best experimental evidence that supports the early observations of Darwin about how earthworms affect soil erosion comes from studies in New Zealand pastures (Sharpley and Syers, 1976; Sharpley *et al.*, 1979). Duplicate plots for measuring surface runoff were established and earthworms were removed from one plot using carbaryl, a broad-spectrum biocide. Elimination of earthworms led to significant reductions in the sediment load of surface runoff. The total sediment in surface runoff was 1120 kg/ha/yr in plots with earthworms and only 290 kg/ha/yr in plots from which earthworm populations had been eliminated. As a result, the loss of particulate phosphorus in surface runoff was greater when earthworms were present. However, because earthworms increased water infiltration threefold and decreased the total amount of surface runoff twofold, the total loss of phosphorus and nitrogen was much lower in plots with earthworms (Sharpley *et al.*, 1979). Hopp (1973) reported that erosion of soil particles and surface runoff from cropland were reduced greatly in the presence of earthworms and were related directly to the numbers of earthworms present.

Earthworms can also influence erosion by removing surface litter and creating bare spots on the soil surface. Plant litter on the soil surface protects soil from the erosive impact of rain droplets and deters the movement of soil particles downslope. Earthworms can interfere with this protective barrier by removing surface litter. This is particularly true of species such as *L. terrestris*, that gather leaves around the mouths of their burrows, creating bare patches in the spaces between burrows. Hazelhoff *et al.* (1981) investigated the influence of these bare patches on erosion in forest soils in Luxembourg. They reported that erosion of soil particles increased in areas where *L. terrestris* had removed the surface litter and created patches of bare soil. Thus, earthworms can contribute to soil erosion in two ways:

1. by removing surface litter and, thereby, exposing the underlying soil to the impact of rain; and
2. by producing surface casts that can be carried downslope during heavy rains.

These processes are probably influenced by soil type, plant cover and degree of slope, and must be considered against the overall beneficial effect of earthworms in reducing total runoff and increasing water infiltration.

10.3 EARTHWORMS AS BIOINDICATORS OF SOIL TYPE

Several workers have proposed that the species of earthworms that occur in a soil can be bioindicators of the soil type and its basic properties (Müller, 1878; Bodenheimer, 1935; Lee, 1959; Saussey, 1959; Volz, 1962). Ghilarov (1956a,b, 1965) was the chief modern proponent of this theory and his work has been comparatively successful. Many other attempts to diagnose soils by the earthworms they support have been unsuccessful, probably because those ecological factors that favor multiplication of earthworms, such as moisture capacity, pH, organic matter content, etc. are not always those properties that are linked directly to soil type. Obviously, certain species of earthworms are associated with extreme soil types, but we need a much more thorough knowledge of the distribution of earthworms in different soils, and corresponding and very thorough soil analyses of the same soils, before we can hope to diagnose soils successfully on the basis of their earthworm populations. ·

10.4 EFFECTS OF EARTHWORMS ON PLANT PRODUCTIVITY

It is certain that earthworms have beneficial effects on soils and many workers have attempted to demonstrate that these effects increase plant growth and yields of crops. Some of the effects of earthworms on soil structure and nutrient availability take much too long to produce detectable effects on plant growth, in experiments lasting only a few months or even years. Many investigations of the influence of earthworms on plant growth have used pots, under controlled conditions, and it is questionable whether the results of these experiments are applicable to plants growing under natural field conditions. There is also the difficulty in distinguishing between the effects of living earthworms on soil conditions and nutrient content, and those due to the release of nitrogenous compounds and other nutrients from decaying worms that die during the course of experiments.

Many workers have obtained inconclusive results from pot experiments because of these errors (Wollny, 1890; Chadwick and Bradley, 1948; Baluev, 1950; Joshi and Kelkar, 1952; Nielson, 1953). However, Russell (1910) accounted for release of nutrients from dead worms, by adding the same number of dead worms to his control soils as he added live worms to test soils. When he added 0.5 g (live weight) of earthworms to each kilogram of soil, he obtained increased dry matter yields of the order of 25%, and he attributed this to improvements in the physical condition of soils. For instance, there was less evaporation from pots that contained earthworms, because the soil surface was covered by earthworm casts. Hopp and Slater (1948, 1949) reported that herbage plants grown in a poorly structured soil to which live earthworms were added at a rate of 120 per m² (together with organic matter) yielded 3160 kg/ha, whereas when dead earthworms were added instead, the yield was only 280 kg/ha. These workers also tested the influence of four species of earthworms on overall soil fertility, and concluded that they consistently increased yields of millet, lima beans, soybeans and hay, and that adding live earthworms increased yields much more than adding dead ones. The growth of soybeans and clover in soils with poor structure was stimulated more than that of grass and wheat.

When large numbers of earthworms were added to soil cultures they doubled the dry-matter yield of spring wheat grown in the soil, and increased grass yields four times and clover yields 10 times, although pea yields were decreased (van Rhee, 1965). Kahsnitz (1922) claimed that the addition of live worms to a garden soil increased yields of peas or oats by 70%, but that the numbers of earthworms that had to be added were very large. *Lumbricus terrestris* and *L. rubellus* inoculations increased yields of barley in heavily manured soils in garden frames (Uhlen, 1953), and winter wheat yielded more in plots to which live earthworms were added (Dreidax, 1931). Ribaudcourt and Combault (1907) also reported that the addition of worms to small field plots increased yields, but they did not account for the effects of dead worms. Aldag and Graff (1974) compared the growth of oat seedlings in 800 g of brown podzol soil to which 18 *E. fetida* had been added, for 8 days, with growth in the same soil without earthworms. The dry matter yield of the oat seedlings was 8.7% greater in the soil with earthworms, and the total protein yield was 21% more. In a series of pot and box experiments, Atlavinyte and her co-workers (Atlavinyte *et al.*, 1968; Atlavinyte, 1971, 1974; Atlavinyte and Vanagas, 1982) demonstrated clearly that there is a strong correlation between the number of earthworms (usually *A. caliginosa*) and the growth of barley. For instance, addition of 400–500 individuals of *A. caliginosa* per m^2 increased the biological productivity of barley in 1 m^2 field plots by 78–96%. Increased yields were proportional to the numbers of earthworms that had been added.

When individuals of *A. caliginosa* were kept for 8 weeks in a soil that contained 30 g of dry dung/kg of soil, and were then removed, and ryegrass planted in the soil, yields doubled in the earthworm-worked soil (Waters, 1951). Another worker showed that addition of straw with earthworms increased barley yields more than addition of earthworms only (Atlavinyte, 1971). In recent greenhouse experiments in Australia, Stephens *et al.* (1995) reported that *A. trapezoides* and *A. rosea* increased the shoot and root growth of wheat plants, and also increased the concentration of many nutrients in the wheat shoots. These experiments showed conclusively that earthworms can increase yields, at least in small-scale experiments. Other workers have confirmed this in field experiments.

There is evidence from field experiments that the addition of earthworms to sown pastures can increase crop yields (Waters, 1955; Fig. 10.2). In some cases, introductions of European species of earthworms into New Zealand fields resulted in an initial increase in above-ground productivity of 70%, which settled down to a 20–30% increase after several years (Stockdill, 1982; Syers and Springett, 1983). The soils were usually acid, so lime was added to counteract this, then colonies of about 25 individuals of

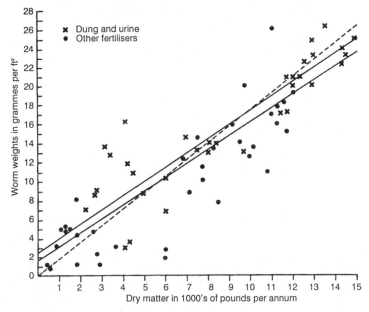

Figure 10.2   Relation between weight of worms and productivity of pasture (after Waters, 1955).

A. caliginosa were added on a 10 m grid pattern, and 4 years later, around each inoculation point, there was a greener and more densely covered area, several meters in diameter. After 8 years, the areas of earthworm activity had spread as far as 100 m from the initial inoculation points (Hamblyn and Dingwall, 1945; Richards, 1955; Stockdill, 1959). There are many instances of increased grass yields caused by earthworm inoculations into soils from different parts of New Zealand. For instance, *A. caliginosa* increased grass production by between 28 and 100%, and yield increases in mixed swards ranged from 28 to 110% (Nielson, 1953; Stockdill and Cossens, 1966; Stockdill, 1982). Waters (1955), Barley (1961), Barley and Kleinig (1964), Kleinig (1966) and Noble *et al.* (1970) reported similar yield increases (Fig. 10.2).

There have been relatively few studies on the effect of earthworms on the growth of forest trees. Live earthworms, when added to pots, increased the growth of 2-year-old seedlings of oak (*Quercus robur*) by 26%, of green ash (*Fraxinus pennsylvanica*) by 37% (Zrazhevskii, 1957), and black spruce (*Picea mariana*) seedlings increased in weight significantly when earthworms were added to the soil in which they were grown (Marshall, 1971). *A. tuberculata* increased the growth of birch (*Betula pendula*) seedlings in laboratory culture, with increases observed in the growth of shoots, roots and stems (Haimi and Enbrok, 1992). Wolters and Stickan (1991) examined the allocation of carbon and nitrogen in beech

(*Fagus sylvatica*) seedlings, in two different beech forest soils, incubated either with one individual of *Octolasion lacteum* or without worms. The production of stems and total incorporation of nitrogen were greater in containers with earthworms than in control soils with no earthworms. The earthworms did not affect total root production, but increased the proportion of large roots relative to small roots and stimulated the transfer of carbon and nitrogen to above-ground parts of the plants.

The effects of earthworms on growth of roots of apple trees was demonstrated clearly by van Rhee (1977). He reported an increase of 70% in aggregate stability and up to 140% increase in density of roots <0.05 mm diameter, and also an increase in densities of thicker roots, when worm-free polders were inoculated with *A. caliginosa* and *L. terrestris* at a rate of 500 per m^2. These changes resulted in average crop yield increases of 2.5%.

Earthworms can be very important in affecting the growth and yield of crops planted into soil that has not been cultivated. Edwards and Lofty (1975b) demonstrated clearly that earthworm populations were much larger in soil that was not cultivated and had crops drilled directly (no-till) with a special slit drill. In box experiments (Edwards and Lofty, 1976), they showed that addition of typical field populations of *L. terrestris*, *A. longa*, *A. chlorotica* and *A. caliginosa* to soil dug out as intact profiles and not cultivated, greatly increased the emergence and growth of barley. In other experiments, Edwards and Lofty (1980) investigated the influence of earthworms on the growth and yield of barley in field soils that had been direct-drilled for 6 years. They fumigated the soil to eliminate earthworms and then inoculated small replicated plots with either *L. terrestris* and *A. longa* or *A. caliginosa* and *A. chlorotica*. The growth of barley in the inoculated plots was compared with that in plots to which no earthworms had been added. There were significant differences among treatments in the yield of barley, and the growth rate of the barley was greater in plots inoculated with worms. Moreover, the production of roots and their penetration into deeper soil layers was significantly greater in inoculated plots, particularly in response to inoculation with the deep-burrowing species of earthworms.

Lavelle (1992) inoculated selected earthworm species into several low-chemical-input farming systems at La Mancha, Mexico, Lamto in the Ivory Coast and Yurimaguas, Peru. Earthworms increased crop growth and yields significantly, in 10 out of 20 cropping cycles at all three sites. At La Mancha, when *Pontoscolex corethrurus* was introduced, growth of grain crops increased by 10–30%. At Yurimaguas, grain yields increased by 145%.

The influence of earthworms on plant growth can be affected by soil type. Doube *et al.* (1995) had a series of greenhouse experiments in which they added *A. trapezoides* or *A. rosea* to three different soil types (sandy loam, loam and clay). In one experiment, both earthworm species produced a significant increase in the growth of wheat seedlings in the sandy loam soil, but had no effect in the loam or clay soils. In another experiment, both earthworm species increased the growth and yield of barley in the sandy loam soil, but had no effect in the loam soil, and decreased the growth and grain yield in the clay soil. MacKay and Kladivko (1985), who reported that earthworms did not increase corn root or shoot growth, attributed their results to the high amounts of nitrogen and phosphorus and low bulk density of the soil at the outset of their experiment, which may have overridden the beneficial effects of earthworms. Buse (1990) also found that the growth response of plants to earthworms depended on soil type. He reported that earthworms increased herbage yields in soil cores from improved pastures but not in soil from unimproved pastures (improved pastures had received additions of lime and nitrogen 5 years previous to the study).

Different earthworm species can have different influences on plant growth. For instance, Spain *et al.* (1992) found that the growth of *Panicum maximum* increased significantly after inoculation of soil in which it grew with two species of eudrilid earthworms, *Chuniodrilus zielae* and *Stuhlmannia porifera*, but was unaffected by inoculation with the acanthodrilid worm, *Millsonia anomala*. James and Seastedt (1986) failed to show a positive effect of earthworms on the shoot growth of big bluestem (*Andropogon gerardii*) in laboratory microcosms, but one of the three earthworm species tested had a positive effect on root growth.

There have been suggestions that earthworms produce plant growth substances. For instance, Nielson (1965) claimed that he detected such substances in eight species of lumbricids and two megascolecids, and stated that they were secreted into the alimentary tract and voided with the feces. This would not be surprising because many micro-organisms produce such substances. Other researchers also suggested that earthworms release substances beneficial to plant growth (Hopp and Slater, 1949; Springett and Syers, 1979; Krishnamoorthy and Vajranabhaiah, 1986), although the relationships of these substances to plant growth are still not fully understood (Tomati *et al.*, 1988). This topic is discussed in more detail in Chapter 11.

More studies are needed using inoculations of different combinations of earthworm species and test plants under a variety of conditions, if we are to improve our understanding of the relationship between earthworm activity and plant growth. Investigations of the effect of earthworms on plant growth have been limited to a relatively small number of earthworm and plant species. All previous studies of the influence of earthworms on yields of crops have been based on a method using addition of live or dead worms to worm-free soil. Another approach would be to compare the yield in plots with natural populations, with those from which worms have been removed either with electrical stimuli (this might, however, kill worms and leave their carcasses behind), or formalin extraction. Dobson and Lofty (1956) used chemicals to kill earthworms in moorland soils, but the chemicals used may have affected other soil animals and microbes, making the interpretation of results less clear.

Earthworms as pests and benefactors

11

It has generally been assumed that the majority of the activities of earthworms benefit soil structure, fertility and productivity by their influence on organic matter breakdown and nutrient cycling. They can also have other benefits such as providing plant growth stimulation. However, there are instances where earthworms are pests and their influence is adverse, such as through their effects on transmission of pathogens and on soil erosion.

11.1 EARTHWORMS AS PESTS

11.1.1 EARTHWORMS AS PESTS OF CROPS

In the nineteenth century, prior to Darwin's book on earthworms (1881), most books on agriculture and natural history regarded earthworms as pests of plants; this was reviewed by Graff (1983) who reported that even after the publication of Darwin's book, numerous reports of earthworms as pests continued to be published.

Some habits of earthworms qualify them as potential pests of crops. The behavior of certain species of earthworms in seizing fallen leaves and pulling them down through the mouth of their burrows, means that if they seize the leaves or seedlings of growing plants in this way, they damage the plant, sometimes sufficiently to kill it. Zicsi (1954) reported such damage to various crops, and Edwards and Lofty (1977) reported that earthworms destroyed a large part of a lettuce crop in this manner when soil containing large numbers of *L. terrestris* was transported into a greenhouse. Stephenson (1930) mentioned several reports from India where *Malabaria paludicola, Aphanascus oryzivorous* and *Criodrilus* sp. were stated to attack plants by feeding on the roots of paddy rice, and *Perionyx* spp. were reported to burrow in the stems of cardamom plants.

As earthworms tunnel through the surface soil, they can damage small and delicate seedlings by uprooting them (Walton, 1928), and they sometimes damage the roots of delicate plants in pots or flower beds, by tunneling through their root systems (Stephenson, 1930). Only one study has confirmed that earthworms can feed on unhealthy plant roots (Cortez and Bouché, 1992), although various authors have speculated that they might damage healthy plants in this way. Such damage is becoming potentially much more important in agriculture with the advent of precision drilling of crops, with seeds such as sugar beet drilled at optimal spacing, so that any loss or damage cannot be compensated during thinning. Valuable and delicate crops such as tobacco have been reported as being damaged by earthworms in Bulgaria (Trifonov, 1957), vegetables have been attacked in India (Patel and Patel, 1959) and rice has been injured in the Philippines (Otanes and Sison, 1947), in China (Chen and Liu, 1963) and in Japan (Inoue and Kondo, 1962). Damage to vegetable crops was also reported by Puttarudriah and Sastry (1961). Heungens (1969a) considered that earthworms could affect the growth of azaleas adversely by their effects on aeration and moisture availability. There have been reports of worm casts being deposited to such a height in cereal harvest stocks that it was difficult to obtain clean grain when the crop was threshed. Similar effects have been reported in pastures and hay meadows in Switzerland, where thick layers of casts on the soil surface interfere with cutting and harvesting of hay. Damage to crops that can be attributed indirectly to earthworms, has occurred when moles tunnel through arable land in search of earthworms for food. In South Africa, the large casts of the giant microchaetoid earthworms, which in the past made farming almost impossible in some areas, have to be destroyed by the use of bulldozers in order to make crop production feasible. The overall conclusion from these reports is that although earthworms can occasionally damage healthy plants, they are more likely to attack moribund plants and there is no reason to regard them as pests of plants in any significant way.

11.1.2 AS PESTS OF GRASSLAND AND TURF

Grassland usually has very large populations of earthworms, and when these are species which excrete casts on to the surface, they can often be serious pests. These can ruin the appearance of ornamental lawns, and the casts are considered to be intolerable on the surface of many sports grounds and golf greens because the projections render the turf unsuitable for golf, tennis or bowling. Schread (1952) estimated that as much as 18 tonnes of casts may be brought up to the surface of golf greens annually. Many hundreds of thousands of UK pounds are spent annually in keeping earthworm populations under control in such sporting facilities.

Weed incidence has been reported to increase through the activities of earthworms. There have been many reports of seeds being ingested by *L. terrestris* and other earthworm species (beginning with Darwin, 1881). McRill (1974) reported that L. *terrestris* consumed the seeds of 14 common British grassland plants. The seeds of some species did not appear in the casts of the earthworms and the germination success of those that did was affected. He reported that the seeds of many weed species occurred in the casts of *L. terrestris* and concluded that this increased the spread of weeds. Corrall (1978) demonstrated that *L. terrestris* could bury weed seeds, and hence increase their germination. Grant (1983) showed that *A. longa* could also ingest and distribute weed seeds. Both *L. terrestris* and *A. longa* choose different seeds preferentially; in choice experiments the seeds of *Poa pratensis* and *Poa trivialis* were preferred to those of *Lolium perenne* and *Festuca rubra*; most seeds (75–90%) were still viable after passage through the earthworm guts. Moreover, small seeds that lack awns or hairs may be selected preferentially by earthworms (Thompson, 1987). Shumway and Koide (1994) concluded that although size influences the relative acceptability of seeds by earthworms, they tend to prefer seeds with smooth coats over those with rough coats, and select these preferentially. About 20% of a total grassland seed pool was transported downwards in the soil profile by earthworms to a depth greater than 4 cm.

11.1.3 RELATIONSHIPS WITH PLANT PATHOGENS

Earthworms have been incriminated in transmitting many diseases of plants. They can spread soil fungi, including pathogens, throughout the soil by dispersing spores and hyphal fragments. For instance, Hutchinson and Kamel (1956) showed that *L. terrestris* could consume the spores of several species of fungi from soil and thereby greatly increase their rates and extent of spread through soil. The spores of many species of fungi can pass through the earthworm gut without harm; for instance, there is circumstantial evidence that dwarf bunt (*Tilletia controversa*) may be spread extensively by earthworms, because the teleospores produced by this pathogen are ingested and do not lose their viability during passage through the earthworm's gut (Hoffman and Purdy, 1964). Thornton (1970) isolated the viable spores of 15 species of phycomycetes from the guts of lumbricid earthworms. The pathogenic fungi, *Fusarium* and *Pythium*, can also be transmitted through soil by earthworms (Baweja, 1939; Khambata and Bhatt, 1957; Edwards and Fletcher, 1988). Rao (1979) concluded that the megascolecid earthworm, *Megascolex insignis,* fed upon the decaying roots of papaya (*Carica papaya*) and spread viable spores. It has been suggested that earthworms could carry the zoospores of the potato wart disease (*Synchytrium endobioticum*) (Hampson and Coombes, 1989).

Earthworms can also dispense beneficial disease-biocontrol agents. Take-all disease is possibly the most serious root pathogen of wheat (*Gaeumannomyces graminis* var. tritici (Ggt)). Various bacterial and fungal biocontrol agents for this pathogen have been identified (Ryder and Rovira, 1993), but the effectiveness of such organisms is limited by their very slow rate of unaided dispersal. Doube *et al.* (1994c) showed that the earthworms *A. trapezoides* and *A. rosea*, two of the most common species in cereal soils in southern Australia, could disperse *Pseudomonas corrugata* strain 2140R, a biocontrol agent for take-all (Ryder and Rovira, 1993), a distance of 10–20 cm through tubes of pasteurized soil in the laboratory after 8 days. The biocontrol agent was introduced to the tubes in a mixture with sheep dung on which both species of earthworm fed. The larger earthworm, *A. trapezoides*, moved the bacteria further and more quickly than did *A. rosea*. Similar densities of *P. corrugata* were present in casts of both earthworm species and in the soil associated with the walls of earthworm tunnels (10^4–10^6 bacteria/g soil), and the numbers of bacteria recovered from casts were not affected by the clay content of the soil in which the worms were living.

Stephens *et al.* (1994b) showed that when *P. corrugata* 2140R was mixed with pea and cereal straw and applied to the surface of soil in pots, the earthworm *A. trapezoides* dispersed the bacteria through the soil, resulting in bacterial colonization of the roots of wheat seedlings. After 18 days, there were c. 103 *P. corrugata* per centimeter of root at 3–9 cm soil depth in the presence of the earthworm *A. trapezoides*, while none was recovered at this depth in the absence of the earthworm. By comparison, seed inoculation was a poor method of achieving root colonization by 2140R. At 3–9 cm soil depth, root colonization was about 100 times greater when the bacteria were applied in earthworm food than when they were applied on the seed. Thus, at least two species of earthworms can disperse biocontrol bacteria from inoculated earthworm food to the surrounding soil and promote colonization of wheat roots by the biocontrol agent.

The earthworm *A. trapezoides* can decrease substantially the symptoms caused by *Rhizoctonia* in wheat seedlings (Stephens *et al.*, 1993; Doube *et al.*, 1994c). In southern Australia, *Rhizoctonia solani*, the causal agent of bare-patch disease, is a common fungal root pathogen of wheat, colonizing living roots. This pathogen was introduced into soil on wheat chaff, and the presence of *A. trapezoides* caused a significant reduction in the severity of the symptoms of the disease at a density equivalent to 470 earthworms/m^2.

Earthworms can also assist in the dispersal of plant parasitic nematodes. Ellenby (1945) demonstrated that earthworms affect greatly the infectivity of cysts of the potato root eelworm (*Heterodera rostochiensis*) (Table 11.1). Different numbers of earthworms (20, 40, 60 and 100 individuals of *A. longa* per 25 cm diameter plant pot) were added to soil heavily

Table 11.1   Numbers of potato root eelworm larvae emerging from cysts ingested by earthworms (from Ellenby, 1945)

No. of earthworms per pot	100	60	40	20	0 (control)
Mean no. of larvae emerging per cyst	158.9	85.4	83.3	71.6	51.8
Standard error	±16.6	±19.7	±22.0	±18.3	±13.8

infected with potato root eelworm cysts. The more earthworms that were introduced to a pot, the more eggs hatched from the cysts; and the more rapidly they hatched, the more viable larvae each cyst produced. Ellenby also collected earthworm casts and examined the infectivity of cysts present in these, as compared with the infectivity of cysts from the soil in which the worms lived. The number of larvae produced by 52 cysts that had passed through earthworms was 1974, and the number from 71 cysts that had not was only 1148.

Ellenby explained that there were three kinds of cysts: those that would have hatched without passing through an earthworm; those that would not hatch unless they passed through an earthworm; and those that would not have hatched anyway. He was unable to prove conclusively how passage through the worm affected the hatchability of cysts, but suggested that probably some digestive enzyme was responsible, although it is also possible that some substance in the casts influenced the hatching of cysts. Yeates (1981) demonstrated that nematode populations, including plant pathogens, decreased by 37–66% when lumbricid earthworms were inoculated into three New Zealand pasture soils; however, this probably was not a significant decrease in terms of potential plant damage. It was demonstrated that *L. terrestris* and *A. trapezoides* could transmit the entomophagous nematode *Steinernema carpocapsae*, thereby enhancing its potential for biocontrol (Shapiro *et al.*, 1993).

11.1.4 TRANSMISSION OF ANIMAL PARASITES

Many animal parasites are transmitted from host to host by earthworms, which may be either essential intermediate hosts or merely reservoir hosts to the parasites, transmitting them without any direct influence on the parasites' life cycles. For instance, earthworms have been reported as vectors of an animal virus, the foot-and-mouth disease of domestic animals (Dhennin *et al.*, 1963). Poinar (1978) distinguished four kinds of associations between nematodes and earthworms. These ranged from passive transmissions to the earthworm being an essential intermediate host.

Earthworms are essential intermediate hosts to a number of tapeworms (Cestoda) and nematodes, parasitic in birds and mammals. The tapeworms transmitted include *Dilepis undula* Schrank to birds and rodents (Ryšavý, 1964), *Amoebotaenia cuneatus* Linstraw (Grassi and Rovelli, 1892; Magalhaes, 1892, Meggitt, 1914; Monnig, 1927), *Amoebotaenia lumbrici* Villot (Villot, 1883) and *Paricterotaenia paradoxa* (Genov, 1963) to chickens. Nematodes that have been reported to be transmitted to animals by earthworms are: *Metastrongylus elongatus* Duradin, *Metastrongylus pudentotectus* Wostowkow, *Metastrongylus salami* Gedoelst, *Hystrichus tricolor* Dujardin and *Stephanurus* sp. to swine (Schwartz and Alicata, 1931; Tromba, 1955; Breza, 1959; Ryšavý, 1969); *Capillaria annulata* Molin and *Capillaria causinflata* Molin to chickens (Ryšavý, 1969); *Capillaria plica* Rudolph, *Capillaria putorii* Rudolph, *Capillaria mucronata* Molin and *Thominx aerophilus* Creplin to small predators (Skarbilovic, 1950; Ryšavý, 1969); *Porrocaecum crassum* Delongchamps, *Porrocaecum ensicaudatum* Zeder, *Syngamus trachea* Montagu (gapes), *Syngamus skrjabinomorpha* Ryzikov, *Syngamus merulae* Baylis, *Cyathostoma bronchialis* Muhling and *Spiroptera turdi* to birds and chickens (Ryzhikov, 1949; Mozgovoy, 1952; Ryšavý, 1969); and *Dioctophyma* sp. and *Stephanurus* sp. to mammals (Woodhead, 1950; Karmanova, 1963). For the four species of tapeworms listed, the invasive phase that occurs in earthworms is the cysticercus; without such a stage in the earthworm, the life cycle of the tapeworm can proceed no further. Some parasitic nematodes have their larval stages in the bodies of earthworms, and many of these cannot develop to maturity until they reach their final host. For instance, the lung worms (*Metastrongylus*) can only develop to an invasive larval stage in earthworms, but when they are then eaten by pigs they can develop to maturity in the pigs' lungs.

Some parasites are transmitted by earthworms without the stage in the earthworm being essential for the completion of their life cycle, although the earthworm is necessary for them to reach their final host; in such a situation the earthworm is termed a reservoir host. These parasites are taken into the body of the worm with food and soil. They accumulate in the earthworm's body cavity, and often remain there for its whole life, which, for some species, may be as long as several years. Thus the earthworm enables the parasite to remain protected, in a situation in which it can retain its capacity for further development, and still be in a position to eventually infect its final hosts. This is particularly true for those lumbricids which migrate from the surface to deeper soil in adverse conditions, and return to the surface soil in warm, moist weather, when they are eaten by various species of birds, which then become infected with the parasites. A further factor is that, while the parasite is in the body of the earthworm, it is protected against control measures used by man. Typical examples of parasites transmitted in this way are the nematodes (gapeworms) such as *Syngamus trachea* and *Cyathostoma bronchialis* which parasitize the tracheae of various species of birds.

Earthworms are also important as passive (or phoretic) agents in the distribution of parasites. They may ingest the eggs of parasites with their food, carry them down to deeper soil, and protect them from harm that may be caused by adverse physical and environmental factors. The eggs can pass through the intestines of earthworms without losing any of their infectivity. For instance, the eggs of *Ascaris suum* and *Ascaridia galli* passed through the intestine of individuals of *L. terrestris* without damage (Bejsovec, 1962). In this way, the eggs are spread throughout the soil, which facilitates infection of domestic animals and birds. The virus that causes foot-and-mouth disease can persist in the muscle tissue of earthworms for 7–8 days and remain virulent. It has also been suggested that earthworms can infest human beings. *Microscolex modestus* has been reported from a human fistula, *E. fetida* in human urine, and *L. terrestris* and *O. lacteum* from human feces and the female vagina (Stolte, 1962).

11.1.5 ADVERSE EFFECTS ON SOIL

It is generally assumed that earthworms improve soil structure, aeration and drainage, as was discussed in Chapter 10, but a few workers have concluded that soil structure and plant growth can sometimes deteriorate as a result of the activities of earthworms. For instance, Agarwal *et al.* (1958) reported that a species of *Allolobophora*, by excreting a waxy fluid, made soil in parts of India cloddy and unproductive; this has not been substantiated and seems doubtful, because *Allolobophora* species are among the most common in fertile agricultural soils (Barley, 1959a,b). Puttarudriah and Sastry (1961) claimed that the earthworms *Pontoscolex corethrurus* and *Pheretima elongata* had adverse effects on soil structure and the growth of some plants in India. They based their conclusions on the large numbers of worms that were present so that the fine earth of their castings, which become mixed with intestinal secretions, mucus and other excreta, had caused the soil to become compacted and cement-like. This resulted in the soil becoming very poorly drained and heavy, forming large clods. A number of test crops were grown, and although maize (*Zea mays*) and ragi (*Eleusine coracana*) grew well, most vegetable crops such as carrots, radishes and beans did not do well, and had only very short tap roots.

Several workers have commented on the way that earthworms can contribute to soil erosion by bringing very finely divided soil to the surface. Darwin (1881) carefully observed the movement of earthworm casts down slopes, and Barley (1959b) observed that this might eventually lead to soil erosion. Evans and Guild (1947c) and Lee (1959) have reported that when there is excessive casting due to very large populations of earthworms, soil may be 'poached' (deep muddy impressions made by livestock). The effects of earthworms on erosion are discussed in section 10.2.5.

11.2 EARTHWORMS AS BENEFACTORS OTHER THAN IN SOIL FERTILITY

The role and use of earthworms in soil fertility and soil improvement, and their importance in breaking down organic matter and releasing the nutrients it contains were discussed extensively in Chapter 8. However, there are other beneficial contributions of earthworms which will be discussed here.

11.2.1 INTERACTIONS WITH ORGANISMS THAT PROMOTE PLANT GROWTH

Micro-organisms in soil affect plant growth in a variety of ways. Many beneficial micro-organisms promote plant growth by mutualistic or symbiotic relationships (e.g. nitrogen-fixing bacteria, mycorrhizal fungi and plant-growth-promoting rhizobacteria), and by fixing nitrogen (e.g. free-living nitrogen-fixing bacteria). One of the major constraints on effective root colonization by beneficial soil bacteria is their minimal capacity for unaided dispersal through soil. Except when carried by water movement, by earthworms (Doube et al., 1994a,c) or on roots growing from inoculated seed (Bull et al., 1991), these bacteria are essentially sedentary and root colonization occurs mainly when roots grow into soil colonized by them. A variety of methods have been used to inoculate beneficial micro-organisms into soil (e.g. seed inoculation, soil inoculation) but these procedures, when used in the field, commonly inoculate only a small portion of the total soil volume that is available to plant roots.

The activity of earthworms has been shown to promote the dispersal through soil of a variety of types of beneficial soil micro-organisms, including pseudomonads, rhizobia and mycorrhizal fungi. Madsen and Alexander (1982) reported that the earthworm L. rubellus enhanced the vertical transport of Pseudomonas putida and Bradyrhizobium japonicum in soil in the presence of percolating water, although less than 5% and 1%, respectively, of the recovered viable bacteria were present below 2.7 cm soil depth. Rouelle (1983) reported that Rhizobium japonicum could pass unharmed through the guts of E. fetida and L. terrestris and produce nodules on the root systems of soybeans. Buckalew et al. (1982) reported that an earthworm (species unspecified) was capable of acting as a vector of Rhizobium trifolii. Rouelle (1983) indicated that the presence of L. terrestris was associated with improved distribution of nodules on the root system of soybean, and suggested that this resulted from earthworm dispersal of Bradyrhizobium japonicum. Reddell and Spain (1991a) reported the spread of spores and hyphal fragments of mycorrhizal fungi in undigested root fragments in the gut of the pantropical earthworm Pontoscolex corethrurus. Doube et al. (1994a) demonstrated the capacity of earthworms to disperse Rhizobium through soil, to increase levels of colonization of legume roots by *Rhizobium*, and to increase levels of root nodulation. One set of experiments by Stephens *et al.* (1994a) showed that the earthworms *A. trapezoides* and *Microscolex dubius* dispersed *Rhizobium meliloti* strain L5-30R through pots of soil. The bacterium was inoculated into a variety of worm foods (including sheep dung and Ezimulch[R]) and placed on the soil surface. Eighteen days later >10^4 cells/g soil were recovered at 9 cm soil depth. A second experiment, using Ezimulch[R] as the carrier, showed that, in the presence of *A. trapezoides*, about 10^4 *R. meliloti* per centimeter of alfalfa root were detected at a depth of 3–9 cm below the soil surface, while none were detected at that root depth in the absence of *A. trapezoides*. A further experiment by Doube *et al.* (1994b) examined the capacity of seven species of earthworms (*A. trapezoides, A. rosea, A. caliginosa, A. longa, O. cyaneum, Gemascolex walkeri* and *E. fetida*) to disperse *R. meliloti* inoculated into Ezimulch[R] and placed as a pellet either on the soil surface or buried at a depth of 7 cm. Only *A. trapezoides, A. longa* and *E. fetida* consumed significant amounts of the Ezimulch[R] pellets, and more buried material than surface-applied material was consumed. Only *A. trapezoides* and *A. longa* were able to disperse *R. meliloti* through the vertical soil column; *E. fetida* remained in the vicinity of the food pellet and did not disperse the micro-organisms. It is possible that the other species would have dispersed *R. meliloti* if different carriers of suitable food had been used.

A further experiment examined the effect of *A. trapezoides* on root nodulation by *R. trifolii* in subterranean clover (*Trifolium subterraneum*) seedlings (Doube *et al.*, 1994a). When *R. trifolii* was inoculated into sheep dung and placed on the soil surface, three times as many root nodules developed when *A. trapezoides* was present as when this earthworm was not present. Furthermore, there was a highly significant increase in the number of root nodules on the main root at 2–8 cm soil depth in pots with earthworms compared with plants from pots lacking earthworms. The presence of earthworms was associated with increased shoot and root biomass, but the additional nodulation did not affect plant growth significantly. From these results, it appears that at least three common species of earthworms have the capacity to disperse *Rhizobium* through soil from an inoculated food source.

11.2.2 PRODUCTION OF PLANT-GROWTH-PROMOTING SUBSTANCES

The production of substances that promote growth by earthworms was first recorded by Gavrilov (1963). He processed extracts of the tissues, mucus, coelomic fluid and casts of *L. terrestris* and showed that they contained plant growth factors, probably produced by the coelamoebocytes.

The presence of plant-growth-promoting substances in the tissues of *A. caliginosa, L. rubellus* and *E. fetida* was first proposed by Nielson (1965). He killed the earthworms in water at 45 °C and homogenized them in distilled water. This was followed by extraction with various solvents to isolate several indole substances, which were then assayed for their effects on plant growth using the Went pea test. He compared the increases in plant growth due to the earthworm extracts with those from indole 3-acetic acid (IAA) and he reported significant effects for extracts of all these species. He also extracted similar substances that stimulated plant growth from *A. longa*, *L. terrestris* and *D. rubida*.

Springett and Syers (1979) compared the growth of roots of ryegrass seedlings in the presence of casts of *A. caliginosa* and *L. rubellus* to the growth without casts, and concluded that the earthworms must alter nutrient availability, alter the plant's ability to take up nutrients or affect the growth mechanisms of the plants. At all dissolved inorganic phosphorus (DIP) levels there was a marked tendency for the plant roots to be negatively geotropic, so that they grew laterally and upwards into the casts of *L. rubellus*, but they did not do so into the casts of *A. caliginosa*. This observation led the authors to conclude that *L. rubellus* casts probably contain an auxin-like substance, or some substance that modifies the effects of the plant's auxins, and that casts of *A. caliginosa* do not contain such substances.

Graff and Makeschin (1980) tested the effects of substances produced by *L. terrestris*, *A. caliginosa* and *E. fetida* on dry matter production of ryegrass in Germany. They added eluates from pots containing earthworms to pots containing no earthworms. They concluded that yield-influencing substances were released into the soil by all three species, but did not speculate on the nature of the substances.

Tomati *et al.* (1983) tested composts produced from organic waste by the action of earthworms (Chapter 13) as media for growing ornamental plants. They concluded that the growth patterns were too great to be explained purely on the basis of the nutrient content of the composts. These included stimulation of rooting, dwarfing, time of flowering and lengthening of internodes. They compared the growth of petunia, begonia and coleus after adding vermicompost, to that achieved by adding auxins, gibberellins and cytokinins to the soil. They considered that the growth patterns that resulted provided circumstantial evidence of hormonal effects, produced by earthworm activity (Tomati and Grappelli, 1984). They extended these studies to cereals and mushrooms (Tomati *et al.*, 1988) and considered that these data supported their earlier conclusions. Edwards and Burrows (1988) described data on the growth of a wide range of plants in media produced by the processing of organic waste by *E. fetida*. They speculated that the success of these media, even at high dilutions, and the growth patterns of the plants, could be explained best by postulating a hormonal effect on plant growth.

11.2.3 PRODUCTION OF BIOLOGICALLY ACTIVE MATERIALS BY EARTHWORMS

The production of vitamins and similar substances by earthworms and their release into soil has been investigated. For instance, Gavrilov (1963) reported evidence of secretion of B-group vitamins, and Zrazhevskii (1957) attributed the production of some provitamin D to earthworms. Gavrilov extracted tissues of *L. terrestris*, their casts and soil in which they had been cultured with water, and tested the effects of the extracts on growth of the yeast *Saccharomyces cerevisiae*, and on the productivity of cultures of the free-living protozoan *Paramecium*. He compared the results with controls with and without the added vitamin of the B group. Increased production of the yeast and the protozoan in the presence of the earthworm and earthworm-affected extracts occurred and was comparable to that in the controls to which B-group vitamins were added.

Atlavinyte and Daciulyte (1969) measured the vitamin B_{12} content of several soils in pots to which they added the earthworms *A. caliginosa*, *A. rosea*, *L. rubellus* and *L. terrestris* at six population levels. The amounts of vitamin B_{12} increased from 4 to 12 months after introduction of the earthworms, reached a maximum after 12–24 months and decreased, but remained higher than that in the controls for 3 years. The total populations of micro-organisms in the pots with earthworms were twice to three times greater than those in control pots after 6 and 12 months. Atlavinyte *et al.* (1971) confirmed similar increases in vitamin B_{12} and showed increases of up to 3–5 times in microbial populations, proportional to the numbers of earthworms that were added. Eitminavicuite *et al.* (1971) reported correlations between soil vitamin B content and microbial populations, and between microbial numbers and earthworm populations. Probably these increased amounts of B-group vitamins in soils are not due directly to the earthworms, but rather to the micro-organisms, which are stimulated in activity and number by the earthworms.

11.2.4 OTHER BENEFICIAL USES

Earthworms can benefit humans in other ways. They are widely used as bait for fish, and large commercial farms that produce earthworms for fish bait are found in the United States, as well as some in Great Britain and other European countries. The most common species used as bait is *L. terrestris*, but many other species, including particularly *A. caliginosa*, *A. chlorotica*, *E. fetida*, *L. rubellus*, *O. lacteum* and *Pheretima* spp., have also been used. The sale of *L. terrestris*, collected from the field for fish bait, in Canada exceeds $50 million per annum (Tomlin, 1983).

Earthworms have also been used for human food. They are regarded as a delicacy by the Maoris in New Zealand. In Japan, earthworm pies have been made, and there is even a specialty restaurant with earthworm-based dishes, and there have been reports from South Africa of fried earthworms being eaten (Ljungström and Reinecke, 1969). Primitive natives from New Guinea and parts of Africa have been reported to eat raw earthworms, and earthworms are believed to be a source of human food in South America.

It has been suggested that, with suitable processing, easily bred earthworms such as *E. fetida* might be suitable for human or animal feeding; the protein content and quality of earthworms exceeding that of most meat animals (McInroy, 1971; Edwards and Niederer, 1988). This is discussed further in Chapter 13.

There have been many reports of earthworms being eaten or applied to humans to alleviate or cure such ills as stones in the bladder, jaundice, piles, fever and smallpox. Earthworm ashes have been used as a tooth powder in primitive societies or applied to the head to make hair grow (Stephenson, 1930). Earthworms have been eaten to cure impotency or to enable mothers to nurse their children. Earthworm poultices have been used to draw out thorns, scientists have isolated a bronchial dilating substance from earthworms (Reynolds and Reynolds, 1972), and it has been suggested that earthworms might contain a substance effective in curing rheumatism (Weisbach, 1962).

Earthworms have been used in testing pregnancy; urine from human females (concentrated, according to Zonek 5:1) is injected into earthworms subcutaneously, and smears taken from their seminal vesicles, both before and after injection, to assess spermatogenesis. An accuracy of 90% was claimed for this method of pregnancy testing (Hasenbein, 1951).

A method of testing substances for carcinogenic properties was described by Gersch (1954) who found that benzopyrine (0.5%), dimethylbenzanthrene (0.5%) and other compounds, when applied to *L. terrestris* for several weeks, induced tumors.

11.3 CONTROL OF EARTHWORMS

The species of earthworms that are considered to be pests of crops and grass are usually those species that burrow close to the surface, come to the surface to cast, or come on the soil surface at night to feed. Hence, a toxic pesticide sprayed on the soil surface is usually an effective control agent, and several chemicals that were originally developed as insecticides, control earthworms effectively (Chapter 14). The most commonly used of these has been the persistent organochlorine insecticide chlordane, which was applied at a rate of 12 kg active ingredient of dust/ha to control earthworms in golf greens and sports grounds. The use of this insecticide has been restricted or banned in many countries in recent years. Heptachlor and endrin are other organochlorine insecticides that can be used to control earthworms. Unfortunately, all of these chemicals are very persistent in soil and, since they are lipophilic, they can become concentrated from soil into the bodies of earthworms, from where they can pass into the tissues of vertebrate predators such as birds and moles, when these feed on earthworms. Their use has been restricted in many parts of the world, and for this reason less persistent chemicals are now used. Two pesticides that are currently available in most countries are carbaryl and phorate, both of which give good control of earthworms at soil applications of between 6 and 12 kg/ha. However, because they are less persistent than the organochlorine insecticides, more frequent applications of the chemical are necessary, although one treatment per year is usually adequate. Other chemicals toxic to earthworms are listed in Chapter 14 and Appendix A.

Earthworms in environmental management

12

It has been proved so conclusively that earthworms aid soil fertility, that many attempts have been made to inoculate them into poor soils that have no earthworms (Chapters 8 and 10) or to encourage the build-up of earthworm populations by addition of organic matter or fertilizers. Farming enterprises that produce earthworms for inoculation into poor land have been set up in both Europe and the United States.

12.1 EARTHWORMS IN LAND AMELIORATION AND RECLAMATION

The success of land reclamation by conventional techniques is often limited by poor soil structure and low inherent soil fertility, and even in productive soils, a marked deterioration in the botanical composition of the sward can occur within a number of years (Hoogerkamp *et al.*, 1983).

A number of studies indicate that earthworms play an important part in improving reclaimed soils. However, natural colonization and establishment occur relatively slowly, and there is increasing interest in accelerating these processes by deliberate introductions. Curry (1988) reviewed the relationships between build-up of earthworm populations and rates of land improvement. However, some amelioration projects are foredoomed because little attention is paid to introducing the most suitable species of worms. Not all worm species are active in soil formation, and often those species that can be bred most easily and rapidly are the least useful in land reclamation. For example, *E. fetida* is easy to breed and is sold in large numbers by breeders, but this species is basically a manure- or compost-dwelling species and cannot benefit, survive in or thrive in field soils, especially those with little organic matter (Grant, 1955a). Any increases in yield after adding *E. fetida* to soils are usually short lived, and are due more to the decomposition of dead earthworms rather than to the earthworm activity in soil. It is also important to recognize that any given area of soil can support only a certain limited earthworm population, and adding more earthworms beyond that maximum capacity will not increase fertility, unless more organic food is supplied and soil structure and conditions are improved.

Adding earthworm casts to soil can improve its structure and fertility greatly. Casts usually have a higher pH (Fig. 12.1), and more organic matter, total ammonium and nitrate nitrogen, total and exchangeable magnesium and available phosphorus than soil, and have a good base capacity and a high moisture equivalent (Lunt and Jacobson, 1944) (Table 8.2). Adding lime to acid soil usually favors the build-up of earthworm populations; for instance, the addition of 1 tonne of lime/acre to New Zealand soils resulted in an increase of 50% in the numbers of *A. caliginosa* (Stockdill and Cossens, 1966). This process has been accelerated by inoculating such sites a second time when soils are moist, after adding lime.

12.1.1 INTRODUCTION OF EARTHWORMS INTO POOR PASTURES IN NEW ZEALAND AND AUSTRALIA

Earthworms of European origin, notably the peregrine species *A. caliginosa*, have become established gradually in many New Zealand pastures. However, there are still considerable areas of upland pastures from which lumbricids are totally absent (Stockdill, 1966). Improved production, and

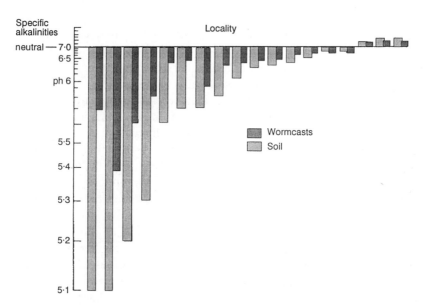

Figure 12.1 Average pH of the soil and earthworm casts from 18 localities (adapted from Salisbury, 1925).

an increase in the proportion of ryegrass and clover in such pastures, was noted following the introduction of *A. caliginosa* in transplanted turfs to pasture with no earthworms (Hamblyn and Dingwall, 1945; Nielson, 1953). The estimated rate of spread of introduced worms from an inoculation point was 10–13 m/yr. Stockdill (1959, 1966) reported that rapid disappearance of the surface organic mat followed introduction of earthworms to hill pastures. There was a marked improvement in soil-structure, aeration and distribution of nutrients. Dry matter production increased initially by as much as 70%, but in older sites, as earthworms became established, the mean dry matter increase was 29% (Stockdill and Cossens, 1966). A turf cutting and distributing machine suitable for large-scale introduction projects was developed, and complete earthworm colonization of a field could be expected from turfs placed at 10 m intervals within 6–7 years under New Zealand conditions (Stockdill, 1982). The expected return on the investment in earthworm introduction projects was calculated to be more than 300%.

Barley and Kleinig (1964) and Noble *et al.* (1970) described the successful introduction, build-up and establishment of earthworm populations in excess of 300 per m² in irrigated sown pasture on sandy loam soil in New South Wales, Australia. This resulted in noticeable improvements in soil structure with a reduction in C:N ratio, a decline in bulk density and a breakdown of the surface organic mat of undecomposed litter, correlated with the earthworm population densities.

12.1.2 INTRODUCTION INTO DUTCH POLDER SOILS

The addition of earthworms to soils to accelerate their productivity seems particularly promising in reclaiming flooded areas that are drained subsequently and put into cultivation, as in the Dutch polders (van Rhee, 1969a,b, 1971). For instance, the earthworms *A. caliginosa* and *L. terrestris* were introduced to polder soils in which fruit trees were grown, at a rate of about 800 worms/tree. More roots grew on trees located in the earthworm-inoculated soils than in those without earthworms (van Rhee, 1971). Earthworms multiplied rapidly in these polder soils; for instance, populations of *A. caliginosa* increased from 4664 individuals to 384 740 in 3–4 years and those of *A. chlorotica* from 2588 to 12 666 in the same period (van Rhee, 1969b). Van Rhee also reported natural earthworm population densities exceeding 200 per m² in soils on Dutch polders 26 years after reclamation, but concluded that the rate of natural spread had been only 4–6 m/yr and that the rate of earthworm establishment could be accelerated greatly by deliberate introduction of earthworms. This author (van Rhee, 1963, 1969a,b, 1977) conducted a series of earthworm inoculation experiments in recently reclaimed polders and recorded eventual population densities of up to 750 per m² in grass areas, but the lower densities of

140–250 per m² were recorded in orchard soils which had received heavy applications of pesticides annually. *Allolobophora caliginosa* was a particularly successful early colonizer and played an important role in the early stages of soil formation in the polder. Other species which became established included: *L. rubellus, A. rosea, A. longa, A. chlorotica* and, to a limited extent, *L. terrestris*. Increased plant litter incorporation, improved aggregation, better aeration and water relationships and the development of mull characteristics were observed following the establishment of earthworms. These improvements in soil structure were confirmed by soil morphological studies (Rogaar and Boswinkel, 1978).

More recent inoculations of earthworms into polder grassland were described by Hoogerkamp *et al.* (1983), who estimated the rates of population build-up with the aid of an infra-red line scanning method to detect the earthworm distributions. This technique detected different patterns of heat exchange, depending on whether or not a surface organic mat was present. When earthworms were well established, the surface mat disappeared, resulting in better heat exchange between soil and air, and diminished daily temperature fluctuations at the surface. The mat was ingested progressively and fully incorporated into the soil within 3 years of earthworm inoculation, and a dark-colored A₁ soil horizon began to develop, increasing gradually to a thickness of 5–8 cm in 8–9 years. This was accompanied by an increase in the C:N ratio of the topsoil. Earthworm activity increased the pore volume, aeration and available moisture at field capacity, and increased the infiltration capacity. These better soil conditions, in turn, influenced root growth and distribution positively; this was reflected in much better sward attachment to the soil and fewer tufts that were detached by grazing cattle. After 10 years, each earthworm-inoculated plot generally had a much better botanical composition, with a higher proportion of *Lolium perenne*, and grass yields were on average 10% higher in earthworm plots.

12.1.3 INTRODUCTION INTO POOR MINERAL SOILS

Some efforts to improve impoverished soils by stimulating earthworm activity have had encouraging results. Large increases in earthworm numbers and improvement in soil structure occurred in poor vineyard soils which had been fertilized with organic manures (Bosse, 1967). Huhta (1979) reported encouraging results from experimental introductions of *A. caliginosa* into limed coniferous forest plots in Finland, and indigenous *L. rubellus* and *E. rubida* populations also benefitted from liming and the addition of deciduous litter. Langmaid (1964) reported considerable soil improvement in afforested virgin podzols in New Brunswick once earthworm populations had become established.

Tréhen and Bouché (1983) studied the role of earthworms in the improvement of impoverished heathland soils in central Brittany, after organic refuse had been spread. Surface-dwelling species such as *E. fetida* and *D. octaedra* played an important part in the initial phases of colonization. Earthworms have also been introduced successfully to newly established areas of artesian irrigation in Uzbekistan, former USSR, in order to improve the rate of soil formation (Ghilarov and Mamajev, 1966)

12.1.4 COLONIZATION OF MINING WASTES SITES

Acidity, metal toxicity, unfavorable moisture conditions and lack of suitable food are among the factors likely to inhibit earthworm establishment on recently reclaimed mining soils. Even under favorable conditions, natural colonization at such sites by earthworms is slow. Dunger (1969, 1991) reported the survival of earthworms after 5 years on reclaimed coal mine spoil in East Germany, but it took 10 years for earthworm populations to become fully established, and on more acid sites, earthworms had not become established even after 15 years. However, once earthworm populations had become established under *Alnus glutinosa*, plant litter incorporation was greatly accelerated and the soil structure changed rapidly from moder to mull. Armstrong and Bragg (1984) reported that it took 10–20 years for earthworm populations to become fully established in the restoration of open-cast coal-mining sites in the UK. Standen *et al.* (1982) reported significant correlations between site age, shade, moisture content and earthworm numbers and biomass in restored open-cast coal-mining sites in Co. Durham, UK. Earthworm numbers were still low (<33 per m^2) 11 years after restoration of a site for agricultural use, but up to 162 *L. rubellus* per m^2 were found in one 14-year-old colliery spoil heap, colonized naturally by scrub woodland. Significant numbers of *L. terrestris* were found in a few sites that were more than 50 years old. Surface organic amendment of newly reclaimed sites favored earthworm establishment considerably, both by increasing food supply for earthworms and by stabilizing temperature and moisture conditions in the surface soil (Dunger, 1969; Luff and Hutson, 1977). This conclusion was supported by a study by Westernacher-Dotzler and Dumbeck (1992) who surveyed earthworm populations in 36 lignite mining sites which were being reclaimed. They reported that addition of poultry manure and the use of minimal tillage greatly encouraged the build-up of earthworm populations and the rates of reclamation of the site.

L. terrestris has been introduced successfully into reclaimed coal mine spoil sites, ranging in pH from 3.5 to over 7 in the USA (Vimmerstedt and Finney, 1973). Earthworm activity accelerated the incorporation of leaf litter, and increased exchangeable cations and available phosphorus. After 12 years, earthworm populations had become dispersed widely throughout an area of afforested calcareous spoil, with earthworm numbers and biomass being greater under *Alnus glutinosa* than under *Robinia pseudoacacia*, where the litter was less palatable (Hamilton and Vimmerstedt, 1980). Earthworms buried or consumed the equivalent of 5 metric tonnes of leaf litter/ha. Ten *L. terrestris* per m² introduced into such a spoil site increased to about 60 per m² over an area of 700 m² in 2 years. The area about 15 m from the point of earthworm introduction had the maximum numbers of earthworms, but the population was still increasing at greater distances from the center (Vimmerstedt and Finney, 1973).

In a field experiment in Australia, the thickness of the organic mat at the base of the grasses was reported to be related inversely to earthworm numbers (Noble *et al.*, 1970). In plots that were inoculated with earthworms the mat tended to disappear quite rapidly, soil bulk density was reduced, and the carbon/nitrogen ratios of the surface soil decreased.

12.1.5 RECLAMATION OF OPEN-CAST MINING SITES BY EARTHWORMS

The acceleration of the rehabilitation of former open-cast coal-mining land was reviewed by Stewart and Scullion (1988). Reports of the rates of recolonization of open-cast mining sites by earthworms is somewhat contradictory. It ranges from very slow (Blenkinsop, 1957) to rapid recolonization (Scullion, 1984). Stewart and Scullion (1988) investigated the effects of different forms of management on earthworm recolonization of mining sites. They concluded that grass-cutting systems of management inhibited the recovery of earthworm populations but sheep-grazing increased populations of *A. caliginosa*, *A. chlorotica* and *L. rubellus*. Top-dressing with poultry manure encouraged earthworm colonization further, as did the promotion of site drainage by regular subsoiling. In later work (Scullion and Mohammed, 1991) they demonstrated that regular subsoiling increased earthworm population densities significantly, the increases in numbers of *A. caliginosa* being greatest. Addition of semi-organic fertilizer further favored the build-up of populations of *L. rubellus*.

12.1.6 EARTHWORM POPULATIONS IN RECLAIMED PEAT

Earthworm populations in virgin peat are limited by high moisture levels, low pH and the poor quality of organic litter, but peat soils drained and reclaimed for agriculture can support moderate densities of earthworm populations. Guild (1948) found 9 of 10 common earthworm species, and population densities ranging from 50 to 100 per m², in improved hill pastures on peaty soils in Scotland, and Baker (1983) recorded 15 earthworm species, with up to 197 individuals/m² in grassland on reclaimed fen peat in central Ireland. A survey of 24 sites in Ireland showed that earthworms were scarce in newly reclaimed peat, and that population densities exceeded 100 per m² only in sites that were more than 25 years old (Curry and Cotton, 1983). Surface-dwelling species, typical of raw humus habitats, such as *D. octaedra*, *D. rubida* and *L. rubellus*, were common in young sites, whereas in the more mature sites, typical soil-dwelling earthworm species such as *A. longa*, *A. rosea*, *L. terrestris* and *O. cyaneum* occurred. The two most abundant species in mature sites were *A. caliginosa* and *A. chlorotica*; these were early colonizers and were distributed widely in young sites, but at only low population densities.

Currently, large areas of cutover peat, consisting of a residual peat layer about 50 cm deep, overlying relict soils derived from limestone till, are available for reclamation in central Ireland following industrial peat extraction. Preliminary studies have indicated that earthworm colonization could be accelerated greatly by deliberate introductions of earthworms (Curry and Cotton, 1983). Their studies showed that well-managed cutover peat can eventually support earthworm populations comparable in density and biomass with those found in fertile mineral soils, and that population development can be accelerated greatly by the deliberate introduction of beneficial species. When shallow-working species such as *A. chlorotica* and *A. caliginosa* are inoculated, they become established fairly quickly, whereas deep-burrowing species such as *L. terrestris*, with low reproduction and dispersal rates, require several years to reach significant population densities.

There are many potential benefits from inoculation of earthworms into reclaimed peat. Curry and Bolger (1984) calculated that a population of *L. terrestris* with a mean biomass of 100 g/m² would consume 365 g dry weight of *Salix* litter/m²/yr in 'energy forest' plantations on cutover peat, thereby considerably accelerating nutrient cycling. Deep ploughing results in an admixture of mineral soil and peat, and soil-ingesting earthworms could greatly facilitate the intimate mixing of mineral and organic fractions and so enhance the rate of soil structural development. Curry and Bolger (1984) estimated that 100 g *L. terrestris*/m² could ingest 1.3 kg soil/m²/yr and that the entire top 20 cm soil layer could be ingested and turned over in 45 years. Ultimately, beneficial results on soil fertility should be reflected in enhanced productivity, and the beneficial results reported by these workers indicate significant increases in grass growth in response to earthworm inoculations and addition of animal manure.

12.2 EARTHWORMS AS INDICATORS OF ENVIRONMENTAL CONTAMINATION

The concerns for understanding the impact of toxic and hazardous materials on the soil environment have multiplied in recent years and there has been an increased effort to find methods of evaluating these environmental effects. Most current methods of environmental assessment of chemicals have measured these impacts primarily in terms of human health (Callahan, 1988). There is an equally important need to assess the long-term toxic impacts of chemicals on the terrestrial environment and its dynamic processes.

A toxic chemical or pollutant can be defined as a substance that has a deleterious effect on organisms and is introduced into the environment by human activity (Moriarty, 1983). The effects of pollutants on the environment are often assessed through their individual impact on particular organisms, and this is difficult to interpret in broad terms, because the severity of the effects of a particular chemical depends upon the dose, formulation, how it reached the environmental compartment, and the importance of the test organisms (see Chapter 14 for further discussion).

12.2.1 THE USE OF EARTHWORMS AS BIOINDICATORS

It has been suggested that earthworms are excellent **bioindicators** of the relative health of soil ecosystems (Kuhle, 1983). Earthworms possess a number of the qualities needed in animals used for biomonitoring of terrestrial ecosystems. They are large, numerous, easy to sample and easily identified. They are widely distributed and relatively immobile, they are in full contact with the substrate in which they live and they consume large volumes of this substrate. They could be considered to be the terrestrial equivalent of aquatic filter-feeders (Morgan *et al.*, 1986). Not only are earthworms killed by toxic chemicals, but their growth rates, reproduction and behavior are also affected; they also accumulate some chemicals into their tissues at levels higher than in the medium in which they live (Chapter 14). This means that low levels of contamination of soil can be detected by periodically monitoring numbers of earthworms and the chemical residues that they contain. Helmke *et al.* (1979) concluded that analyses of earthworm tissue could provide an excellent index of bioavailability of heavy metals in soils.

A standardized bioassay which exposes the earthworm *E. fetida* to chemicals mixed in a carefully standardized artificial soil was developed and tested extensively by Edwards (1983b,c, 1984). This has been adopted by the European Economic Union and the Organization for Economic Development and Cooperation. In this bioassay contaminated soils are mixed with a standard uncontaminated soil to make a range of dilutions of the contaminated soils. Earthworm assays can be performed using suitable containers to which the diluted mixtures and worms are added. Mortality should be assessed after 14 and 28 days and quantities of organic chemicals and heavy metals that have accumulated in the tissues of the earthworms analyzed (Edwards and Bohlen, 1992).

The main problem in using earthworms for biomonitoring contaminated soils, or other materials, is that the nature of the polluting chemical or mix of contaminants in the soil is often unknown (Callahan, 1988). However, a modification of the standardized artificial soil earthworm bioassay can be used to determine the toxicity of such contaminated soils. For assessment of potential hazards from contaminants in sites used to dispose of materials dredged from rivers and estuaries, similar earthworm assays have been used in the field. Such data supplement field data from contaminated sites where earthworms are collected, counted and analyzed to assess the degree of contamination and its bioavailability (Simmers et al., 1983; Rhett et al., 1988; Callahan and Linder, 1992). Such laboratory and field bioassays have proved to be extremely useful in the evaluation and prediction of contaminant levels and their mobility in dredged river sediments disposed on upland sites. They have been used successfully to assess the suitability of dredged materials for land disposal. Venables et al. (1992) described a suite of biomarkers in earthworms that could be used as indicators of ecotoxicity. These include bioavailability of the chemical, and reproductive/developmental, neurological and immunological markers that could be interpreted as potential indicators of hazards to organisms other than earthworms.

12.2.2 EARTHWORMS AND RADIOACTIVITY IN SOILS

Earthworms have also been used as bioindicators of radioactive soil pollution (Krivolutsky et al., 1982) with radioactive materials including strontium-90 (Krivolutsky et al., 1972, 1982) and cesium-137 (Crossley et al., 1971). Earthworms proved to be sensitive to a radiation background of ^{226}Ra. Some radiation levels were toxic and others affected the reproduction of earthworms and caused changes in the structure of the epithelium of the midgut. Exposed individuals had fewer enzyme-producing cells and fewer regenerating cells, and an increase in mucous cells in the epithelium of the outer integument. Earthworms can take up radioactive cesium into their tissues readily and retain a portion of it (Crossley et al., 1971) and are relatively sensitive to γ-irradiation (Edwards, 1969).

12.2.3 EARTHWORMS AND HAZARDOUS WASTES

It has been suggested that the earthworm E. fetida could be used to determine the extent and severity of environmental contamination of hazardous waste sites, using methods similar to those used to assess the toxicity of dredged river sediments. Such laboratory and field assays could be used to assess contamination by both organic contaminants and heavy metals.

If earthworms are used in these ways for environmental biomonitoring, it has been suggested that the same species should be used consistently at different sites (Morgan et al., 1986). Although E. fetida has been used widely in this way (Simmers *et al.*, 1983; Greig-Smith *et al.*, 1992) some workers (Kruse and Barrett, 1985) suggested *L. rubellus* as a suitable species and other workers have proposed *A. caliginosa* as another alternative. Morgan (1986) emphasized the value of earthworms in assessing the potential biological significance of heavy metal pollutants. He stressed that direct measurement of heavy metal concentrations in sediments and soils has relatively little value and is not very sensitive, whereas uptake into earthworms is a much more sensitive and meaningful criterion. Beyer *et al.* (1987) also stressed the value of using earthworms to assess heavy metal contamination.

12.2.4 EARTHWORMS AND POLYCHLORINATED PHENOLS

The potential of earthworms for monitoring levels and effects of polychlorinated phenols and their metabolites in soils near sawmills was discussed by Knuutinen *et al.* (1990). Similarly, the use of earthworms in evaluating contamination by polychlorinated biphenyls (PCBs) was discussed by Kreis *et al.* (1987), who reported that the uptake of PCBs from soil was usually in direct proportion to the amounts present. Earthworms were used successfully as bioindicators of soil contamination by dioxin (2,3,7,8-tetrachlorodibenzo-*p*-dioxin) after an explosion at a chemical plant near Seveso, Italy, in 1976 (Martinucci *et al.*, 1983).

Earthworms in organic waste management

13

In recent years, the disposal of organic wastes from domestic, agricultural and industrial sources has caused increasing environmental and economic problems. These wastes include: sewage sludges and solids from wastewater (Neuhauser et al., 1988); materials from the brewery (Butt, 1993), processed potato (Edwards, 1983a) and paper industries (Butt, 1993); wastes from supermarkets and restaurants; wastes from poultry, pigs, cattle, sheep, goats, horses (Edwards et al., 1985; Edwards, 1988) and small domestic animals such as rabbits (Allevi et al., 1987); as well as horticultural residues from dead plants and the mushroom industry (Edwards, 1988). In the US, such organic wastes can represent more than 50% of the total waste stream, and are major components of landfills; so if earthworms can be used to process them, this would be a valuable contribution to a major environmental problem.

Many of these wastes contain considerable amounts of inorganic and organic contaminants, such as heavy metals, pesticides, aromatic hydrocarbons and sulfur compounds, as well as human and animal pathogens. In landfills, these pollutants may drain into groundwater and contaminate it, or produce dangerous levels of methane. If disposed of on urban or agricultural land, they may cause serious contamination problems.

Since 1978, there has been increasing interest in possible methods of processing many of these wastes using earthworms. This interest resulted in a series of conferences aimed at assessing and developing this potential. The first, entitled 'Utilization of Soil Organisms in Sludge Management' (Hartenstein, 1978) was held in Syracuse, New York, USA, and focused on the processing of sewage sludge by earthworms. The second, a 'Workshop on the Role of Earthworms in the Stabilization of Organic Residues' (Appelhof, 1981) was held in Kalamazoo, Michigan, USA. This was followed by an 'International Symposium on Agricultural Prospects in Earthworm Farming' (Tomati and Grappelli, 1984), held in

Rome, Italy. The largest, a 'Symposium on the Use of Earthworms in Waste Management and Environmental Management' (Edwards and Neuhauser, 1988) was held in Cambridge, UK. Additionally, sessions on earthworms and waste management were held at international symposia in Bologna, Italy in 1985 (Bonvicini-Pagliai and Omodeo, 1987); Avignon, France in 1990 (Kretzschmar, 1992); and Columbus, Ohio, USA in 1994.

Research and commercial projects have been set up in many countries including England, France, Germany, Italy, the United States, Japan, the Philippines, India and other parts of South-East Asia, Australia, Cuba, the Bahamas and many countries in South America. The major research in the USA concentrated mainly on the utilization of sewage sludges and solids (Neuhauser *et al.*, 1988) and that in the UK on processing animal, vegetable and industrial wastes (Edwards, 1988).

13.1 MANAGEMENT OF SEWAGE SLUDGE BY EARTHWORMS

Research into the use of earthworms to manage sewage sludge began at the State University of New York (SUNY), Syracuse (Hartenstein, 1978). In this program, Mitchell (1978) demonstrated that aerobic sewage sludge that is ingested and egested as casts by the earthworm *E. fetida* is decomposed, and thus stabilized (i.e. rendered innocuous) about three times as fast as noningested sludge, apparently because of the enhancement of microbial decomposition in the casts; he found that, relative to non-ingested sludge, objectionable odors disappeared much more quickly and there was a marked reduction in populations of the pathogenic *Salmonella enteriditis*, *Escherichia coli* and other Enterobacteriaceae. Most of the sludge produced in sewage plants is anaerobic, and when fresh is toxic to *E. fetida*. Collier (1978) used a large sample of sun-dried anaerobic sludge and showed that although it was lethal to *E. fetida* when first moistened, and that with daily watering it remained toxic until the tenth day, it had lost its toxicity after 14 days.

Kaplan (1978) suggested that mixing sewage sludge and other materials, e.g. garden wastes, paper pulp sludge or other lignin-rich wastes, and composting the mixture with earthworms might accelerate their decomposition due to maceration and mixing of such materials into earthworm casts. A variety of possibilities exists for utilization of timber mill wastes and other materials, with simultaneous disposal of municipal sewage sludge. Neuhauser *et al.* (1979) called the use of earthworms in sludge management **vermicomposting**, or **vermistabilization** (Loehr *et al.*, 1984). Because vermistabilization has not been well studied, many fundamental factors still need to be evaluated to assure the technical and economic success of such a process. These factors include:

1.   how earthworms affect and are impacted by sludge characteristics;

2. the comparative ability of different earthworm species to grow and reproduce in sludge;
3. processing sludge to make it acceptable to earthworms; and
4. the effect of mixed earthworm species in a vermistabilization system.

Neuhauser *et al.* conducted research along two complementary lines:

1. basic studies to identify fundamental factors that affect the performance of the vermistabilization process; and
2. applied studies to determine the design and management relationships.

For earthworms to be useful in stabilizing sludge, they must increase the stabilization rate. This can be demonstrated best if the presence of earthworms in sludge causes an increase in the rate of volatile solids reduction. Maximum reduction of volatile solids is a goal of any sludge-stabilization system.

It is well established that the earthworm, *E. fetida*, can increase the rates of sludge volatile solids destruction when present in aerobic sludge. This increase in the volatile solids destruction rate reduces the probability of putrefaction occurring in the sludge due to anaerobic conditions. The more rapid degradation of the organic matter is probably due to increased aeration and other factors brought about by the earthworms. Both excessive and insufficient moisture can impact the growth of earthworms adversely. A series of experiments were conducted to determine the most desirable moisture content for media in vermistabilization units. Such information is important since it identifies the conditions that may be necessary for successful vermistabilization.

Neuhauser *et al.* (1988) used aerobically digested sludge dewatered to a moisture content of 75% (total solids = 25%) without the aid of chemicals. The drier sludge was mixed with a more liquid sludge to provide a range of moisture contents of the sludge. Only one species, *E. fetida*, and one temperature, 25 °C, were tested. Although there was some variation at the various sludge concentrations evaluated, a definite pattern was observed. Worm growth was hindered at the lower and higher solids concentrations. Worm growth at the low and high total solids levels (6.3–7.9 and 17.9–25.1% solids) was statistically different, at the 1% level, from the worm growth at the middle range of solids (9.3–15.9% solids). This indicates that optimum worm growth occurred over a range of total solids, on a wet basis, from about 9% to 16% in the media to which the worms were exposed (84–91% moisture content).

It must be emphasized that the appropriate range of total solids identified is the total solids or moisture contents to which the earthworms are exposed. More liquid wastes can be applied to the vermistabilization process so long as the liquid drains readily from the media, the organics are retained, and aerobic conditions are maintained. The desirability of maintaining the total solids content of the media in the range noted does not mean that sludge slurries cannot be processed with vermistabilization. Thus, the upper range of the appropriate total solids content was about 16% in the absence of stressed conditions. In an operational context, it might be necessary to add moisture to materials in a vermistabilization process if the media should become too dry. Such could be the case in hot and/or dry climates. When earthworms are used to stabilize sludges, it is important to know when in the stabilization process earthworms can be most useful. It is important to relate rates of earthworm growth to the period of time after a sludge is removed from an aerobic reactor. Sludge age in these experiments refers to the time after the sludge was removed from the reactor and immediately dewatered. Neuhauser *et al.* (1988) showed that as sludge aged, its nutritive value to earthworms decreased rapidly, after about 12 weeks removal from the digester. The ash content of the sludge increased with time, a further indication of sludge stabilization.

Loehr *et al.* (1984; 1988) conducted a project to explore the practical feasibility of using earthworms to stabilize wastewater treatment sludges. They investigated the performance of what they called a vermistabilization process using stabilized and unstabilized sludges, in an attempt to develop a commercial design for a vermistabilization process. They concluded that a liquid sludge vermistabilization process was feasible and provided data on the rates of stabilization and the physical and chemical characteristics of residual stabilized solids resulting from vermistabilization. This conclusion was supported by Pincince *et al.* (1980).

There have been various attempts to use earthworms in sewage sludge utilization in the USA (Green and Penton, 1981; Pincince *et al.*, 1981), in the UK (Edwards and Neuhauser, 1988) and in Italy (Tomati and Grappelli, 1984), but full-scale successful vermiprocessing of sewage sludge has not yet been achieved. A major factor that has to be overcome is to avoid exposure of the vermistabilization process to toxic chemicals that might enter the wastewater stream.

13.2 MANAGEMENT OF ANIMAL, VEGETABLE AND INDUSTRIAL ORGANIC WASTES BY EARTHWORMS

Research began at Rothamsted Experiment Station, UK in 1980, on using earthworms to break down animal manures, such as pig and cattle solids and slurries, wastes from laying chickens, broilers, turkeys and ducks, horse manure and rabbit droppings (Edwards, 1983a). This was extended later to studies using earthworms to break down vegetable wastes, including those from the mushroom industry, processed potato industry, brewery industry and paper pulp. This work began as fundamental research on a laboratory scale but was extended to development research on a field scale and eventually commercialization.

This research had two main aims:

1. to turn animal and other agricultural and industrial wastes into useful materials that could be added to agricultural land to improve soil structure and fertility, and also have considerable potential in horticulture as a plant growth medium or component of commercial potting composts; and
2. to harvest worms from the worked waste and process them into a highly nutritious protein supplement for fish, poultry and pigs (see later in this chapter).

To accomplish these aims, a complex network of collaborative research was set up involving six research stations, six university or college departments and eight commercial organizations, with a research program coordinated by an interdisciplinary research committee. The following areas of research and development were studied:

1. A laboratory screening program into the suitability of five different earthworm species and 10 different waste materials to assess their biological and economic potentials, and study the biology and ecology of these worms in organic wastes. This research included studies of the population dynamics, life tables and the influence of environmental factors on the growth, reproduction and survival of these different species (Edwards, 1988).
2. Microbial studies to assess the source of nutrition of the earthworms that live on organic wastes, the relative importance of bacteria, protozoa, fungi and nematodes in their diet, and whether particular microorganisms were essential to their survival (Edwards and Fletcher, 1988).
3. Evaluation of productivity of worm biomass and processed waste in relation to type of waste, worm stocking rate and environmental factors (Edwards *et al.*, 1985).
4. Development of systems of production of worm protein and processed waste (section 13.5.3) (Edwards, 1985).
5. Harvesting and processing of worms into animal feed protein, animal feeding trials and toxicological tests (section 13.5.3) (Edwards, 1985).
6. Production and preparation of plant growth media processed by worms from organic wastes; plant growth trials and amendment of media to produce maximum plant growth (section 13.4) (Edwards, 1983; Edwards *et al.*, 1985).

The five earthworm species that were identified as candidate species for breakdown of organic wastes included *E. fetida*, *D. veneta* and *L. rubellus*, of temperate climatic origin and two, *E. eugeniae* and *P. excavatus*, from the tropics. The survival, growth and reproduction of these species was studied in the laboratory in pig, cattle, duck, turkey, poultry, potato, brewery and paper wastes, and compared with that in activated sewage sludge. All the species grew satisfactorily and survived well in a wide range of organic wastes, but some were very prolific, some grew rapidly and some attained a large biomass; all characters contributing to the practical usefulness of the earthworms.

Most organic wastes can be broken down by earthworms, but some need to be pretreated and not all wastes support earthworm growth equally well. Suitable wastes include:

Cattle solids

These are the easiest animal waste in which to grow worms. They usually contain no materials that are unfavorable to growing earthworms. Solids have to be separated from slurries, as with pig wastes, before they can be used to grow worms.

Horse manure

This is an excellent medium for growing earthworms and needs very little modification.

Pig solids

These are probably the most productive wastes for growing earthworms. If the waste is in the form of a slurry, the solids must be separated mechanically or by sedimentation. Pig waste tends to contain some ammonia and inorganic salts, and may have to be composted for about 2 weeks prior to inoculation with worms. Pig wastes tend to have a content of heavy metals, particularly copper (Edwards et al., 1988; Wong and Griffiths, 1991).

Poultry wastes

Poultry wastes, including chicken, duck and turkey manures, contain significant amounts of inorganic salts and ammonia, which may kill worms in fresh waste. After removal of these materials through composting, washing or aging, earthworms grow well in them and the compost produced is high in nutrients.

Potato waste

This waste, in the form of peel from the processed potato industry, is an ideal growth medium for earthworms and needs no modification in terms of moisture content or other preprocessing.

Paper pulp solids

Paper pulp solids, produced by mechanical separation or sedimentation of solids from the press washings, are an excellent material for the growth of earthworms.

Brewery waste

This needs no modification in terms of moisture content to grow earthworms. Worms can process it very quickly and grow and multiply rapidly in it.

Spent mushroom compost

This is a good medium for growing earthworms, which are able to break down the straw it contains into small fragments and produce a finely structured material.

Urban wastes

Urban wastes, including grass clippings and leaves and food wastes from supermarkets and restaurants, are all good growth media for earthworms, particularly when first macerated and thoroughly mixed (Huhta and Haimi, 1988).

Systems of growing earthworms range from very low-technology simple systems, such as windrows, waste heaps or boxes, through moderately-complex, to completely automated reactors (Jensen, 1993). The engineering details of such systems are dealt with in a later section in this chapter, but the basic principle of all of the breeding systems is to add wastes frequently in small, thin layers and allow the earthworms to process successive aerobic layers of wastes. The earthworms will always be concentrated in the upper 15 cm of waste. Many of the operations can be mechanized, a suitable balance being needed between costs of mechanization and savings in labor. The key to maximum productivity is to maintain aerobicity and optimal moisture and temperature conditions, and to avoid excessive amounts of ammonia and salts. Hence, for year-round production in a temperate climate, the processing must be done under cover, although heating is not usually necessary, because the decomposing waste maintains warm conditions in the beds. Zharikov *et al.* (1993) reported successful production of vermicompost from a mixture of organic wastes from microbiological plants, consisting of husks, growth media and bacterial preparations.

13.3 SPECIES OF EARTHWORMS SUITABLE FOR WASTE MANAGEMENT

13.3.1 SELECTION OF BEST SPECIES

There are many species of earthworms with potential for waste management, but relatively few have been used on a widespread scale. They include *E. fetida* (brandling or tiger worm) and *E. andrei* (red tiger worm), *E. eugeniae* (African night-crawler), *D. veneta, P. excavatus* and *L. rubellus* (red worm).

Extensive studies on the potential of *E. fetida, E. eugeniae, P. excavatus* and also *P. hawayana* to grow in sewage sludge were made by Neuhauser *et al.* (1988). They concluded that all these species have optimum temperatures between 15 and 25 °C. In their studies, cocoon production was restricted by temperature more than growth; all five species produced most cocoons at 25 °C.

The individual live weight of the five species was followed over a period of 20 weeks; *P. excavatus* had the lowest increase in weight and *E. fetida* the next smallest increase. Both *P. hawayana* and *E. eugeniae* reached their peak weight at about 10–12 weeks. *Eudrilus eugeniae* began to lose weight after 14 weeks. *Dendrobaena veneta* achieved the largest increase in weight of the five species, but took 16 weeks to achieve its maximum weight. In terms of cocoon production for the five species of earthworms over the 20 week period, *P. excavatus* produced the largest number of cocoons, while *D. veneta* produced the lowest number of cocoons and did not start producing cocoons for 10 weeks. The other three species, *E. fetida, E. eugeniae* and *P. hawayana*, all produced a similar number of cocoons, peaking in numbers at the middle of the experiment (10 weeks) and decreasing gradually toward the end of the experiment (Neuhauser *et al.*, 1988).

Since not all earthworm cocoons hatch, and it is possible to have more than one young earthworm per cocoon, it is important to evaluate the number of live young that would be obtained from the cocoons of each species. Cocoons from five species of earthworms, *D. veneta, E. fetida, E. eugeniae, P. excavatus* and *P. hawayana*, were collected and allowed to hatch. Individual cocoons were kept under nonstressed conditions, at 25 °C, and were checked twice per week to determine the number of cocoons that hatched and the number of worms that were produced per hatched cocoon.

It was concluded from Neuhauser's data that *E. fetida* produced 6 cocoons/worm/week (19 young worms); *D. veneta*, 5 cocoons (19 young worms); *E. eugeniae*, 11 cocoons (20 young worms); *P. excavatus*, 24 cocoons (13 young worms); and *P. hawayana*, 10 cocoons (9.5 young worms) per parent worm. These workers also studied the growth of different combinations of these species in polyculture. Although the total earthworm biomass tended to be greater in polyculture, the results were not clear-cut (Neuhauser *et al.*, 1988).

Edwards (1988) studied the life cycle and optimal conditions for growth and survival of four earthworm species in animal wastes (Tables 13.1 and 13.2). The species were *E. fetida, D. veneta, E. eugeniae* and *P. excavatus*. All these species of earthworms required the waste in which they lived to be aerobic, and as soon as anaerobicity developed the worms moved out of the waste. Each of the four species differed very consider-

Table 13.1   Maximum reproduction rate of earthworms in animal and vegetable wastes (from Edwards, 1988)

Species	No. of cocoons	% Hatch	No. of hatchlings	Net reproductive rates per week
Eisenia fetida	3.8	83.2	3.3	10.4
Eudrilus eugeniae	3.6	81.0	2.3	6.7
Perionyx excavatus	19.5	90.7	1.1	19.4
Dendrobaena veneta	1.6	81.2	1.1	1.4

Table 13.2   Productivity of earthworms in animal and vegetable waste – length of life cycle (from Edwards, 1988)

Species	Time for cocoons to hatch (days)	Time to sexual maturity (days)	Time from egg to maturity (days)
Eisenia fetida	32–73	53–76	85–149
Eudrilus eugeniae	13–27	32–95	43–122
Perionyx excavatus	16–21	28–56	44–71
Dendrobaena veneta	40–126	57–86	97–214

ably in terms of their response to and tolerance of different temperatures. The optimum temperature for *E. fetida* was 25 °C, and it had a temperature tolerance from 0 to 35 °C (Fig. 13.1). *Dendrobaena veneta* had a rather lower temperature optimum and rather less tolerance of extreme temperatures. The optimum temperatures for *E. eugeniae* and *P. excavatus* were about 25 °C, but they died at temperatures below 9 °C and above 30 °C. Optimum temperatures for cocoon production were much lower than for growth for all species. These species also differed in their optimum moisture requirements from those of *E. fetida* (Fig. 13.2), but not greatly. Some, such as *D. veneta*, were able to withstand a much wider range of moisture than others, such as *P. excavatus*.

All earthworms are very sensitive to ammonia and do not survive in organic wastes containing high levels (e.g. fresh poultry litter). They also die in wastes with large amounts of inorganic salts. Both ammonia and inorganic salts have very sharp cutoff points between toxic and nontoxic, i.e. <0.5 mg/g of ammonia and <0.5% salts. However, wastes that have too much ammonia become acceptable after this is removed by a period of composting, or when both excessive ammonia and salts can be washed out of the waste. Earthworms are relatively tolerant with regard to pH, but when given a choice in a pH gradient, they move towards the more acid material, with a pH preference of 5.0. The optimal conditions for breeding *E. fetida* are summarized in Table 13.3. These do not differ too much from those suitable for the other species (Edwards, 1988).

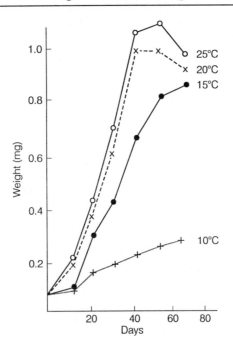

Figure 13.1   Growth of *E. fetida* at different temperatures (Edwards, 1988).

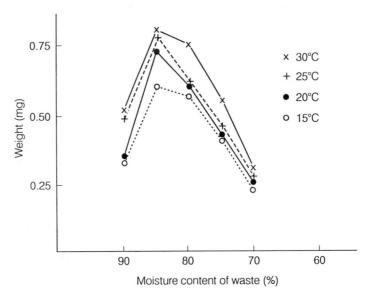

Figure 13.2   Growth of *E. fetida* at different moisture contents (Edwards, 1988).

Table 13.3 Optimal conditions for breeding *E. fetida* in animal and vegetable wastes (from Edwards, 1988)

Condition	Requirements
Temperature	15–20°C (limits 4–30°C)
Moisture content	80–90% (limits 60–90%)
Oxygen requirement	Aerobicity
Ammonia content of waste	Low: <0.5 mg/g
Salt content of waste	Low: <0.5%
pH	>5 and <9

13.3.2 CHARACTERISTICS OF DIFFERENT SPECIES

As can be seen from the preceding section, different earthworm species have different requirements for optimal development, growth and productivity in organic wastes.

Eisenia fetida

The species most commonly used for management of organic wastes is *E. fetida*, or the closely related *E. andrei*. There are several reasons why this species is preferred. It is ubiquitous, and many organic wastes become naturally colonized by this species. It has a wide temperature tolerance and can live in organic wastes with a good range of moisture contents. It is a tough worm, readily handled, and in mixed cultures usually becomes dominant, so that even when field systems begin with other species, they often end up with a large proportion of *E. fetida*. *Eisenia fetida* and *E. andrei* (the 'tiger worm' and 'red tiger worm') have been distinguished as separate species (Sheppard, 1988), but their overall reproductive performances and requirements do not differ (Reinecke and Viljoen, 1991). These species have both been distributed widely throughout temperate regions of the world, and are the species most commonly used in commercial vermiculture and waste reduction. Graff (1974), Vail (1974), Watanabe and Tsukamoto (1976), Tsukamoto and Watanabe (1977), Hartenstein *et al.* (1979), Kaplan *et al.* (1980), Edwards (1988) and Neuhauser *et al.* (1979) all investigated the productivity, growth and population biology of *E. fetida*, when fed on animal manures or sewage sludge.

In surveys of commercial earthworm farms in the US and Europe by Edwards, and Australia by Buckerfield and Baker, the earthworms sold under the name *L. rubellus* were all *E. fetida* or *E. andrei*. Data on the biology and ecology of *E. fetida* are given in the previous section.

Eudrilus eugeniae

This is a large worm that grows extremely rapidly and is reasonably prolific, commonly known as the African night-crawler. It is cultured extensively, and under optimum conditions it would be ideal for protein production. Its main disadvantages are poor temperature tolerance and poor handling capabilities, so that it can be difficult to harvest. *Eudrilus eugeniae* has high rates of reproduction (Bano and Kale, 1988; Edwards, 1988), and is capable of decomposing large quantities of organic wastes rapidly and incorporating them into the topsoil (Neuhauser *et al.* 1979, 1988; Edwards, 1988; Kale and Bano, 1988). *Eudrilus eugeniae* has a preference for higher temperatures and cannot tolerate extended periods below 16 °C (Viljoen and Reinecke, 1992), and it does not survive below 10 °C. Its use in outdoor vermiculture may therefore be limited to tropical and subtropical regions, unless winter temperatures are controlled.

Dendrobaena veneta

This species is a large worm with potential for use in vermiculture that can also survive in soil (Satchell, 1983), but it is not very prolific and does not grow very rapidly (Edwards, 1988; Viljoen *et al.*, 1992). Of the species that have been studied in detail, it is probably the least suitable for organic waste breakdown.

Perionyx excavatus

This tropical earthworm is extremely prolific and is almost as easy to handle as *E. fetida,* and very easy to harvest. Its main drawback under temperate conditions is its inability to withstand adverse temperature conditions, but for tropical conditions it is an ideal species. It has an extremely high reproductive rate (Kale *et al.,* 1982; Edwards *et al.,* 1988; Neuhauser *et al.,* 1988). It is a very common species in Asia and is used extensively in vermiculture in the Philippines and Australia.

Polypheretima elongata

This species has been tested for use in reduction of organic solids, including municipal and slaughterhouse wastes; human, poultry and dairy manures; and mushroom compost in India. A project in India using this species claimed to have a commercially viable facility for the 'vermistabilization' of 8 tons of solid wastes/day. They have developed a 'vermifilter', packed with vermicompost and live earthworms, which produces reusable water from sewage sludges, manure slurries and organic wastewaters from food-processing. *Polypheretima elongata* appears to be restricted to tropical regions, and may not survive temperate winters.

Lumbricus rubellus

This is a common species of earthworm that is found in moister soils, particularly those to which animal manure or sewage solids have been applied (Cotton and Curry, 1980a,b). It is believed to be suitable for organic waste breakdown, but this has yet to be substantiated.

13.4 EARTHWORM COMPOSTS AS PLANT GROWTH MEDIA

Earthworm composts can be produced from almost all organic wastes with suitable preprocessing and controlled processing conditions. They can also be used as structural additives for poorer soils.

13.4.1 CHARACTERISTICS OF PLANT GROWTH MEDIA PRODUCED BY EARTHWORMS FROM ORGANIC WASTES

The final physical structure of the plant growth media or vermicomposts produced from organic wastes depends very much on the original material from which they were produced. Some source materials, such as cattle, pig and poultry manures from indoor systems and spent mushroom compost, contain straw, which takes longer for the earthworms to fragment, than more particulate materials, such as animal solids and slurries, brewery waste, paper pulp and similar materials. However, the final product is still usually a finely divided peat-like material with excellent structure, porosity, aeration, drainage and moisture-holding capacity (Edwards, 1982, 1993a). It has the appearance and many of the characteristics of peat.

The nutrient content of a vermicompost differs greatly, depending on the parent material. However, when the nutrient content is compared with that of a commercial plant growth medium to which inorganic nutrients have been added, it usually contains more of most mineral elements,

Table 13.4 Major plant nutrient elements in worm-processed animal wastes (from Edwards and Burrows, 1988)

Waste material	Element content (% dry wt)					
	N	P	K	Ca	Mg	Mn
Separated cattle solids	2.20	0.40	0.90	1.20	0.25	0.02
Separated pig solids	2.60	1.70	1.40	3.40	0.55	0.03
Cattle solids on straw	2.50	0.50	2.50	1.55	0.30	0.05
Pig solids on straw	3.00	1.60	2.40	4.00	0.60	0.05
Duck solids on straw	2.60	2.90	1.70	9.50	1.00	0.10
Chicken solids on shavings	1.80	2.70	2.10	4.80	0.70	0.08
Commercial plant growth medium	1.80	0.21	0.48	0.94	2.20	0.02

Table 13.5   Effect of worm activity on nutrients in organic wastes (from Edwards and Burrows, 1988)

Organic waste	Nitrate nitrogen (p.p.m.)	Readily soluble P (% d.m.)	Exchangeable (% d.m.)		
			K	Ca	Mg
Cattle waste (unworked)	8.8	0.11	0.19	0.35	0.05
Cattle waste (worm worked)	259.4	0.18	0.41	0.59	0.08
Pig waste (unworked)	31.6	1.05	1.49	1.56	0.45
Pig waste (worm worked)	110.3	1.64	1.76	2.27	0.72
Potato waste (unworked)	74.6	0.19	1.94	0.91	0.24
Potato waste (worm worked)	1428.0	0.22	3.09	1.37	0.34

although there is often a deficiency of magnesium (Table 13.4). An important feature is that during the processing of the wastes by earthworms, many of the nutrients they contain are changed to forms more readily taken up by plants, such as nitrate or ammonium, nitrogen, exchangeable phosphorus and soluble potassium, calcium and magnesium (Table 13.5). Moreover, many organic wastes tend to be on the alkaline side of neutral (>7.0) whereas most plants prefer a growth medium on the acid side of neutral (e.g. with a pH of 6.0). The processing by earthworms does not change the pH of the material appreciably, so for ideal plant growth, worm-worked wastes benefit from some acidification or by mixing with an acid medium such as peat.

13.4.2 GROWTH OF PLANTS IN WORM-WORKED ORGANIC WASTES

There is good evidence that materials produced by earthworms can promote the growth of plants. Fosgate and Babb (1972) grew earthworms in cattle wastes and reported that the material produced was equal to greenhouse potting mixes for production of flowering plants. Reddy (1988) reported increased growth of *Vinca rosea* and *Oryza sativa* after addition of cast material from *Pheretima alexandri*. Buchanan *et al.* (1988) suggested that vermicomposts had higher levels of available nutrients than parent materials. Edwards (1988) reported that samples of vermicomposts had high levels of available nitrogen. Handreck (1986) reviewed the utilization of vermicomposts as horticultural potting media. He concluded that they could supply most of the trace element needs of plants, but many vermi-composts may not have sufficient nitrogen to supply all the needs of the plants. However, it seems difficult to justify this conclusion since many organic wastes have excess amounts of nutrients and only a small propor-tion is lost during vermicomposting (Edwards *et al.*, 1985).

A research program at Rothamsted Experimental Station involved an extensive series of trials testing the growth of a wide range of plants in a variety of worm-worked wastes (Edwards and Burrows, 1988). There is no space here to report the results from these extensive trials in detail, but it would be appropriate to give some general data and results on particu-lar aspects and uses of worm-worked composts.

A wide range of plants has been grown successfully in both undiluted wastes and a number of mixes, including 3:1 or 1:1 ratios of worm-worked wastes to peat, pine bark or Kettering loam. Growth has usually been better than in recommended commercial growing media, such as Levington compost, and seed germination has been found to be more rapid for most species of plant in worm-worked wastes.

Plants tested for growth in vermicomposts by Edwards and Burrows (1988) included vegetables, such as aubergine, cabbage, capsicum, cucum-ber, lettuce, radish and tomato; bedding plants, such as alyssum, antir-rhinum, aster, campanula, calceolaria, cineraria, coleus, French marigold, plumose asparagus, polyanthus, salvia and sweet pea (Edwards and Burrows, 1988); and ornamental shrubs such as *Eleagnus pungens, Cotoneaster conspicua, Pyracantha, Viburnum bodnantense, Chaemaecyparis lawsoniana* and *Juniperus communis* (Scott, 1988).

The waste, after working by the worms, is at about 75% moisture con-tent. Some form of partial sterilization, such as heating to between 60 and 80 °C for 24 hours, or passing through a flame-sterilizer, may be used to kill residual earthworms and their cocoons, to kill insects, and to lessen pathogen problems. If the waste is likely to contain human pathogens, some precomposting for 3–4 days may be advisable. Usually, a magne-sium sulfate supplement is added to rectify the usual deficiency in mag-nesium, and the pH is adjusted in some way, such as by adding acid peat, to bring the medium to a pH of about 6.0.

A wide range of tests of seedling emergence of pea, lettuce, wheat, cab-bage, tomato and radish were made in small pots and trays using the standard EEC recommended seedling emergence test. The emergence of tomatoes, cabbage and radish seedlings tended to be as good, and usually better, in worm-worked animal wastes than in a commercial plant growth medium, and much better than in composted animal wastes with no earthworms. Similarly, the early growth of seedlings of ornamentals up to the stage when they were transplanted into larger pots or outdoors, was as good or better in the worm-worked animal wastes mixed with peat as in a commercial plant growth medium.

The growth of the ornamentals listed earlier was also good in worm-worked animal wastes and mixtures of these with peat, and often better than in a commercial plant growth medium. Aubergines, dahlias, coleus, capsicum and polyanthus did particularly well in such worm-worked waste/peat mixtures. Some of the ornamentals, particularly chrysanthemums, salvias and petunias, flowered much earlier in worm-worked waste mixtures, and this could possibly be due to a hormonal effect resulting from microbial action. A number of other species of plants followed significantly different growth patterns in vermicomposts, another indication of a possible hormonal effect.

Tomatoes and cabbages germinated well and grew better in mixtures of worm-worked wastes with peat and sand than in a commercial plant growth medium, or in 100% peat or a peat/sand mixture. Worm-worked paper waste was one of the best materials tested and there were no problems in processing it or producing a standardized material.

The effects of dilution of worm-worked animal wastes, with a commercial plant growth medium, on growth of ornamentals at a range of levels were quite dramatic (Scott, 1988). For instance, in growth trials of three ornamental plants, when a 50/50 mixture of pig and cattle worm-worked animal wastes was diluted at a range of levels, ranging from 5 to 10%, with a commercial plant growth medium, all of the mixtures grew *Chaemaecyparis lawsoniana*, *Pyracantha* sp. and *Viburnum bodnantense* better than the recommended commercial medium itself. The most surprising result was that addition of even 5% of worm-worked animal waste to the worm-worked waste/commercial medium mixture produced a significant improvement in the growth of the plants. However, all dilutions of the different mixtures also tended to grow better plants than the 100% worm-worked wastes, which had a tendency to dry out more rapidly than the different mixtures. These results, where a small amount of worm-worked waste has a significant effect, indicate that the response is not based only on nutrient content.

Worm-worked animal waste was also used as a 'blocking' material to grow seedlings for transplanting into the field. Cabbages were grown in machine-compressed blocks made from either a commercial seedling medium or from worm-worked cattle or pig wastes. The seedlings were transplanted into the field in the same blocks and their subsequent growth was followed and measured. Those seedlings grown in blocks of worm-worked pig waste were significantly larger and more mature at harvest than those grown in the commercial blocking material (Edwards and Burrows, 1988)

Thus, although much more research is needed, there is evidence that good worm-worked organic wastes, mixed with peat and other materials, make excellent growth media for a variety of purposes. They need standardization to ensure an adequate nutrient status, possible addition of deficient nutrients, adjustment of pH and sterilization to kill insects and pathogens, etc., but have a considerable commercial horticultural potential.

The potential commercial outlets for vermicomposts vary greatly between countries, and so do the possible economic returns on the sale of these media. In general, for the high-value market for plant growth media, the product must be produced to within relatively small tolerances, as a standard material varying little in consistency or nutrient content. For such a product, uniform sources of organic wastes must be available and, if more than one kind of waste is used, the mixture and additives must be in constant proportions. Additionally, there must be batch analyses to ensure the standardization of the product. When the product is produced with lower technology and more variable sources of wastes, so its value decreases, but so also does the cost of production, processing and packaging for these less valuable materials.

13.5 THE USE OF EARTHWORMS AS A FEED PROTEIN SOURCE FOR ANIMALS

The first workers to suggest that earthworms contained sufficient protein to be considered as animal food were Lawrence and Millar (1945), and this potential has been confirmed only in the past 15 years, when full analyses of the body tissues of earthworms have been available to show the kinds of amino acids they contain and the nature of the other chemical body constituents. The first successful animal feeding trials were by Sabine (1978); but subsequently there have been a number of other such trials by various workers.

13.5.1 FOOD VALUE OF WORMS

Lawrence and Millar (1945) reported a high protein content for earthworms. The first analyses of the constituents of the tissues of different species of earthworms was by McInroy (1971), but there have been various other analyses since this (Schulz and Graff, 1977; Sabine, 1978; Yoshida and Hoshii, 1978; Mekada et al., 1979; Taboga, 1980; Graff, 1982). The results of all these studies have shown clearly that the overall composition of earthworm tissues does not differ greatly from that of many invertebrate and vertebrate tissues. The essential amino acid spectrum for earthworm tissues, as reported by these different authors, compares well with those from other currently used sources of animal feed protein. There is good evidence that the mean amounts of essential amino acids recorded are very adequate for a good animal feed, when compared with the recommendations of FAO/WHO, particularly in terms of lysine and the combinations of methionine and cysteine, phenylaniline and tyrosine, all of which are important components of animal feeds. In addition, earthworm tissues contain a preponderance of long-chain fatty acids, many of which cannot be synthesized by nonruminant animals, and an adequate mineral content. They have an excellent range of vitamins and are rich in niacin, which is a valuable component of animal feeds, and are an unusual source of vitamin B_{12}. The overall nutrient spectrum of worm tissues has excellent potential as a feed for fish, poultry, pigs or domestic animals.

13.5.2 GROWTH OF WORMS IN ANIMAL, VEGETABLE AND INDUSTRIAL WASTES

The biology of four species of worms that grow and reproduce well in organic wastes has been investigated at Rothamsted in the UK and at SUNY, Syracuse and Cornell University (Ithaca) in the USA (Edwards, 1988; Neuhauser et al., 1988). These species are E. fetida, E. eugeniae, P. excavatus and D. veneta. The growth patterns of individual worms, or whole populations, follow classical sigmoidal growth curves, with rapid initial growth followed by a steadier phase and levelling off. The maximum protein production per unit time could be achieved by inoculating large volumes of animal wastes with relatively small numbers of young worms to take advantage of the initial phase. Dry matter conversion ratios of waste to worms, which range from 10% for cattle and pig waste to 2% for duck waste, have been achieved readily in the laboratory and although rather lower conversion rates have been attained in the field, the higher ratios should be feasible ultimately in large-scale production.

13.5.3 PRODUCTION OF EARTHWORM FEED PROTEIN

The production of feed protein from earthworms has been reviewed by Edwards (1985) and Edwards and Niederer (1988). The efficient productivity of earthworm protein depends mainly upon detailed knowledge of the population dynamics of the appropriate species, maintenance of optimal environmental conditions, and upon the engineering of suitable production methods and particularly development and use of harvesting systems that involve small labor inputs. Our knowledge of all these aspects is sufficient to make worm protein production an economic proposition if non labor-intensive harvesting methods are available.

Earthworms grow best at relatively high moisture levels (80–90% MC), and this raises subsequent harvesting problems, since it is not easy to separate worms mechanically from the finely divided organic matter at such high moisture contents, and some drying of the vermicompost before harvest is usually necessary. There are good methods of separating worms from fully worked organic materials, but an improved method was developed at Rothamsted and the National Institute for Agricultural

Engineering, Silsoe, UK (Phillips, 1988). The efficiency of this machinery, in terms of percentage recovery, is very high. The machine that has been developed currently will separate worms from about 1 tonne of waste/hour and this machine can be automated and scaled up to increase output.

After worms are collected from the separating machinery, they may have small particles of waste attached to their bodies and are likely to contain waste in their guts. Hence, the first stage of all the methods of processing tested is to wash the worms thoroughly and leave them standing in water for several hours, in order to evacuate the residual waste particles from their guts completely. A range of different methods of processing the worms for animal feed has been developed (Edwards and Niederer, 1988). Two of the methods produced a paste product and the other four a dry worm meal; all of the products were acceptable for particular types of animal feeds, and the ultimate choice of a method of processing must depend upon:

1. the type of animal feed required;
2. the cost of production of the protein;
3. minimal loss of dry matter; and
4. minimal loss of nutrient value.

The following methods were all suitable:

Incorporation with molasses

This method involves first blanching the worms in boiling water for 1 minute, then incorporating 30% molasses with them, together with 0.3% potassium sorbate to produce a paste. The molasses lowers the water activity to about A_w 0.90, and the addition of potassium sorbate reduces this further, to approximately A_w 0.65. At this level, the growth of yeasts and molds is inhibited, permitting indefinite storage of the product.

Ensiling with formic acid

A second method involved incorporating 3% formic acid with the earthworms, thoroughly homogenizing this, then allowing it to ensile until eventually a very stable paste or liquid product was produced.

Air-drying

In this method, a dry meal can be produced by blanching earthworms in boiling water for 1 minute then air-drying them and grinding. This needs high ambient temperatures and is a relatively slow process.

Freeze-drying

Another kind of dry meal can be produced by freezing earthworms quickly, then freeze-drying and grinding them. This is efficient, but both slow and expensive.

Acetone heat-drying

A third type of dry meal was produced after killing earthworms by immersing them in acetone for 1 hour, then air-drying and later oven-drying them in a hot oven at 95 °C and finally grinding them into a pan. The acetone was evaporated off and reclaimed.

Oven-drying

A dry meal was produced by killing and drying earthworms in trays in a large commercial oven at 95 °C and then grinding them up into a dry powder.

All of the methods tested produced a good material that could be used as animal feed, but there were variations in the dry matter yield. Killing earthworms in boiling water, then drying them in an oven resulted in the lowest dry matter (11.6%). Freeze-drying produced a meal with 13.5% dry matter. Killing worms in acetone, then drying them in air produced a dry matter of 14.5%, but after subsequent oven-drying, this fell to 12.8%. Killing and drying the earthworms in a hot air oven gave the greatest dry matter yield, 15.2% (Edwards, 1985).

The different processing methods affected the amounts of essential amino acids present slightly; but the differences were very small. Most of the methods had relatively little effect on the total amounts of different amino acids in the product. However, the lysine content was decreased slightly by incorporation with molasses, using formic acid and by freeze-drying, compared with the other methods. Clearly, a stable protein feed can be produced by any of the methods and the choice of method must depend mainly on the use to which the protein is to be put, the animal to be fed, and the cost of the processing method in relation to the value of the protein.

13.5.4 ASSESSMENT OF THE VALUE OF WORM PROTEIN AS ANIMAL FEED

The main outlets suggested for utilization of earthworm protein have been in fish farming and as protein supplements in poultry and pig feeds, and all of these have been tested experimentally (Sabine, 1978, 1981; Edwards, 1985; Edwards and Niederer, 1988).

Fish-feeding trials

The first trials using earthworms as a protein source in fish feed were by Tacon et al. (1983). The growth of trout fed only on E. fetida, A. longa and L. terrestris was compared with that of fish fed on a commercial ration. Fish fed frozen *A. longa* and *L. terrestris* grew as well as or better than fish fed on commercial trout pellets. Trout did not grow so well on a whole diet of freeze-dried *E. fetida*, although they grew much better on *E. fetida* that had been 'blanched' in boiling water before freezing (Stafford and Tacon, 1988) than on unsterilized worms. However, dried earthworm meal derived from *E. fetida* which had not been blanched could satisfactorily replace the fishmeal component of formulated trout pellets at the normal levels of inclusion, between 5 and 30%, without affecting growth of trout adversely. The overall conclusion was that all earthworms have good potential both as a complete feed or a protein supplement for trout or other fish. Hilton (1983) reported that trout did not grow well on another earthworm species, *E. eugeniae*, but there are some doubts about his experimental techniques, because other forms of protein supplement currently used commercially, such as blood meal, would also have been unsuccessful if used in the same way as he used worm protein. Guerrero (1983) reported that *Tilapia* fish grew better on diets containing earthworm protein supplements from *P. excavatus*, than those with other fishmeal supplements. Velasquez *et al.* (1991) reported that earthworm meal produced from *E. fetida* produced satisfactory growth of rainbow trout, with a significant increase in lipid content.

Chicken-feeding trials

The first trials to assess the growth of chickens on earthworm protein were reported by Harwood and Sabine (1978) and Sabine (1981). They compared the use of earthworm meal as a protein supplement for chickens with that of meat meal, and found no significant difference in growth of chickens between the two diets. Similar results were reported by Taboga (1980) and Mekada *et al.* (1979), and the latter workers reported that when earthworms were fed to older birds, egg production was maintained. Jin-you *et al.* (1982a) reported that chickens fed on earthworms put on weight faster than those given other diets (including fishmeal); they had more breast muscle and consumed less food.

These conclusions have been confirmed by Fisher (1988). In his experiments, chickens grew well, had a good weight gain per unit of food, and had an excellent nitrogen retention when fed on diets with levels of worm meal from 72 to 215 g/kg (Fig. 13.3).

Pig-feeding trials

Two trials reporting the growth of pigs on earthworm protein supplements (Harwood and Sabine, 1978; Sabine, 1981) showed that in feeding trials with both starter and grower pigs, young pigs fed on an earthworm protein supplement grew equally well and had similar feed conversion ratios to those grown on commercial rations. Jin-you *et al.* (1982b)

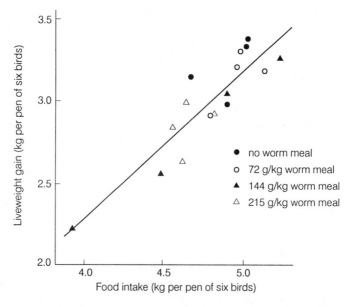

Figure 13.3   Food intake and gain of pens of six chickens given diets containing 0 (●), 72 (○), 144 (▲) or 215 (△) g/kg worm meal. The pooled regression line has the equation $y = -1.22 + 0.875x$ and is not significantly different from separate lines fitted for each treatment (Fisher, 1988).

reported that piglets grew better on earthworm protein supplements than on other protein supplements, and weaning was accelerated, estrus was produced earlier in sows, disease resistance was increased and there was a decreased incidence of white diarrhoea. Edwards and Niederer (1988) also reported good growth of pigs on earthworm protein.

Rat-feeding trials

Ibanez *et al.* (1993) fed rats with earthworm protein and reported good growth and no adverse effects on growth.

Shrimp-feeding trials

Garcia and Jaime (1990) reported feeding earthworm meal produced from *E. eugeniae* to shrimps (*Renaeus schmitti*) and stated that they grew well on this diet.

13.5.5 ECONOMICS OF PRODUCTION OF EARTHWORM PROTEIN

Fieldson *et al.* (1985) made a detailed study of the economics of production of earthworm protein. Their general conclusions were that earthworm production had the best prospects of good profits when it was done by larger farmers producing considerable amounts of animal wastes. The most important criterion is that the earthworm meal must be produced at an economic price, although the value of the compost can also be taken into account as complementary income. Currently, the only labor-intensive part of worm protein production is the harvesting process, and this remains the main barrier to successful commercial production, but could be relieved by engineering of better earthworm harvesting methods.

In a computer analysis of the economic value of earthworm protein based on its amino acid, fatty acid, mineral and vitamin content, it emerged that it is extremely valuable as feed for certain animals, particularly eels and young turkeys, and it had the same value for fish, pig and poultry feed as fishmeal or meat. Sabine (1983) calculated that, on the basis that meat meal for stock food manufacture costs Australian $177/tonne, earthworm meal, on the basis of its proven food value, is an economic substitute in poultry food at prices up to $236/tonne, and may be economic up to about $300/tonne.

Satchell (1983) also discussed the costs of production of earthworm protein. He quoted Tomlin (1983), who calculated the price of *E. fetida* in Canada as $21/kg ($21 000/t) and that of dressed tenderloin beef as $16/kg ($16 000/t). He calculated, based on the amino acid, mineral, energy and water content of earthworms, that in England, earthworm meal (1983 prices) was worth £230/tonne (dry weight) compared with £160/tonne for meat bonemeal and £230/tonne for fishmeal; the value of earthworm meal (wet weight) on this basis would be £53/tonne. Thus, earthworm feed protein still cannot compete economically with other commercial animal feeds. However, as a by-product of a vermicomposting process it could be economically viable.

13.6 ENGINEERING OF WASTE MANAGEMENT BY EARTHWORMS

Methods of managing organic wastes by earthworms range from simple outdoor windrows to complex continuous reactors (Price, 1987; Phillips, 1988). It is outside the scope of this book to deal with the technology of worm production in detail, but it has been found that growing worms in beds, filled with animal waste to a depth of about 50 cm in successive shallow layers at regular intervals by automatically operated gantries, is an easy system and not labor intensive. In colder climates, some form of insulated housing is necessary. Different forms of batch production in crates or boxes are useful and systems of continuous processing of wastes with automatic addition and removal of wastes have been developed. The technology to maximize worm and/or compost production is still evolving, but holds considerable promise for development into well-engineered, highly productive, nonlaborious systems. A rapid rate of growth and multiplication of earthworms, and the efficient conversion of organic wastes into earthworm tissue protein would mean that systems could be developed using a minimum of labor or sophisticated technology.

Organic wastes range from almost liquid slurries to relatively dry and finely dispersed solids. All of these wastes must be brought to a suitable moisture content and temperature and the ammonia and salt contents reduced to acceptable levels by leaching, composting or some other method, before the worms can be grown successfully and productivity maximized.

Processing of each kind of organic waste involves different problems (Edwards and Neuhauser, 1988). Animal wastes, such as pig and cattle wastes, can be processed by earthworms when they are either straw-based or solids separated from slurries using a commercially available mechanical slurry separator. Cattle solids become acceptable to earthworms a few days after collection and/or separation, but it may take up to 2 weeks before the worms will enter pig solids and feed satisfactorily. Duck, turkey or chicken wastes with straw or wood-shavings pose an even greater problem, since they contain considerable amounts of ammonia and, until this falls to below 0.5 mg/g, the earthworms will not enter or utilize this material. The key to successful processing of poultry wastes is to monitor adverse factors closely and use different techniques to accelerate their acceptability for vermistabilization. Once processed, poultry wastes, which are rich in nutrients, can be diluted with other materials to make excellent plant growth media. Industrial wastes such as paper wastes, brewery wastes, processed potato wastes, restaurant wastes and garden wastes are much easier to preprocess and make acceptable to earthworms.

The techniques available for processing organic wastes with earthworms are of three main types. These differ in cost and complexity; the simplest costing the least initially, and the high-technology systems costing much more to set up, but being much less labor intensive and more efficient in operation repayment. Wastes are processed very much faster in the high-technology automated systems. The relative economic performances of the three systems depend upon land availability, labor availability and types of waste to be processed.

13.6.1 LOW-COST FLOOR BEDS

Outdoor windrows or beds with simple walls are the simplest type of process. The size of such beds is flexible, but the width of the beds should not exceed 8 feet (2.4 m), which allows the entire bed to be inspected easily, without the need to walk on the bed, and is also compatible with the sizes of many suitable covering and construction materials. The length is less important and depends on the area available. They can be laid on soil which is freely draining and not subject to waterlogging. Concrete areas are ideal for earthworm-processing systems since they provide a firm surface for tractor operations. However, it is essential for precautions to be taken to prevent too much water from entering the beds and to allow excess water to drain away from the bed. Usually, such floor beds are covered and the covers removed only for watering and addition of new waste materials.

13.6.2 GANTRY-FED BEDS

An important principle in improving the efficiency of organic wastes by earthworms is to be able to add the wastes to the beds in thin layers, 1–2 cm thick, at frequent intervals. This can be done most readily by adding the wastes by means of an overhead gantry running on wheels on the walls of the beds. This gradual addition of waste minimizes the generation of heat through composting and ensures that earthworms are continually processing the fresh wastes near the surface.

13.6.3 CONTAINERS OR BOXES

Edwards (1988) discussed batch vermicomposting in stacked boxes or containers and suggested that it was too labor intensive, since batches had to be moved to add more wastes or water. Grappelli *et al.* (1985) described a modular container system.

13.6.4 RAISED GANTRY-FED BEDS

Earthworms are usually confined to the top 10–15 cm of a bed of organic wastes. The efficiency and rate of processing of the wastes can be increased considerably by raising the bed on legs above the ground. If the bed has a mesh base, the worm-processed organic matter can be passed through this by some mechanical means, such as a breaker bar, and collected on a moving belt or with a slurry scraper. If waste is added to the top of the bed in thin layers from a mobile gantry daily and collected through the base, continual processing of the waste can be achieved. Such a system can become relatively sophisticated through complete mechanization of the waste addition and collection systems. Such automated continuous-processing reactors have operated successfully for as long as 2 years.

13.6.5 COMPLETE RECYCLING SYSTEMS

Bouché has developed a vermiculture-based complete urban waste recycling system (SOVADEC) in France. This involves putting the waste through a selectory which breaks up plastic and removes it, followed by manual sorting, sorting of rolling objects such as bottles, and separation of ferrous metal objects with magnets. The waste is then composted for 30 days, followed by vermicomposting.

13.7 ROLE OF EARTHWORMS IN PROCESSING ORGANIC WASTES APPLIED TO AGRICULTURAL AND OTHER LAND

Traditionally, animal wastes applied to agricultural land were the source of nutrients for crop production. However, with the progressive separation of animal and crop production in most countries in Europe and North America, animal wastes are produced in quantities much too large for land disposal. Similarly, sewage and urban organic wastes are not applied extensively to land and are often disposed of in landfills. However, in recent years, there has been a movement back towards land disposal of organic wastes and there is increasing evidence that earthworms play an important role in breaking these down and incorporating them into soils (Chapter 8).

For instance, Curry (1976) investigated the effects of application to pastures of slurries of cattle waste, pig waste and poultry waste on earthworm populations in Ireland. Slurries were applied for 3–4 years, at rates of about 28 tonnes/ha/yr and 56 tonnes/ha/yr. These rates of application for cattle slurry corresponded to additions of the order of 98–252 kg/ha nitrogen, 14–56 kg/ha phosphorus and 98–280 kg/ha potassium; for pig slurry, 112–252 kg/ha nitrogen, 56–168 kg/ha phosphorus and 14–84 kg/ha potassium; and for poultry slurry, 196–560 kg/ha nitrogen, 98–252 kg/ha phosphorus and 56–224 kg/ha potassium. Earthworm populations, relative to those in untreated controls, were 41% higher in plots with added cattle slurry, 53% higher in plots treated with pig slurry, and 40% higher in plots treated with poultry slurry. Earthworm populations were assessed over a 2 year period: after treatments with cattle slurry, they increased relative to untreated control plots by 5% and 31%, with increases in biomass of 33% and 95%, in response to a single application of 56 tonnes/ha and six applications of 112.5 tonnes/ha, respectively.

Derby *et al.* (1982) compared soil pH, and NH_4^+, NO_2^- and NO_3^- content of litter and A_1 horizons of an acid forest soil from a larch (*Larix*) plantation, treated with a surface application of pig slurry, in the presence and in the absence of lumbricid earthworms. In the presence of earthworms, the pH fell to about 4.0, while in the absence of earthworms, the pH fell to about 5.0. The rapid pH fall in soils with earthworms was accompanied by a rapid decline in NH_4^+ and NO_2^-, and a corresponding rapid increase in NO_3^- relative to soil without earthworms. It was concluded that the end result of the presence of earthworms was to improve the supply of nitrogen (as NO_3^-) that was available to plants. The results of experiments by Curry (1976) and Cotton and Curry (1980b) suggest that there is an upper limit to the amount of organic nitrogen that will result in a population increase. At very high slurry application rates there were changes in the earthworm community structure, including a reduction in proportion of immature worms, as well as changes in proportions of the species present. Curry (1976) showed that cattle and pig slurries were initially very toxic to earthworms that were confined in small containers in a soil peat moss medium. They suggested that the earthworm toxicity declines relatively quickly, hence it would be preferable to apply aged animal slurries to land rather than fresh slurries. It has been suggested by van Rhee (1975) that copper residues in some pig slurries could be toxic to earthworms, but this remains to be confirmed.

The application of anaerobically and aerobically digested sewage sludges to land in the UK and other parts of Europe is becoming increasingly common. Dindal *et al.* (1977) compared the response of earthworm populations to municipal wastewater irrigation in grassland, spruce forest and a mixed oak/hardwood forest. Earthworm numbers and biomass increased in response to all the rates of the treatments he applied, but effects were not consistent between sites. Ferreira and Cruz (1992) reported that adding a compost produced from municipal wastes by earthworms had excellent effects on the yield of corn (*Zea mays*) if used together with a low level of inorganic fertilizer.

Edwards and Lofty (1979) reviewed the effects of adding cattle, pig and poultry manure, municipal sewage wastes, and industrial wastes from breweries, potato processing, and paper production to agricultural land. They concluded that all of these organic amendments increase earthworm populations, which, in turn, accelerate the release of nutrients from these wastes. Marquenie and Simmers (1988) and Rhett *et al.* (1988) described the use of earthworms in a bioassay to assess the suitability of sediments dredged from US rivers for land disposal. We need more information on the overall impact of animal and sewage sludges and slurries applied to agricultural land and, in particular, the role of earthworms in nutrient turnover and soil structural changes on treated land.

Effects of agricultural practices and chemicals on earthworms 14

We have seen in earlier chapters that earthworms are important in providing soil fertility and improving soil structure (Chapters 8 and 10). Consequently, it is important to know how earthworm populations are affected adversely or favorably by different agricultural practices. The four main management inputs into any farming system are **cultivations, cropping patterns, fertilization** and **crop protection** (Edwards, 1989a). Each of these four inputs interacts strongly with all the others, so it is often difficult to extrapolate the overall effects from the results of component research, which investigates the influence of only one or another of these inputs at a time, on earthworm populations. Additionally, in arid and semi-arid regions, **irrigation** is a further major input that can also have an important influence on the other inputs as well as direct effects on earthworm populations. Earthworm populations can also be affected by industrial pollutants, atmospheric depositions, radioisotopes and heavy metals, so we shall also consider the effects of these factors on earthworms in this chapter.

14.1 THE EFFECTS OF CULTIVATIONS

Darwin (1881) considered the earthworm to be nature's plough, but he did not consider the converse, i.e. how ploughing affected earthworm populations. There is good evidence that many natural ecosystems tend to contain more earthworms than ploughed agricultural systems, although repeated, minimal or no-tillage systems allow earthworm populations to build up progressively.

It is now well established that grassland usually contains more earthworms than arable land in most parts of the world (Table 5.1) (Hopp and Hopkins, 1946b; Evans and Guild, 1948b; Ponomareva, 1950; Graff, 1953a; Barley, 1961; Dzangaliev and Belousova, 1969; Zicsi, 1969; Atlavinyte,

1974). The decreased numbers of earthworms that occur in cultivated arable land could be due to mechanical damage during cultivation, to the loss of the insulating layer of vegetation, to a decreased supply of food as the organic matter content gradually decreases with repeated cultivations, or to predation by birds when earthworms are brought to the surface during cultivating. Many earthworm researchers have considered that these differences in earthworm populations are due mainly to mechanical damage during cultivation. This opinion has become common because, when old grassland is ploughed, the number of earthworms in it decreases steadily with time after ploughing with repeated cultivations (Graff, 1953a). In one such study, 5 years after grass was ploughed, the earthworm population had declined by 70%, although the population was unchanged by the first ploughing of the sward, so it is unlikely that mechanical damage was a primary or initial cause of the decreased numbers of worms (Evans and Guild, 1948b). Low (1972) reported that after 25 years of regular cultivation, the numbers of earthworms were only 11–16% of those in old grassland, but Hopp and Hopkins (1946b) reported that the ploughing of arable land in late spring did not decrease earthworm numbers. It would be surprising if mechanical damage to earthworms by ploughing was a major factor influencing earthworm populations, because the plough merely turns the soil over, and probably has little effect on those earthworm species with deep burrows; however, predation after ploughing may be a factor. Preparation of seed beds by rotary cultivation, harrowing, disking or rolling can be expected to damage more of the surface-dwelling species of earthworms than ploughing, but only a few would be killed outright and the regenerative powers of individuals and populations of earthworms is considerable.

Edwards and Lofty (1969a) investigated the effects of annual maximal and minimal cultivation of grass plots on earthworm populations. They compared earthworm populations in plots that were ploughed and cultivated in spring with others that were left unploughed for 2 years. The more that the soil was cultivated during the first two seasons, the greater were the number and weight of earthworms in the soil (Table 14.1). However, repeated cultivations over successive seasons after the first two depressed the earthworm populations progressively. Edwards and Lofty (1978) discussed the influence of cultivation on earthworms further and concluded that there are more important indirect effects of heavy cultivation on earthworm populations, due to incorporation of surface litter into soil, the loss of the insulating cover on the soil surface that becomes buried by ploughing, and the redistribution of the organic food sources of the earthworms through the soil profile. Mechanical damage did not appear to decrease populations in this experiment. Zicsi (1958a) stated that cultivations after harvest had considerable effects on earthworm populations, with a mortality of 16.1% after stubble-stripping, 39.3% after summer

Table 14.1   Earthworm populations in cultivated plots (from Edwards and Lofty, 1969a)

Treatment	Earthworm populations			
	wt (g/m²)	% control	no./m²	% control
Control (unploughed)	6.28	100	4.18	100
Minimal cultivation (once spring ploughed)	6.35	101	5.06	121
Minimal cultivation (twice spring ploughed)	6.50	104	4.37	104
Maximal cultivation (once spring ploughed)	8.13	129	5.30	127
Maximal cultivation (twice spring ploughed)	7.35	117	5.55	133
Maximal cultivation (once autumn ploughed)	8.66	138	6.39	153
Maximal cultivation (twice autumn ploughed)	10.30	164	6.88	164

ploughing and 67.2% after cultivating further, but these were effects on only surface-living species of earthworms. In later work, Zicsi (1969) reached the conclusion that moderate cultivation (i.e. disk cultivation) favored earthworms by loosening the soil, and he concluded that by careful selection of cultivating machinery, earthworm numbers can actually be increased by cultivation. Some workers in the United States have postulated that decreased numbers of earthworms in regularly cultivated arable land could be prevented by leaving some remains of the crop lying on the soil to form an insulating layer (Hopp, 1946; Hopp and Hopkins, 1946b; Slater and Hopp, 1947). The evidence that they presented was fairly conclusive, but winters are usually very cold in the north-central United States, and these conclusions may not be valid for more temperate areas.

Stubble-mulch farming consists of leaving crop residues on the soil surface and working the soil underneath this layer with subsurface tillage implements called 'subtillers'. Teotia et al. (1950) reported that there were 3–5 times more earthworms in subtilled and stubble-mulched plots than in ploughed plots and this conclusion was confirmed by McCalla (1953). Graff (1969) pointed out that mulching usually favors the build-up of earthworm populations linked with the deposition of casts on the surface of arable land.

Edwards and Lofty (1979) compared the effects of four methods of handling straw residues from cereal crops, on earthworm numbers and biomass. The four treatments, used for three successive years before sampling, were:

1. straw chopped and spread evenly on soil surface;
2. straw baled and removed, stubble not burnt;
3. straw burnt in rows; and
4. straw spread evenly and burnt.

The large, deep-burrowing *L. terrestris* was common in the unburnt plots and almost completely absent in burnt plots; *A. longa*, also a deep-burrowing species, was unaffected; but numbers of *A. caliginosa*, a shallow-burrowing species, increased, more or less compensating for the decrease in *L. terrestris*. The overall effect was a considerable change in the relative proportions of the different species, a small effect on total numbers, and virtually no effect on total biomass. In the long term, continued removal of straw, by whatever means are used, must lead to eventual decreases in soil organic matter, with consequent decrease in earthworm populations. The resulting changes in relative proportions of earthworm species are likely to lead to the deterioration of soil structure, macroporosity and water infiltration. Different mulches can also cause changes in the species composition of earthworm communities. Kuhle (1983) reported that mulching with wood chips resulted in maximum earthworm abundance (540 per m², wet weight 208 g/m²), and to increased *A. caliginosa* populations (154 per m², 82.9 g/m²), compared with green grass-mulching (272–362 per m², 94–114 g/m²) and bare soil surface treatment (up to 64 per m², 9.6 g/m², with no *A. caliginosa*). *Lumbricus terrestris* populations were favored particularly by green-mulching when it was applied to a grass ground layer, whereas *Helodrilus antipae* was favored by green-mulching with grass cuttings, and was absent from plots receiving only a mulch of wood chips.

There seems to be little doubt that growing grass over several years favors increases in earthworm populations, and that the best way of maintaining a large earthworm population in regularly activated agricultural land is by including ley farming, or at least by the inclusion of some leys in the crop rotation.

Many of the beneficial effects that earthworms produce in arable soils, such as turnover of soil, breakdown of plant organic matter, nutrient cycling, aeration, drainage and general improvements in soil structure can be substituted artificially by the use of inorganic fertilizers and suitable cultivations. However, since 1973, the practice of killing the crop residue and weeds with a suitable herbicide that is inactivated when it reaches soil, as a substitute to cultivation, has been adopted increasingly in Europe and particularly in North America, as well as in parts of Africa and in South America. The crop is then sown with a special drill that cuts a slot into which the seed is inserted; this is called direct drilling in Europe and no-till in North America. After the general adoption of these practices and also in relation to the wide range of minimal or conservative cultivations that are being adopted widely in agriculture, the activities of earthworms are becoming increasingly more important in removing organic matter from the soil surfaces as well as in maintaining soil fertility and structure, and it has been demonstrated clearly that earthworm burrows aid root growth greatly, as well as improving aeration and drainage (Edwards, 1975; Wilkinson, 1975; Edwards and Lofty, 1976). Ehlers (1975) noted that the number and percentage volume of earthworm channels in untilled cropped soil doubled in 4 years in the A_0 horizon. Populations of *L. terrestris* are favored particularly by direct drilling or no-till, numbers increasing up to five times and biomass by up to eight times of those in ploughed soil; similarly, the activities of *L. terrestris* are very important in improving plant growth. If such conservation tillage practices become universally or widely adopted, much more attention will have to be given to ensuring the maintenance of earthworm populations in long-term programs of minimal cultivation.

The kinds of cultivations used for crop production have changed dramatically over the past 30–40 years, from deep, mold board ploughing to various forms of conservation tillage, including chisel ploughing and ridge tillage, and eventually to no tillage (direct drilling). No-till, direct drill and most of the other conservation tillage practices, such as chisel ploughing and ridge tillage, all favor the build-up of earthworm populations, to levels limited usually only by the availability of food (Edwards and Lofty, 1977, 1978, 1982; Barnes and Ellis, 1979; Gerard and Hay, 1979). A survey of seven cereal fields that had been direct drilled for several years and seven cereal fields that had been ploughed on the same farm and soil type showed significant increases in populations of *L. terrestris* in the direct-drilled fields (mean 28 per m^2 direct drilled v. 4.5 per m^2 ploughed), *A. longa* (mean 10 per m^2 direct drilled v. 3.2 per m^2 ploughed) and *A. chlorotica* (29 per m^2 direct drilled v. 9 per m^2 ploughed), but not of *A. caliginosa* (Edwards and Lofty, 1982).

Barnes and Ellis (1979) also showed that earthworm populations in wheat and barley croplands in England increased in direct drilled relative to ploughed plots, and that the effect increased with successive years with repeated treatments, but they were not able to show any corresponding increase in total crop yields.

Edwards and Lofty (1977, 1978, 1980) compared the effects of direct drilling (no-till) with those of ploughing on earthworm populations and on growth of barley, through six successive years of crops. They showed that overall, populations and biomass were much higher in direct-drilled (no-till) than in ploughed plots, but the effects of cultivations differed for various earthworm species. *Lumbricus terrestris* populations were 1.5–6.0 times higher and their biomass was 1.7 to 6.7 times higher in direct-drilled than in ploughed plots, whereas populations of *L. castaneus, A. caliginosa, A. longa, A. rosea, A. chlorotica* and *Octolasion cyaneum* were only 1.1–2.6

times higher and their biomass 1.2 to 3.1 times higher in direct-drilled than in ploughed plots. In longer-term experiments, earthworm populations were estimated for up to 8 years in plots or fields that were ploughed, chisel ploughed or direct drilled, and sown to cereals each year (Edwards and Lofty, 1982). Deep-burrowing species (*L. terrestris* and *A. longa*) increased in direct-drilled plots to up to 17.5 times the numbers in ploughed plots, while the shallow-burrowing species (*A. caliginosa* and *A. chlorotica*) were only 3.4 times as numerous. In chisel-ploughed plots, populations were intermediate between those in direct-drilled and those in ploughed plots.

In a comparison of earthworm populations in ploughed and reduced-tillage stubble cultivations, *L. terrestris* and *D. rubida* populations increased in the reduced tillage, but those of *A. caliginosa* did not differ (Nuutinen, 1992). The deep-burrowing species, *L. terrestris*, *A. longa* and *A. nocturna* benefitted most from the minimal cultivations. Chisel ploughing, shallow tining, harrowing and disking all seem to have relatively small effects on either deep-burrowing or shallow-working species. Thus, the increases in earthworm populations that occur under long-term conservation tillage can be large. For example, Edwards and Lofty (1982) reported a 30-fold difference between earthworm populations in ploughed and no-tilled soils after 8 years. In Nigeria, the effects of cultivations on earthworm populations were very great in one experiment; as many as 2400 earthworm casts/m^2 were counted in no-tillage plots, compared with fewer than 100 casts/m^2 in ploughed soil (Lal, 1974). There seems little doubt that adoption of no-till or conservation tillage practices is an important key to building up and maintaining earthworm populations.

14.2 THE EFFECTS OF CROPPING

It seems probable from the data available that the most important factor controlling earthworm populations in arable land is the amount of organic matter that is available as food for earthworms. For instance, it has been shown conclusively that the availability of food can limit the numbers of earthworms in grassland and arable land (Satchell, 1955a; Waters, 1955; Jefferson, 1956) (Table 7.4), and Evans and Guild (1948a) attributed the decrease of numbers of worms in arable land to the gradual decrease in the availability of food. Large populations of earthworms can be maintained in arable land if farmyard manure or other organic inputs, such as sewage sludge, are applied regularly, but exhaustive cropping without adding any organic manure usually decreases earthworm populations to a very low level, especially if the crop residues are removed (Morris, 1922; Evans, 1947b; Tischler, 1955; Barley, 1961).

It is clear that cropping can influence the numbers of earthworms in arable land considerably. Hopp and Hopkins (1946b) reported that alfalfa–grass-cropped plots contained more earthworms than lespedeza–grass-cropped plots, and that grass-cropped plots in orchards contained more earthworms than plots with timothy grass. When land was cropped on a rotation that included a pasture ley, the numbers of earthworms changed every year according to the phase of the rotation. Ponomareva (1950) measured the numbers of earthworms during each phase of a wheat, rye and 2 year grass rotation in the former USSR, and reported the largest numbers of earthworms in the second year of the ley. In a similar experiment in Australia, with a fallow, wheat and 2 year grass rotation, Barley (1959a) reported that the weight of earthworms after the fallow was only 25% of that occurring in permanent pasture, but by the end of the ley, the weight was at least 70% of that in permanent pasture.

Hopp (1946) reported that earthworm populations in the United States differed greatly under different crops, being least under row crops and greatest in plots growing winter cereal and summer legumes (Fig. 14.1). The more often row crops were grown, the smaller were the earthworm populations (Fig. 14.2). Other studies (Hopp and Hopkins, 1946b) reported larger earthworm populations under continuous maize than under continuous soybeans, and even bigger populations under continu-

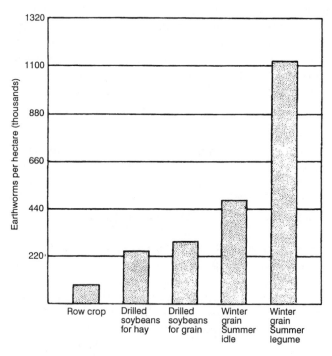

Figure 14.1   Earthworm counts in February 1946 under different kinds of annual cultures. (After Hopp, 1946.)

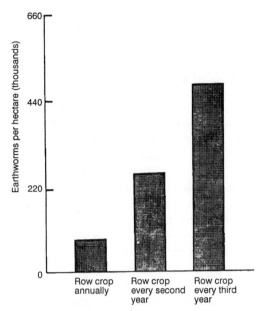

Figure 14.2 Earthworm count in February 1946, after row cropping in 1945, in plots planted to row crops annually, every second year, and every third year (after Hopp, 1946).

ous winter cereals. Earthworm numbers were as large as those in pasture, in fields with winter cereals followed by legume hay. Probably, one of the more important factors affecting the influence of cropping on earthworm populations is the proportion of the plant material that is returned to the soil after harvest. In plots at Rothamsted Experimental Station in England that had grown the same crops since 1843, the largest earthworm populations occurred under continuous cereals, with much lower populations under root crops, and the lowest of all under fallow (Edwards and Lofty, 1977). In one of the few detailed investigations of the influence of cropping on earthworm populations, the supply of organic matter to the soil was the most important factor favoring the build-up of earthworm populations. Straw residues ploughed into the soil and the growing of short-term hay crops increased earthworm populations greatly. In general, the inclusion of crops such as cereals, that leave considerable residues, encourages the build-up of earthworm populations much more than growing crops like soybeans and other legumes, which decompose quite rapidly and leave little residue (Edwards, 1983a). Perennial crops such as alfalfa or clover, when included in a rotation, are particularly beneficial to the build-up of earthworm populations, partly because of the absence of tillage and partly because of the high protein content of the residues of these crops. Root crops, where most of the crop is removed, or periods of fallow both tend to discourage the build-up of earthworm populations. Much more research is needed on the influence of cropping patterns on earthworm populations.

14.3 THE EFFECTS OF FERTILIZERS

Nearly all cropped soils in developed countries are treated with either organic or inorganic fertilizers. The effects of these fertilizers on earthworms may be direct, for instance, by changing the acidity of soil or through toxicity (e.g. the ammonium radical), or indirect, by changing the form and quantity of the vegetation that ultimately turns into decaying organic matter that provides food for earthworms. For instance, application of superphosphate and lime to a pasture caused a dense clover sward to develop, and this in turn increased the weight of earthworms in the soil about fourfold (Johnstone-Wallace, 1937). Marshall (1977) reviewed the overall effects of organic fertilizers on earthworm populations and reported that most of them increased earthworm abundance. Curry (1976) and Anderson *et al.* (1983) reported that addition of farmyard manures and animal slurries increased earthworm populations quite rapidly. The addition of a broad range of organic manures from sources such as cattle, pigs, and poultry, municipal sewage wastes, and industrial wastes from breweries, paper pulp or potato processing can all have a considerable influence on the build-up of earthworm populations in agricultural land (Edwards and Lofty, 1979). Additions of some organic materials to soils can double or triple earthworm populations in a single year. However, some liquid organic manures that have not been aged or composted can have short-term adverse effects on earthworm populations (Chapter 13) (Curry, 1976), due to their ammonia and salt contents; but populations usually recover quickly and increase thereafter.

There is good evidence that most inorganic nitrogenous fertilizers favor the build-up of large numbers of earthworms, probably due to increased amounts of crop residues being returned to the soil (Zajonc, 1975; Barnes and Ellis, 1979; Edwards and Lofty, 1979; Lofs-Holmin, 1983). Large amounts of nitrochalk applied to pastures increased the earthworm populations in them, but this was probably due to greatly increased grass production and resultant organic residues (Watkin, 1954). Jacob and Wiegland (1952) also reported increased numbers of earthworms in grassland, after application of different forms of inorganic nitrogenous fertilizers; they estimated that there were 128 earthworms per m^2 in plots without nitrogen, and 176 per m^2 in plots to which nitrogenous fertilizers had been added. Heath (1962) reported a linear correlation between the amounts of inorganic nitrogenous fertilizers used on grass leys and the weight of earthworms in the soil in these fields (Fig. 14.3). Applications of nitrochalk and nitrate of soda to grassland also resulted in increased

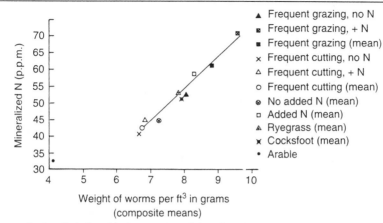

Figure 14.3   Relationship between weight of earthworm populations and amount of nitrogen mineralized on incubation of soil samples (after Heath, 1962).

earthworm populations (Escritt and Arthur, 1948). Hendrix *et al.* (1992) reported that earthworm numbers in inorganically fertilized meadows were, on average, nearly twice those in unfertilized meadows on the Georgia Piedmont in the south-eastern US. This was probably due indirectly to the increased plant growth, resulting in increased organic matter inputs, on the fertilized pastures.

Conflicting results have been reported from a long-term experiment on Park Grass, Rothamsted (Edwards and Lofty, 1975a). This is a long-term experiment with plots that have been treated regularly with a range of different fertilizers since 1856. After this extremely long exposure to a range of doses of nitrogen (0, 48, 97 and 145 kg/ha) populations of earthworms decreased to levels that were inversely proportional to the dose of nitrogen applied, but this may have also been related to the kind of fertilizer used (Edwards and Lofty, 1975a). Other workers have reported similar effects of fertilizers on enchytraeid worms (Huhta *et al.*, 1969; Axelsson *et al.*, 1973) and earthworms (Zajonc, 1970). Zajonc reported that increasing doses of nitrogen (0, 100, 200 and 300 kg/ha) decreased numbers of earthworms progressively, but that addition of phosphorus or potassium could, to some extent, ameliorate that effect.

Lime seems to be beneficial to earthworm populations (Richardson, 1938; Crompton, 1953; Jefferson, 1955; Hanschko, 1958; Marshall, 1977) and this is probably because most species of earthworms tend to avoid acidic soils, probably because they may have a need for calcium (Chapter 5). Moreover, many species of earthworms have been reported to increase in numbers after application of lime to Australian fields (Buckerfield and Doube, 1991). The introduced species *A. trapezoides* increased significantly in numbers after liming, but a native species, *Gemascolex walkeri*, was unaffected. Liming of forest soil in a mature stand of Norway spruce in

Germany caused a considerable increase in the density of earthworms, particularly *L. rubellus* (Ammer and Makeschin, 1994). The enormous increase in earthworm activity in limed plots caused a drastic change in humus forms, with much of the A_0-horizon being converted into earthworm casts after 5 years. Probably, lime is effective in providing earthworm build-up only in soils that have a pH below 4.5–5.0, because above this level, acidity does not seem to influence the size of populations of many species of earthworms very much. Crompton (1953) believed that the way that lime influenced earthworm populations was by helping to decompose organic matter into a form that earthworms could eat, but this conclusion is dubious, if earthworms are to be considered as primary litter breakdown organisms.

It is not clear what the effects of inorganic superphosphate fertilizers on earthworms are, and these effects probably vary with the soil conditions (Bachelier, 1963). Doerell (1950) and Zajonc (1970) concluded that superphosphate was very beneficial to earthworms, but Escritt and Arthur (1948) reported that this fertilizer decreased the numbers of earthworms in grass plots and so did Gerard and Hay (1979).

Most other inorganic fertilizers have little effect on earthworm populations. For instance, Uhlen (1953) could find no significant influence of commercial inorganic fertilizers on earthworms. However, small increases in numbers of worms, in plots treated with inorganic fertilizers, have been reported occasionally (Ogg and Nicol, 1945; Jacob and Wiegland, 1952).

There is good evidence that sulfate of ammonia is antagonistic to earthworm populations (Escritt and Arthur, 1948; Rodale, 1948; Jefferson, 1955; Wei-Chun, *et al.*, 1990). Slater (1954) stated that the use of sulfate of ammonia on grass plots, for 3 years in succession, decimated the earthworm population, and Richardson (1938) reported that annual treatments with sulfate of ammonia completely eliminated earthworms from plots on Park Grass at Rothamsted (Table 14.2). Earthworms react negatively to ammonia but the reason that sulfate of ammonia is unfavorable to earthworms is probably that it gradually makes soils more acid, its effect being greatest in soils that are already rather acid; this was certainly so on Park Grass at Rothamsted. It may also be due to the toxic effect of the ammonium radical. Sulfate of ammonia is sometimes so drastic in its effect on earthworms that Slater (1954) suggested it could even be used as a means of eliminating earthworms from golf courses.

14.4 THE EFFECTS OF CHEMICALS

The chemicals that reach soils include pesticides, heavy metals, polychlorinated biphenyls (PCBs) and acid precipitation. The degree of exposure of earthworms to such chemicals in soils depends upon a wide range of

Table 14.2 Wormcasts on grassland (Park Grass, Rothamsted) (from Richardson, 1938)

Treatment	Number of (thousands per acre)	No./m^2
Sulfate of ammonia + minerals	0	0
Sulfate of ammonia + minerals + lime	127	31
Nitrate of soda + minerals	153	38
Nitrate of soda + minerals + lime	161	40
Unmanured	245	60
Unmanured	294	72
Unmanured + lime	276	68
Dung	423	104
Dung + lime	337	83

variable factors that may be associated not only with the chemical, the route of exposure and the soil type, but also the environmental conditions, and the species and behavior of the earthworms.

Chemical characteristics that affect their impact upon earthworms include: their route of exposure, water solubility, volatility, lipid/water partition coefficient, adsorptive capacity and persistence in soils. Various soil characteristics can also influence the toxicity of chemicals to earthworms. A soil with a regularly low moisture content can decrease the toxicity of some chemicals through competitive adsorption on to clay colloids or organic matter, while water-soluble compounds can be leached readily from a wet soil when the adsorption is reversible. Different chemicals can bind reversibly or irreversibly with the particles of clay and organic matter in the soil, and binding of a chemical to soil will depend on the proportions and amounts in which these two soil components occur. Clay and organic matter can affect the cation exchange capacity of the soil, which also influences the binding of chemicals. The pH of the soil can influence the ionic state of the test chemical, as well as the adsorptive capacity of the soil, which in turn will affect the amount of chemical that is available to the earthworms. Soil temperature can affect not only the vapor pressure of a chemical but also its solubility in soil water, and thereby can influence the rate at which compounds are taken up into the tissue of earthworms. The loss of chemicals from soils as a vapor, or their breakdown through the action of soil micro-organisms, is also strongly dependent upon soil temperature. Micro-organisms are often instrumental in detoxification of chemicals, yet the range and activity of micro-organisms present in the soil at an experimental site is often difficult to determine and standardize.

Several aspects of the behavior of earthworms can affect the assessment of the toxicity of chemicals to earthworms in the field. The horizontal distribution of earthworms in the soil is limited by various aspects of the physico-chemical environment, such as their reactions to temperature, moisture content and pH of the soil, the availability of food and the capacity of the earthworms to reproduce and disperse from breeding sites (Chapter 6). The vertical distributions of earthworms are also affected by changes in the soil conditions, such as temperature and moisture and the ability of the earthworms to respond to these changes. Juvenile earthworms may be unable to burrow deeply into soil and can therefore be affected more severely than the adults by toxic chemicals, through greater exposure to chemicals on the soil surface. By contrast, adult earthworms can move rapidly into the safety of the deeper soil layers using their semi-permanent, mucus-lined burrows.

Different species of earthworms can be exposed to chemicals to quite different degrees and in very different ways. For instance, *L. terrestris* is often exposed to a high concentration of pesticides, because this species moves over and feeds at the soil surface. *Allolobophora caliginosa* also lives in the superficial layers of the soil and the adults may move over the soil surface in wet weather, thereby becoming particularly vulnerable to surface pesticide residues. *Aporrectodea longa* seems to be less susceptible to pesticides than many other species of earthworm, because it can burrow deeply into the soil (Wheatley and Hardman, 1968) and enters an obligatory diapause during the summer (Gerard, 1967).

14.4.1 METHODS OF TESTING EFFECTS OF CHEMICALS

There are several tests for assessing the toxicity of chemicals to earthworms in the laboratory.

Immersion test

Some reports in the literature describe testing the toxicity of chemicals to earthworms by immersing the worms in dilute solutions of chemicals for set periods of time, then transferring them to clean soil to assess mortality (Goffart, 1949; Martin and Wiggans, 1959; Edwards and Lofty, 1973; Stringer and Wright, 1973; Stenersen, 1979). Such tests can give very reproducible results and an LC_{50} for a heavy exposure can be calculated for any species of earthworm, but this dose can refer only to the concentration of a chemical in the test solution and cannot be related to a dose in soil to which earthworms may be exposed.

Topical application test

Several workers have applied chemicals to the surface of earthworms using a paintbrush or microapplicator (Aspock and van der Laan, 1963;

Stringer and Wright, 1976; Ebing and Haque, 1979; Fisher 1984), but the mucus on the earthworm's body can impair the contact between the test chemical and the earthworm. Moreover, the chemical is often sloughed off with the mucus. Such tests produce very variable results, are almost impossible to interpret in terms of field exposure of earthworms to chemicals, and have little relevance as test methods.

Injection tests

A number of workers have tested the effects of chemicals injected in an ethanolic solution into the coelomic cavity of earthworms that are then kept in soil, and subsequently assessed for mortality (Nakatsugawa and Nelson, 1972; Stenersen et al., 1973; Fisher, 1984). Administering the chemicals tends to injure the earthworm and the results from most of these tests have been very inconclusive. Often the effects of the test chemical on the earthworm cannot be distinguished from those of the solvent. The results cannot be related to field exposure to the chemical.

Forced feeding tests

This test procedure was first reported by Stringer and Wright (1976). The test chemical was suspended in 1.5% agar-agar gel and injected into the esophagus of an earthworm which had been anesthetized previously in a 10% aqueous ethanol solution. Data from such a test cannot be interpreted in relation to field exposure.

Feeding on treated food

Leaf tissues treated with chemicals can be offered to *L. terrestris*, or other species, as food (Stringer and Wright, 1973). Discs of apple leaf (1 cm^2) were presoaked in water for 2 days to soften them and then treated with the test chemical in a spray tower. The earthworms were observed for any toxic, behavioral, or antifeedant effect of the test chemical for 2 days. This may have more relevance to field exposure.

Laboratory soil test

Many workers have applied chemicals to soil and exposed earthworms to these toxicants in containers in the laboratory. Such tests have usually been performed in boxes (Hoy, 1955; Edwards and Lofty, 1973) or in pots (Caseley and Eno, 1966; Heungens, 1969a; Stringer and Wright, 1973; Atlavinyte, 1975; Agarwal et al., 1978; Kale and Krishnamoorthy, 1979; Lofs-Holmin, 1980; Hamilton and Dindal, 1989). Soil tests in containers of this kind usually provide more consistent and reproducible results than field tests, because standard numbers of a single species of earthworm can be exposed in a uniform medium, in which the worms are in intimate contact with the medium containing the chemical. However, a wide range of soils, of greatly differing adsorptive capabilities, have been used in such tests, and the conditions, doses and times of exposure have also differed considerably. One group of workers (Karnak and Hamelink, 1982) recommended a common standard loess soil as a test soil, but this is not readily available in many parts of the world.

Contact filter paper test

The basic principle of the contact filter paper test is to treat standard filter papers with chemicals, by dipping them in solutions of known concentrations, or by spraying the chemical in solution on to them with a fine chromatograph spray and then exposing single earthworms, replicated 10 times, to the chemical residues for 48 hours. It is important to obtain a uniform and adequate deposit across the surface of the filter paper, and to achieve this, the test chemical is usually dissolved in water or organic solvents (such as acetone, hexane or chloroform) if it is relatively insoluble in water. This test is an excellent screening technique to assess the relative toxicity of a range of chemicals, but does not factor in the differential adsorption of chemicals that occur in many soils.

Artificial soil test

In response to urgent requests from the European Economic Community (EEC) and the Organization for Economic Cooperation and Development (OECD), a contract from the EEC was provided to Edwards (1983a, 1984; Goats and Edwards, 1988) to develop a standardized earthworm toxicity laboratory test. In order for laboratory tests to be reproducible and to relate to field exposure, it is important that the earthworms are exposed to a test chemical in a medium that closely simulates natural field conditions. To minimize the variability in natural soils, a standardized 'artificial soil' was developed with an adsorptive capacity similar to that of a typical loam soil (25 meq). The artificial soil was prepared from a mixture of the following components:

- 10% finely ground sphagnum peat (pH 5.5–6.0)
- 20% kaolinite clay (containing >50% kaolinite)
- 69% industrial quartz sand (dominant fine sand with more than 50% of particle size 0.05–2.0 mm), ~1% calcium carbonate ($CaCO_3$ –pulverized to bring the pH of mixture to 6.0 + 0.5).

These components were mixed thoroughly and brought to 45% dry weight water content. Ten earthworms were exposed to the chemical which was dissolved in water, or an organic solvent, applied to 500 g dry weight of artificial soil, and then placed in a glass container. The treatments were replicated four times for each dose and kept for 14 days at 20 °C. The worms were then hand-sorted from the soil and their mortality assessed. If a range of doses is used then an LC_{50} can be calculated. If appropriate, earthworms can be weighed at the beginning and end of the experiment and numbers of cocoons in the soil counted to assess effects of the test chemical on growth and reproduction.

14.4.2 EARTHWORM MORTALITY FROM CHEMICALS

There is an extensive literature on the effects of pesticides on earthworms but we know much less about the effects of other chemicals. There have been a number of reviews on the toxicity of pesticides to earthworms (Satchell, 1955c; Davey, 1963; Edwards and Thompson, 1973; Thompson and Edwards, 1974; Edwards and Lofty, 1979; Dean-Ross, 1983; Edwards and Neuhauser, 1988; Lofs-Holmin and Bostrom, 1988), but only one has included the effects of other chemicals (Edwards and Bohlen, 1992). Some workers have tested a range of pesticides and other chemicals for their toxicity to earthworms (Ruppel and Laughlin, 1977; Edwards, 1983b, 1984; Haque and Ebing, 1983; Roberts and Dorough, 1984).

The toxicities of different chemicals and pesticides vary greatly (Appendix A) but some generalizations can be made:

Inorganic chemicals

Most of these, such as lead arsenate (Polivka, 1951; Escritt, 1955) and copper sulfate (Edwards, 1984) were used commonly as pesticides before the Second World War. They were moderately toxic to earthworms, but were potentially harmful in some orchard soils which contained large residues of these chemicals. These chemicals are very persistent, and when they build up in soil, can lead to long-term exposure of earthworms. None of these chemicals seems to be extremely toxic to earthworms based on the relatively sparse data available in the literature.

Organochlorine insecticides

The organochlorine insecticides, most of which are persistent in soils, were used extensively from the 1950s to the 1970s but currently only relatively small quantities are used in developed countries. Most were not very toxic to earthworms. For instance, aldrin had low toxicity to earthworms (Edwards and Thompson, 1973).

Low toxicity of BHC has been reported (Gunthart, 1947; Morrison, 1950; Prisyaznyuk, 1950; Polivka, 1951; Grigor'eva, 1952; Richter, 1953;

Weber, 1953; Lipa, 1958; Ghilarov and Byzova, 1961). Chlordane was extremely toxic to earthworms (Polivka, 1951; Schread, 1952; Hopkins and Kirk, 1957; Doane, 1962; Edwards, 1965; Lidgate, 1966; Long et al., 1967; Legg, 1968).

Many workers have studied the effects of DDT on earthworm populations, and most concluded that normal rates of application of this insecticide did not harm earthworms (Fleming and Hadley, 1945, 1950; Goffart, 1949; Polivka, 1951; Richter, 1953; Hopkins and Kirk, 1957; Barker, 1958; Martin and Wiggans, 1959; Edwards and Dennis, 1960; Ghilarov and Byzova, 1961; Stringer and Pickard, 1963; Edwards, 1965; Edwards et al., 1967a; Thompson, 1970; Perfect, 1980).

Dieldrin is chemically related to aldrin, and its toxicity to earthworms was also low (Polivka, 1951; Hopkins and Kirk, 1957; Luckman and Decker, 1960).

Endrin was reported as very toxic to earthworms (Hopkins and Kirk, 1957; Patel, 1960; Edwards and Lofty, 1973; Thompson, 1970). Endosulfan and isobenzan are moderately toxic to earthworms.

Organophosphate insecticides

In general, most organophosphate insecticides are not very toxic to earthworms. Azinphos methyl was not toxic to earthworm populations (Hopkins and Kirk, 1957), but chlorfenvinphos had slight effects (Edwards et al., 1967b); disulfoton was only slightly toxic to earthworms; and dyfonate had a slight effect on the numbers of earthworms (Edwards et al., 1971). Dursban (Whitney, 1967; Kring; 1969; Thompson, 1970), fenitrothion (Griffiths et al., 1967), malathion (Hopkins and Kirk, 1957; Voronova, 1968) and menazon (Raw, 1965) did not affect earthworm populations. Parathion was moderately toxic to earthworms (Goffart, 1949; Schread, 1952; Weber, 1953; Scott, 1960; van der Drift, 1963; Heungens, 1966). Ethopropos, fonofos and terbufos were all moderately toxic to earthworms. Sumithion and trichlorfon had little effect on earthworms. Phorate was extremely toxic to earthworms (Edwards et al., 1967b; Kelsey and Arlidge, 1968; Way and Scopes, 1968) and has almost eliminated earthworms from many soils, even at normal agricultural rates.

Carbamate insecticides and fungicides

Many workers have reported that benomyl is very toxic to earthworms (Stringer and Wright, 1973; Stringer and Lyons, 1974, 1977; Black and Neely, 1975; Heimbach and Edwards, 1983). Of the other carbamate insecticides tested for toxicity to earthworms, to date, relatively high toxicity has been reported for aminocarb, methiocarb, oxamyl and promecarb, and very high toxicity for aldicarb, bufencarb, carbaryl, carbofuran, methomyl, propoxur and thiofanox (Edwards and Thompson, 1973; Stringer and Wright, 1973, 1976; Ruppel and Laughlin, 1977; Dean-Ross, 1983; Haque and Ebing, 1983; Roberts and Dorough, 1984). It seems reasonable to predict that the majority of carbamate pesticides are toxic to earthworms.

Pyrethroid insecticides

Of the many natural or synthetic pyrethroids that have been tested, none have been demonstrated to be toxic to earthworms (Edwards and Thompson, 1973; Inglesfield, 1984; Roberts and Dorough, 1984).

Contact and fumigant nematicides/fungicides

Fumigant nematicides and fungicides, such as D–D mixture (dichloropropane–dichloropropene), metham sodium and methyl bromide, normally applied to soil to control pathogens and nematodes, are usually broad-spectrum biocides. They permeate the soil as vapors and kill most of the earthworms, even those that live in deep burrows (Buahin and Edwards, 1963). Chloropicrin is also very toxic to all earthworms (Blankwaardt and van der Drift, 1961). The contact nematicide methomyl (11.2 kg/ha), which kills earthworms through contact action, is also very toxic. There seems little doubt that the majority of fumigant and contact nematicides are toxic to earthworms.

Fungicides

None of the fungicides that have been tested were toxic to earthworms (van Leemput et al., 1989; Anton et al., 1990), with the exception of the carbamate fungicides such as benomyl, which is very toxic, and carbendazim, which is moderately toxic (Stringer and Wright, 1973, 1976).

Herbicides

Very few herbicides are directly toxic to earthworms (Edwards, 1989b), although they may exert considerable indirect effects due to their influence on weeds as a source of supply of organic matter on which earthworms feed in soil. There have been several reports that DNOC, chlorpropham, propham, dinoseb and triazine herbicides, such as simazine, have moderate effects on earthworm populations (Edwards and Thompson, 1973). Chio and Sanborn (1978) reported that L. terrestris could metabolize atrazine, chlorambar and dicamba. Haque and Ebing (1980) and Haque et al. (1982) reported that earthworms could take up metabolites of monolinuron, and Reinecke and Nash (1984) found that TCDD accumulated in earthworms.

Polychlorinated biphenyls

These chemicals are common products of many industrial processes and may be a contaminant of sewage sludge and dredged river sediments. There have been few studies on the toxicity of PCBs to earthworms (Edwards and Thompson, 1973). The main publication is by Kreis *et al.* (1987) who studied the bioconcentration of PCBs and reported higher levels in earthworms than in the soils in which they live, and this potential has been confirmed by other authors.

Heavy metals

The toxicity of heavy metals to earthworms is discussed in section 14.6. Edwards and Bohlen (1992) reviewed in detail the toxicity of about 200 chemicals, and those that were most toxic were identified (Appendix A).

14.4.3 UPTAKE OF CHEMICALS INTO EARTHWORMS

Earthworms are not very susceptible to many toxic chemicals and can live in soil containing large amounts of some toxicants that may be taken up into their tissues in large concentrations. These may be lipophilic, and gradually become absorbed from soil into the earthworm tissues, as the worms pass soil through their intestines. This may be very important ecologically, because earthworms are eaten by many species of birds and several species of mammals, and these mammals can further concentrate the pesticide in the higher trophic levels of food chains.

Organochlorine insecticides

One of the first reports on the uptake of chemicals into earthworms was by Barker (1958), who found large residues of DDT and its breakdown products in the tissues of earthworms taken from soil under elm trees that had been sprayed with DDT to control the insect vectors of Dutch elm disease. The largest amounts occurred in the crops and gizzards of the worms. Large residues of DDT in earthworms were also reported by Doane (1962), and subsequently, many other workers have found DDT residues in earthworm tissues (Stringer and Pickard, 1963; Cramp and Conder, 1965; Hunt, 1965; Davis, 1968; Davis and Harrison, 1966; Dustman and Stickel, 1966; Korschgen, 1970; Yadav *et al.*, 1976).

Wheatley and Hardman (1968) investigated the relationship between the amounts of residues of organochlorine insecticides in the tissues of earthworms and the amounts in the soil in which they were living. They found that not all species concentrated these insecticides into their tissues to the same degree, and that the largest concentration factor was found in *A. chlorotica* Savigny, a small species which lives in the surface layers of soil. Smaller concentration factors occurred in the larger species (*L. terrestris, A. longa* and *O. cyaneum* Savigny) than in the smaller species (*A. caliginosa, A. chlorotica* and *A. rosea*), but the amounts concentrated were probably correlated more with the habits of these species than their sizes. The smaller species tend to live mainly in the upper few centimeters of soil, where insecticides are most likely to occur, both in arable and orchard sites. By contrast, *L. terrestris* lives in well-defined burrows, often 1–3 m deep, and feeds mainly on surface debris, so its exposure is less. When large concentrations of residues have been applied to the soil surface, the insecticide residues in *L. terrestris* have approximated those found in other species. Davis (1971) confirmed that the uptake of DDT differs with species.

There is evidence (Wheatley and Hardman, 1968; Yadav *et al.*, 1976) that the amounts of pesticide residues in soil and those in worms are not related linearly, and there is proportionately less insecticide concentrated into worm tissues from soils containing large quantities of residues than from those with small amounts. In these studies, the degree of concentration from soil to worms changed from about five- to tenfold when pesticide residues in the soil were between 0.001 and 0.01 p.p.m., to less than unity when the concentrations exceeded 10 p.p.m. in the soil.

Worms do not concentrate all organochlorine insecticides from soil to the same degree. The concentration factors for dieldrin and DDT and its metabolites ranged from 9.0 to 10.6; whereas for aldrin it was 3.3; for endrin, 3.6; heptachlor, 3.0; and chlordane, 4.0. These agree with Wheatley and Hardman's (1968) results and also with data given by Edwards (1970a,b) and Edwards and Thompson (1973). The concentration factors reported by Gish (1970) tend to be higher than those found by other workers (Fig. 14.4).

Organophosphate insecticides

There is little evidence that earthworms can concentrate many organophosphorous insecticides from soil into their tissues (Thompson, 1973). Edwards *et al.* (1967b) reported that earthworms from plots treated with the organophosphorous insecticide, chlorfenvinphos, did not contain appreciable quantities of this insecticide or its metabolites. However, other data have shown that individuals removed 19 days after pasture plots were treated contained more of the organophosphorous insecticide dasanit and its sulfone than would be expected to occur in the soil in the same study (Thompson, 1970).

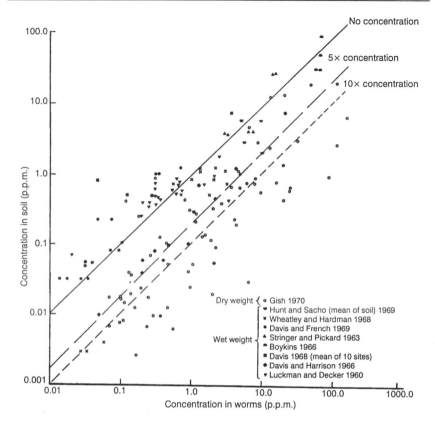

Figure 14.4   Concentration of organochlorine insecticide residues from soil to earthworms. (After Edwards and Thompson, 1973.)

Carbamate insecticides

There is evidence that soluble insecticides such as aldicarb can be bioconcentrated very rapidly into earthworm tissues in large amounts when the insecticide is applied to flooded soils, although it can concentrate only as the parent compound, not as metabolites, which are formed quite rapidly.

14.4.4 CHRONIC AND SUBLETHAL TOXICITY OF CHEMICALS

Many sublethal chronic toxicity symptoms of earthworms exposed to chemicals have been recorded in the literature (Venter and Reinecke, 1988; Edwards and Bohlen, 1992), some of which are serious and short term, others of which are minor but involve long-term effects on the earthworms and their functions.

Malformations

Many workers have reported different types of malformations of earth-worms in response to exposure to chemicals. For instance, the fungicide benomyl had teratogenic effects on the posterior segment regeneration of the earthworm *E. fetida* (Zoran *et al.*, 1986). These effects included an increased frequency of segmental groove anomalies and a variety of mon-strosities, including two tails. Various malformations occurred in response to exposure to captan (Anton *et al.*, 1990). Many carbamates have been reported to produce tumors and swellings along the earthworm's body (Stenersen, 1979).

Haque and Ebing (1983) reviewed the symptoms caused by 23 pesti-cides to earthworms and reported that the most common reaction to chemicals was coiling of the body and longitudinal muscle contraction, after which the body became rigid and sometimes swellings appeared on the body surface. The swellings often burst, creating bleeding sores. These symptoms occurred with propoxur, methidathion, endosulfan, tria-zophos, carbofuran, terbufos and methamidophos. Aldicarb, endosulfan, benomyl and calcium cyanide caused constrictions of the body to occur. Stenersen *et al.* (1973) reported swellings and tumors on earthworms exposed to carbamate insecticides such as carbofuran.

Effects on activity

Hans *et al.* (1990) reported that aldrin, endosulfan, heptachlor and lindane all produced different symptoms of chronic toxicity. Aldrin caused coiling and curling of the worms; endosulfan, excretion of mucus; heptachlor, a lifting of the body and extrusion of coelomic fluid; and lindane, glandular swelling, segmental constriction and white banding.

Earthworms in contact with imazalil showed signs of hyperactivity and became very sensitive to mechanical stimuli (van Leemput *et al.*, 1989). Similarly, low doses of phosphamidon caused hyperactivity of *L. mauritii* (Bharathi and Subba Rao, 1984). By contrast, some pesticides, such as car-bofuran and carbaryl, produced a long-lasting immobility and rigidity in all species tested (Stenersen, 1979). These responses must be related to the mode of action of the chemical.

Effects on nervous and physiological functions

There is an accumulating body of evidence that chemicals can have vari-ous drastic effects on the nervous system of earthworms. For instance, earthworms exposed to benomyl by contact showed sublethal neurotoxic effects (Drewes *et al.*, 1987; Drewes and Callahan, 1988). Locomotory reflexes were impaired by benomyl at doses approximately two orders of magnitude less than the LC_{50} (Zoran *et al.*, 1986). Benomyl was extremely toxic to *L. terrestris* (Stringer and Wright, 1973). However, the toxicity of benomyl is not due to acetylcholinesterase (AChE) inhibition (Stringer and Wright, 1976). Indeed, benomyl is converted rapidly to carbendazim under field conditions; and the latter compound possesses no cholinergic activity (Krupka, 1974). Since neither benomyl nor its conversion product, carbendazim, are cholinergic, the toxicity of benomyl must be due to some mode of action other than through AChE inhibition. Both carbamates and organophosphates have been shown to inhibit AChE activity (Edwards and Fisher, 1991), and von Niklas (1979) reported that carbamates inhibited AChE more than did organophosphates. The organophosphate insecticides phosphamidon, monocrotophos and dichlorvos all inhibited AChE activity in *L. mauritii* (Bharathi and Subba Rao, 1984). However, carbamates were demonstrably more toxic to *L. terrestris* than organophosphate compounds possessing comparable cholinergic activity (Ruppel and Laughlin, 1977; Stenersen, 1979; Roberts and Dorough, 1984). Since the two classes of compounds share AChE inhibition as their primary mode of action, the greater toxicity of carbamates to *L. terrestris* than organophosphate compounds has led to the suggestion that carbamates may affect additional target sites in the earthworm.

Effects on growth

It is difficult to interpret results from experiments on the effects of chemicals on growth of earthworms because laboratory experiments have often tested the effects of chemicals on earthworms without supplying food. Earthworm species differ greatly in their sources of food and growth patterns (Edwards and Lofty, 1977) and any effect on growth is usually most common in rapidly growing species. For instance, loss of weight or slowing of growth in response to chemicals has been reported much more often for the fast-growing *E. fetida* than for the slower-growing *L. terrestris* or *A. caliginosa*.

Neuhauser *et al.* (1984) tested the effects of five different heavy metals, mixed with manure, on the growth of *E. fetida*, and found that all metals reduced growth relative to that in the controls. Cadmium had the greatest effect. Worms that were removed from the contaminated manure and placed in uncontaminated manure for 6 weeks showed compensatory growth, so that by the end of the experiment, their mass was not significantly different from the control worms. These results indicate that sublethal effects of heavy metal contaminants on earthworm growth will disappear once the contaminants are removed. This suggests the possibility of using earthworm growth responses to monitor the remediation of contaminated soils.

Effects on reproduction

There is gradual accumulation of evidence that sublethal doses of chemicals can have considerable effects on earthworm reproduction. Cocoon production and hatching of E. *fetida* were influenced by pentachlorophenol at 10 mg/kg in soil, but not by copper and dichloroaniline (van Gestel *et al.*, 1989). Copper at low concentrations increased cocoon production by D. *rubida* at 100 mg/kg. Bengston *et al.* (1986) and van Leemput *et al.* (1989) reported that copper increased cocoon production of earthworms. Neuhauser *et al.* (1984) tested the effects of five heavy metals on cocoon production in E. *fetida* and found that cocoon production decreased with increasing concentration of all the metals. Abbasi and Soni (1983) reported that chromium and mercury enhanced the reproduction of O. *pattoni*. Polychlorinated biphenyls (PCB) increased the number of infertile earthworm cocoons at doses above 12 mg/kg (van Leemput *et al.*, 1989). Clitellum development and cocoon production were completely inhibited in earthworms exposed to doses of carbofuran in soil greater than 2 mg/kg (Bouwman and Reinecke, 1987). Cocoon production in E. *fetida* was inhibited by cadmium and zinc at 2.5 p.p.m. or greater (Neuhauser *et al.*, 1984). Ivermectin, a broad-spectrum antiparasitic drug for livestock, the use of which has expanded greatly in the past decade, caused a 56% decline in cocoon production in E. *fetida* in soil containing 4 mg/kg ivermectin (Gunn and Sadd, 1994). This result indicates that ivermectin may have a deleterious effect on earthworm populations in pastureland.

14.5 THE EFFECTS OF RADIOISOTOPES

There has been considerable pollution of soil by radioactive fallout, and it has been suggested that the dispersal of radioactive contaminants in soil is accelerated by the ability of living organisms to accumulate isotopes in their tissues. There is evidence (Peredel'sky *et al.*, 1957; Peredel'sky, 1960b) that worms are important in dispersing radioisotopes (^{66}Co) through soil.

There have only been two detailed experimental studies of the effects of radiation on earthworms; Edwards (1969) studied the effects of treating individuals of A. *chlorotica* with doses of 5, 10, 25, 50, 100 and 200 kilorad (kr) of γ-radiation (Fig. 14.5).

After treatment with 200 kr all worms died within a few hours but all other doses except 100 kr caused little mortality for 3 months, although thereafter the worms began to die quite rapidly. All worms treated with 100 kr were dead within 20 weeks of irradiation. Some of the worms treated with the other doses were still alive after 25 weeks (Lofty, 1972).

Crossley *et al.* (1971) studied the uptake of ^{137}Cs by L. *terrestris* and *Octolasion* sp., and reported that about 12% of the amount consumed was

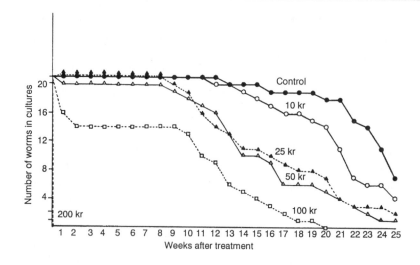

Figure 14.5   Effects of radiation on *A. chlorotica* in cultures (after Lofty, 1972).

assimilated. Maldague and Couture (1971) made similar studies of the uptake of ^{59}Fe by *L. terrestris*, with remarkably good agreement with the data of Crossley *et al.* (1971). Both groups of workers reported that these isotopes did not persist long in the bodies of earthworms, so the movement of isotopes through the soil by the agency of earthworms seems more likely to be a form of passive transport with the soil in the gut. Krivolutsky *et al.* (1982) suggested that earthworms were one of the best indicators of radioactive soil pollution.

14.6 HEAVY METALS AND EARTHWORMS

Environmental pollution by heavy metals is increasing rapidly. The most important pollutants are cadmium (Cd), lead (Pb), copper (Cu), mercury (Hg), zinc (Zn), nickel (Ni), antimony (Sb) and bismuth (Bi), but many other metals can also cause pollution. The environmental pollution can arise from atmospheric contamination from emissions from industrial plants, contamination of roadsides from release of lead in exhaust fumes of motor vehicles using leaded fuel, the spreading on agricultural land of sewage sludge and slurries of animal dung containing copper and other metals, and landfill disposal of industrial and municipal wastes. Hughes *et al.* (1980) reviewed atmospheric heavy metal pollution of terrestrial ecosystems.

14.6.1 TOXICITY OF HEAVY METALS TO EARTHWORMS

Relatively few data are available on the toxicity of heavy metals to earthworms, and much more research is needed (Pietz *et al.*, 1984). Van Rhee (1975) reported that 110 p.p.m. copper and 1100 p.p.m. of zinc in soil were toxic to earthworms. By comparison, copper toxicity resulting from long-term use of copper-based fungicides in orchards completely eradicated earthworms where copper levels in the soil were >80 p.p.m. (van Rhee, 1963, 1975). Large amounts of pig wastes containing >1000 p.p.m. copper, which had been applied to pasture soils in Holland, caused a progressive decline in earthworm populations (van Rhee, 1977).

Soni and Abbasi (1981) reported chromium to be extremely toxic to the earthworm *P. posthuma*. These authors found that mercury was about 20 times more toxic than chromium (Cr) to the earthworm species *O. pattoni* (Abbasi and Soni, 1983).

It has been reported that copper affected earthworm populations at concentrations above 287 p.p.m., cadmium at 33 p.p.m. and lead at 4800 p.p.m. Bengtsson *et al.* (1983) provided comparable data for significant effects on earthworm populations at 78 p.p.m. for copper, 171 p.p.m. for zinc, and 36 p.p.m. for lead. Wentsel and Gueltar (1987) reported a 14 day LC_{50} of 190 p.p.m. for a mixture of 20% copper and 30% zinc (brass).

It is apparent from such variable data that probably the chemical form in which toxic heavy metals are presented to earthworms is an important factor in determining the level at which toxicity appears. Lee (1985) suggested that the differences between the relative toxicity of the compounds tested may explain some of the conflicting data in the literature on the concentrations that have deleterious effects on earthworms. For instance, the very high concentrations of lead reported to influence growth and reproduction of earthworms in some studies, may be due more to the very low solubility of lead compounds that are found in soils and the ability of earthworms to sequester absorbed lead, than to any lower toxicity of lead compared with other heavy metals. Clearly there is a need for much further testing of the toxic levels of the different heavy metals to earthworms.

14.6.2 SUBLETHAL AND CHRONIC EFFECTS OF HEAVY METALS ON EARTHWORMS

Sublethal levels of heavy metals can affect earthworm growth and reproduction. Such sublethal effects may be valuable indicators of the overall impact of metal contaminants on earthworm populations (Ma, 1984; Wentsel and Gueltar, 1987). Malecki *et al.* (1982) examined the effect of metal acetate and chloride salts on the reproduction of a surface-feeding species of earthworm *E. fetida*, and found that its reproduction was inhibited at 50 mg cadmium/kg soil, 400 mg nickel/kg, 2000 mg copper/kg and

>5000 mg/kg zinc. Hartenstein *et al.* (1980) determined the concentrations at which heavy metals added to activated sludge would affect the growth of *E. fetida*. They reported that manganese, chromium and lead were innocuous to earthworm growth even at 52 000 mg/kg and that growth inhibition for ionic forms of copper, nickel, zinc and cadmium started at 1100, 1200, 1300 and 1800 mg/kg, respectively.

Some of the effects of copper, lead, nickel, cadmium and zinc on the growth and cocoon production of *E. fetida* were assessed by Malecki *et al.* (1982) in laboratory cultures. The body weights of *E. fetida* in untreated controls after 4, 6 and 8 weeks were compared with those in cultures to which were added a series of concentrations of acetates, carbonates, chlorides, nitrates, oxides and sulfates of the five metals. Carbonates and oxides were less toxic than the other compounds, possibly because of their relatively low solubility. Cadmium was the most toxic, producing significant decreases in earthworm growth rate at 100 p.p.m. cadmium acetate and chloride, and at 50 p.p.m. for cadmium sulfate; minimum concentrations of the other metals that significantly retarded growth were 200 p.p.m. for nickel (as chloride), 100 p.p.m. for copper (as nitrate), 2000 p.p.m. for zinc (as chloride or nitrate) and 12 000 p.p.m. for lead (as acetate). Cocoon production was inhibited totally by the acetates of these five metals at concentrations of 50 p.p.m. for cadmium, 400 p.p.m. for nickel, 2000 p.p.m. for copper and at ≥5000 p.p.m. for zinc and lead.

Van Rhee (1975) kept *A. caliginosa* for 8½ weeks in soils to which were added various heavy metal contaminants, and reported that copper at 110 p.p.m. did not affect earthworm body weight or mortality, but caused a slight reduction in the rates of cocoon production; zinc at 1100 p.p.m. reduced the body weight by about 50%, stopped cocoon production and inhibited development to maturity; a mixture of 10 p.p.m. mercury and 20 p.p.m. of cobalt caused increases in body weight and rates of development to maturity and reduced the rate of cocoon production by about 65%; a mixture of all four elements at the same rates caused a reduction of about 34% of body weight, as well as cessation of development and cocoon production, and 22% mortality, about the same as that due to addition of zinc alone. These levels of heavy metal contamination are comparable with those in mud containing industrial wastes from the harbor bottom at Rotterdam, which may be put on land.

Neuhauser *et al.* (1984) studied the growth and reproduction of the earthworm *E. fetida* during and after exposure to sublethal concentrations of five metals: cadmium, copper, lead, nickel and zinc. Most concentrations of heavy metals led to a decreased rate of earthworm growth. However, *E. fetida* was capable of compensatory growth which allowed it to recover to approximately the control mass values after the metal was removed from its food source. *Eisenia fetida* appeared able to compensate for growth much more than for reproduction. This study demonstrated that earthworms affected by heavy metals have the ability to resume normal growth and reproductive processes when they are removed from an environment that is highly contaminated with metals and placed in an environment that does not contain excessive metal concentrations. The test procedures exposed the earthworms to concentrations similar to those they might experience under natural conditions at a spill and/or a hazardous waste site with subsequent reclamation and clean-up activity.

Abbasi and Soni (1983) reported enhancement of reproduction in *O. pattoni* in terms of greater production of juveniles in response to exposure to chromium and mercury, at levels ranging from 0 to 25 p.p.m. Clearly, we do not understand the potential sublethal effects of heavy metals on earthworms very well, and much more research on this subject is needed.

14.6.3 ACCUMULATION OF HEAVY METALS BY EARTHWORMS FROM CONTAMINATED SOILS

There is abundant evidence in the literature that earthworms can accumulate heavy metals from polluted soils and other media (Gish and Christensen, 1973; van Hook, 1974; Ireland, 1979, 1983; Ash and Lee, 1980; Beyer, 1981; Beyer et al., 1982; Eijsackers, 1982; Kruse and Barrett, 1985). They probably accumulate much higher levels of metals than most other soil animals (Martin and Coughtrey, 1975; Beyer et al., 1982). Many of the terrestrial investigations of bioaccumulation have concerned heavy metal accumulation into earthworms from mine spoil tips (Ireland, 1979), from land contaminated by the emissions from metal smelting or refining works (Wright and Stringer, 1980; Bengtsson et al., 1983; Ma et al., 1983), from motor vehicle emissions (Ash and Lee, 1980), from municipal waste sites (Fleckenstein and Graff, 1982), from metals deliberately applied to the soil as pesticides (Ma, 1984), from sediments dredged from rivers (Marquenie and Simmers, 1988), from sewage sludge (Beyer et al., 1982) or from animal manures (Edwards and Neuhauser, 1988). The uptake of lead from soil into earthworms seems to be related to that of calcium. It is known that lead can enter calcium metabolic pathways in vertebrates (Morgan, 1986). Cadmium and zinc can concentrate into earthworms from soil more than other heavy metals (Beyer et al., 1982) but less calcium is taken up in the presence of high levels of zinc.

Most studies of heavy metal accumulation by earthworms have involved mainly lead, cadmium, copper and zinc, but there are data available on a wide range of other elements. Analyses have sometimes been made of total heavy metal contents and sometimes of water extracts or acid extracts as an indication of those elements that are available to plants and as some measure of the significance of earthworms in the transfer of heavy metals from contaminated soil into plants and through food chains.

Lee (1985) pointed out that the significance of the concentration of plant-available heavy metals in earthworm tissues is difficult to assess. Earthworms feed on organic debris as well as on soil and, although plant roots can absorb nutrients from dead earthworm bodies and might thus be expected to absorb the available heavy metals, the total amounts involved must be very small compared with the total heavy metal contents of soils (Neuhauser *et al.*, 1995).

On the basis of the total heavy metal content of earthworms relative to soil (Table 14.3), the bioconcentration factor for lead is generally in the range 0.1–1.0 (Ireland, 1975). For cadmium, bioconcentration factors are usually in the range of 11–22 in relatively unpolluted soils, with soil cadmium levels about 0.2–0.8 p.p.m. and earthworm levels about 3.0–9.0 p.p.m., but as soil cadmium levels rise, the concentration factor tends to fall. For zinc, bioconcentration tends to increase with increasing pollution, but not proportionally so, and the bioconcentration factor for many heavy metals in earthworms tends to fall as soil concentrations rise.

There is no doubt that different earthworm species differ in their capability to take up heavy metals in their tissues. For instance, Ireland (1979) compared concentrations of lead, copper, cadmium, zinc, manganese and calcium in the tissues of *L. rubellus*, *D. veneta* and *E. tetraedra* from three soils with varying concentrations of these elements. *Lumbricus rubellus* had the highest concentrations of zinc and manganese, *D. veneta* of cadmium, and *E. tetraedra* of lead. The concentrations of copper, zinc and manganese in earthworms appeared to be regulated more effectively than those of the other metals; cadmium was concentrated more by earthworms than any of the elements tested, especially by *D. veneta*. In other work Ireland (1983) recorded the lead and zinc levels in tissues of *D. rubidus* and *L. rubellus* from a highly contaminated site; for lead, *D. rubidus* had about twice the level that *L. rubellus* had, whereas for zinc the situation was reversed. He suggested that these differences could be due to differences in food selection. Terhivuo *et al.* (1994) reported interspecific differences in the lead concentrations and lead concentration factor in different lumbricid species living near a lead smelter in southern Finland. They also reported intraspecific differences in lead accumulation, indicating that worms living in contaminated soils may become acclimatized to heavy metal pollution. They suggested that species, such as *A. caliginosa*, that accumulate and tolerate high lead concentrations in their tissues, might be the best species for biomonitoring.

Part of the significance of the accumulation of heavy metals by earthworms is the potential for movement into the higher tropic levels in food chains. Earthworms are preyed on extensively by birds, amphibians, reptiles, mammals and other vertebrates, so that heavy metals accumulated in their tissues or contained in their intestines are readily passed on in food chains. The heavy metals that are absorbed by earthworms from

Table 14.3 Selected studies on accumulation of heavy metal by earthworms

Species	Metal	Amount in soils (p.p.m.) (S)	Amount in earthworm (p.p.m.) (E)	Bioconcentration factor (E/S)	Reference
L. rubellus	Cd	0.08	15.05	188.0	Kruse and Barrett (1985)
A. tuberculata		0.6	3.8	6.3	Pietz et al. (1984)
A. tuberculata		1.0	12.0	12.0	Pietz et al. (1984)
L. rubellus		7.3	66.0	9.0	Morgan et al. (1986)
A. tuberculata		0.3	4.4	14.6	Beyer et al. (1987)
Mixed spp.		0.35	5.7	16.2	Hook (1974)
L. rubellus	Zn	111.37	942.36	8.3	Kruse and Barrett (1985)
A. tuberculata		50.0	174.0	3.5	Pietz et al. (1984)
A. tuberculata		72.0	264.0	3.7	Pietz et al. (1984)
L. rubellus		427.0	1187.0	2.8	Morgan et al. (1986)
Mixed spp.		43.0	317.0	7.4	Hook (1974)
L. rubellus	Cu	10.45	11.40	0.9	Kruse and Barrett (1985)
A. tuberculata		22.0	7.0	0.32	Pietz et al. (1984)
A. tuberculata		24.0	6.4	0.27	Pietz et al. (1984)
A. tuberculata		6.1	4.8	0.79	Beyer et al. (1987)
L. rubellus	Pb	14.43	2.54	0.18	Kruse and Barrett (1985)
A. tuberculata		18.0	0.20	0.01	Pietz et al. (1984)
A. tuberculata		13.0	0.44	0.04	Pietz et al. (1984)
L. rubellus		8740.0	696.0	0.08	Morgan et al. (1986)
A. tuberculata		16.0	6.9	0.43	Beyer et al. (1987)
Mixed spp.		27.0	4.7	0.20	Hook (1974)

their food may accumulate, both in earthworms and their predators, if either lacks adequate biochemical or physiological mechanisms to eliminate them rapidly from their bodies.

14.7 ACID DEPOSITION AND EARTHWORMS

Acid deposition from industrial sources has become widespread in many parts of Europe and North America. There are considerable data on what it does to trees but not much on how it affects soils and the organisms that live in them, including earthworms.

Very acidic conditions have been reported to decrease the abundance of many species of lumbricid earthworms, and earthworms tend to occur in low numbers in most naturally acid forest soils (Edwards and Lofty, 1975a; Ma, 1984; Kuperman, 1993). Edwards and Lofty (1975a) reported that not all species of earthworms responded to changes in soil pH in the same way, and none of the species they studied could tolerate a pH below 4.0. They stated that individuals of *L. terrestris* became progressively more numerous as the soil pH increased from 4.0 to 7.5, whereas those of *A. nocturna*, *A. caliginosa* and *A. rosea* tended to have an optimal pH range of 5.0–6.0, decreasing markedly in numbers at either higher or lower pH levels. Satchell (1955a,d) reported the lower pH tolerance limit for *A. nocturna*, *A. caliginosa* and *A. rosea* as 4.6 and that for *L.terrestris* and *O. cyaneum* as 4.1, whereas Edwards and Lofty (1975b) considered that the lower tolerable limit of pH for most species was 4.2.

A single experimental field treatment with sulfuric acid (pH 3.0) decreased the overall earthworm numbers significantly under a pine plantation. However, it was impossible to determine whether the earthworms migrated from the acid-treated plots or simply moved deeper into their burrows. A few species of earthworms are relatively tolerant of soil acidity (Edwards and Lofty, 1977). Huhta *et al.* (1969) reported no effects of experimental manipulations of pH on earthworm populations in coniferous forest soils in Finland, but the number of earthworms in their study sites was low. In laboratory experiments, which tested the effects of acidification on *E. fetida*, 100% mortality occurred at pH levels below 5.0 or higher than 9.0 (Kaplan *et al.*, 1980). The abundance of earthworms decreased significantly in oak–hickory forests which had received large amounts of acidic deposition over the past several decades in southern Ohio (USA), compared with numbers in ecologically similar forests in southern Illinois, that had received much smaller amounts of acidic deposition (Kuperman, 1993). Ammer and Makeschin (1994) reported that frequent applications of simulated acid precipitation (pH 2.7–2.8) to soils of a mature Norway spruce stand caused significant decreases in earthworm populations; *L. rubellus* and *D. rubida* disappeared entirely. Clearly, some species of earthworms can survive at pH levels down to 4.0, but many are favored by higher pH levels and each species seems to have its individual optimal pH range (Chapter 10). Ma *et al.* (1990) have assessed the effects of acidification due to nitrogenous fertilizers, such as mineral ammonium sulfate and sulfur-coated ureas on earthworms, and reported drastically decreased populations, particularly of endogeic species. Much more data are needed before meaningful conclusions on the effects of acidic deposition on earthworms can be made.

Appendix A
Summary of results of laboratory and field experiments on testing the toxicity of chemicals to earthworms

Chemical tested	Rate in field (kg/ha)[a] or laboratory* (mg/kg)	Species tested[b]	Relative toxicity[c]	References[d]
Inorganic chemicals				
Calcium arsenate	424–508	–	0	62,104
Calcium cyanamide	120*	A,B	NS	82
Copper chloride	370	C	NS	110
Copper oxychloride	80–90*	A,D		82,136
Copper sulfate	108	A	0	40,41,63,112,133
Lead arsenate	244–1307	–	0	32,62,63,104,126
Sulfur	678	–	2	63
Mercuric chloride	8.5	–	NS	63
Potassium bromide	393	B	2	40,41,83
Potassium permanganate	169–376	–	0	63,128
Sodium chlorate	150–5700	A,B	0	20,82,113
Biological agents				
Enterobacterin	0.6–60	A,D	0	8,12,144
Mowrah meal	1017–2712	–	0	32,33,63
Rotenone	6.8–508	–	0	32,33,63
Mustard	NR	–	NS	63
Aromatic and organochlorine insecticides				
Aldrin	0.2–33	A,B,D,E,F,L	0	4,27,38,43,47,49,56, 59,79,81,85,86,102, 125, 143
Aramite	9.76	–	NS	89
BHC	0.24–500	A,B,C,D,G,H	0	19,33,38,49,59,64, 80,82,86,88,105,122, 141,145
Chlordane	4.5–19.2	A,B,D,E,F	4	35,38,40,41,49,59,76, 77,83,86,102,104,139, 144

Chemical tested	Rate in field (kg/ha)[a] or laboratory* (mg/kg)	Species tested[b]	Relative toxicity[c]	References[d]
DDE	61*	A	NS	27
DDT	2.2–60	A,B,C,D,E,F	0	9,31,35,38,43,54,56, 59,67,78,86,88,114, 116,126,141,153,154
Dicofol	3.0	–	NS	
Dieldrin	2.25–80	A,B,D	0	32,35,49,59,86,102, 110,143
DNOC	NR	B,G	1	54,64,83
Endosulfan	2.2	A,B,L	2	19,61,81,82,83,138
Endrin	1.0–9	A,B	4	27,59,60,61,90,125, 153,155
Heptachlor	1.4–8	B,L	4	14,38,45,54,59,81,86
Isobenzan	2.2	–	3	38,41,95,121
Lindane	1.0–100*	A,B,L	NS	81,82
Naphthalene	6000*	D	NS	2,4
1-Naphthol	–	B	NS	138
Telodrin	2.2–4.5	–	1	38,94
Tetradifon	2.2	–	0	59
Toxaphene	10.8	B	2	86,102

Organophosphorous insecticides

Acephate	–	B,C	NS	138
Azinphos methyl	0.06*	B	NS	86
Bromophos	270	B	NS	19
Carbophenothion	NR	–	NS	59
Chlorfenvinphos	4.0–4.5	–	1	53,57,59
Chlofmephos	41*	B	NS	19
Chlorpyrifos	0.45–4.48	A,B,C	3	98,127,138,139,153, 154
Chlorpyrifos-ethyl	205	B	NS	19
Demeton-S-methyl	NR	–	NS	34
Dialifos	133	B	NS	83
Diazinon	2.25–8.0	A,D	1	49,53,54,57,59,98, 127,139
Dimethoate	18.2	G	1	13,64
Disulfoton	4.0–8.0	A	1	49,53,58,59,98,139
Ethoprophos	4.0–5.6	A,B	3	115,124,127,140
Ethyl-parathion	4.0–32	A,B,C,D,E,G	3	27,34,53,54,55,59,64, 89,125,138,143,145
Fenamiphos	8–18.6	A,B,D,I	1	118,120,139
Fenitrothion	1.5–4.0	A,D,F	0	49,53,59,79,114
Fensulfothion	2–4.48	A,D	1	69,98,114,121,139, 153,154
Fonofos	NR	A,B,C	2	49,59,138,139
Formothion	200*	D	1	2,7
Isazophos	2.24	D	3	127
Isofenphos	0.45–2.24	–	1	65,127
Leptophos	3.4	–	NS	155
Malathion	NR	A,B,C,D	1	71,116,138
Menazon	NR	A,D,G	0	59,132,134,158,159
Methamidophos	17.3–109*	A,B	NS	82
Methaphenamiphos	318*	B	NS	83
Methidathion	3.6*	B	NS	83
Methyl-parathion	20–37*	–	NS	89
Monocrotophos	NR	B,D,G	0	49,61

Chemical tested	Rate in field (kg/ha)[a] or laboratory* (mg/kg)	Species tested[b]	Relative toxicity[c]	References[d]
Paraoxon	35–200*	A,D,G	NS	145
Phorate	1.0–44.8	A,C,D,J	4	53,57,59,61,113,139, 142,155,158,159
Phosalone	16–150*	A,B,D	1	3,7,19,139
Phosphamidon	16.5–27.0*	K	NS	13
Terbufos	2.8	A,B	2	19,82,139
Tetrachlorvinphos	8.0	–	0	49,60,61
Thionazin	NR	A,D,F	0	49,53,59,79,158
Triazophos	210*	A	NS	82
Trichloronate	2.0–8.36	B,C,D,G	0	69,145,154
Trichlorfon	4.0	D	1	2,3,7,49,53,71,127
Carbamate insecticides				
Aldicarb	4.0–11.2	A,B,C,D,G	4	34,48,60,64,82,99, 138,139,140,145
Aminocarb	0.18	–	NS	22
Bendiocarb	4.48	–	4	127
Bufencarb	1.0–4.48	A	3	98,139,153,154
Carbaryl	0.2–25	A,B,C,D,F G,M,P,U,V	4	2,4,19,27,34,40,41,49, 53,59,66¹,76,77,83,85, 92,93,102,123,127,138, 145,153,154
Carbofuran	1.1–4.48	A,B,C,D,G	3	19,21,24,61,68,75,82, 87,98,99,114,138,139, 140,145,146,153,154
Ethiofencarb	262	B	NS	82
Methiocarb	1.06	A,B,C,D,G	3	9,10,19,49,61,64,83,151[g]
Methomyl	3.4–11.2	A,B,C,D	4	48,49,60,138,139,155
Oxamyl	12.7	A,B,C,D,G	2	118,139,145
Promecarb	16*	A	NS	139
Propoxur	2.0	A,B,C	3	70,84,138,139
Thiofanox	3.5–67*	A,B	3	19,87,99
Synthetic pyrethroids and chitin inhibitors				
Alphamethrin	100*	B	NS	91
Cypermethrin	100*	B	.1	91,138
Diflubenzuron	NR	–	0	50
Fenvalerate	0.1	A,B,D,J	0	108,138
Permethrin	–	B	NS	138
Pyrethrins (unspecified)	5*	B	NS	19
Soil fumigants and nematicides				
Chloropicrin	4.5	D,G,N	4	18,136
Dazomet	364	A,B	2	19,48,49,60
DBCP	0.57–0.75	B	0	84,85,120
Ethylene dibromide	66.2*	B	NS	120
1,2-Dichloropropane	3880–4272*	B,P,U,V	NS	123
Dichloropropane– dichloropropene	400–600	B	4	23,25,59,120
Formaldehyde	60–5050	D,G,N	2	18,49,129
Metham sodium	NR	–	NS	59
Methyl bromide	NR	–	NS	59
Fungicides				
2-Aminobutane	–	A	NS	150

Chemical tested	Rate in field (kg/ha)[a] or laboratory* (mg/kg)	Species tested[b]	Relative toxicity[c]	References[d]
Aniyaline	85.4	B	1	97,137
Benomyl	0.28–360	A,B,D,F,G, H,J,O,P	4	15,16,17,19,30,42,46, 49,61,64,82,83,94,96, 97,106,107,108,125, 138,147,148,149[h],150[g], 155,156,161[g,h],162
Bupirimate	338*	B	NS	83
Captafol	496–800*	A,B	0	82,83
Captan	4.73–18.9	A,B,D,F	1	1,30,49,82,83,101,116, 147
Carbendazim	0.15–0.56	A	3	30,96,107,149[h],150[g], 161[g],162
Chlorthalonil	1.25–64	A,B,D,J	2	97,108,127,137
Dichloran	2.0	–	0	49
Dinocap	2.6*	B	NS	137
Ethazole	42.0*	B	NS	137
Fenaminosulf	83.9*	B	NS	137
Fenarimol	3.02	–	NS	127
Folpet	338–459*	A,B	NS	82
Fuberidazole	–	A	NS	150[g]
Imazalil	–	B	NS	100,101
Mancozeb	136.5	B	0	97,137
Maneb	43·9*	G	NS	64
Propiconazole	3.36	–	NS	127
Quintozene	5.6	–	NS	49
Thiabendazole	15.9–9.5	A	3	17,137,149[h],150[g],161[g,h], 162
Thiophanate-methyl	0.78–9.0	A,B,C,F,G	3	19,30,77,97,137,138, 147,149[h],161[g,h],162
Thiram	6.7	B	0	19,49,137
Triadimefon	0.125–3.02	A,B,D,J	0	81,108,127
Triforine	–	B	NS	36
Ziram	169*	B	NS	19
Herbicides				
Aminotriazole	64–100*	A,D,Q,R,S	0	26,74,115
Asulam	100*	D	NS	115
Atrazine	8.0	A,B,D,Q,R,S	0	26,69,74,82,115
Aziprotryne	1.0–100*	D	NS	115
Bromacil	NR	A,D,Q,T	0	26,115,152
Chlormequat chloride	460	A,B	NS	82
Chloroacetamide	50	A,B,D	4	40,41,42,83,91
Chlorpropham	1.0–6.4*	A,Q	1	11,26
Chlorthiamid	8.96–16	–	2	51,52
Chlortoluron	6100*	B	NS	19
Cyanazine	2.0–4.0	A	0	59,60,61
Cycloate	3.0*	–	NS	103
2,4-D	2.24	A,B,D,R,S	1	74,115,116,127,138
Dalapon	20.0	A,D,Q	0	3,26,56,69,115
Di-allate	1.4*	–	NS	103
Dicamba	0.56	–	NS	127
Dinoseb	RR1	A,G	2	64,160
Diphenamid	16–32*	A,Q	NS	26
Diquat	258*	B	NS	19
Diuron	16–100*	A,D,Q	0	26,115
Endothal	64*	A,Q	NS	26

Chemical tested	Rate in field (kg/ha)[a] or laboratory* (mg/kg)	Species tested[b]	Relative toxicity[c]	References[d]
Glyphosate	NR	D,T	0	115,152
Hexazinone	100*	D	NS	115
Lenacil	4.0*	–	NS	103
Linuron	0.84–4.5	D	0	39,44,115
Maleic hydrazide	6.4	–	0	111
MCPA	1.68–3.0	D	0	39,44,115
MCPB	100*	D	NS	115
Mecoprop	NR	–	NS	56
Methabenzothiazuron	100–215*	D,G	0	64,115
Metribuzin	100*	D	NS	115
Monolinuron	280*	A	NS	82
Monuron	10.0	A,D,R,S	0	69,74,116
Nitrofen	166*	G	NS	64
Oxadiazon	100*	D	NS	115
Paraquat	11.4–200 (RR)	A,B,D,Q,T	0	26,39,42,46,60,138, 152
Pendimethalin	3.36	–	NS	127
Pentachlorophenol	10–75	A,B,F	4	40,41,72,73,76,77,83
Phenmedipham	1*	–	NS	103
Prometryn	100*	D	NS	115
Propazine	100*	D	NS	115
Propham	100*	NS	1	11
Pyrazone	16–48*	G	1	64,103
Sesone	64*	A,Q	NS	26
Simazine	1.68–3.0 (NR)	D,R,S	2	5,39,44,71,72,90,115
Sodium trichloroacetate	NR	D	0	2,4,71
2,4,5-T	–	B	NS	138
2,3,6-TBA	64*	A,Q	NS	26
Terbacil	100*	D	0	115
Trichlopyr	0.56	–	NS	127
Trichloroacetic acid	80	A,D,G,J	1	6,40,41,69,106
Tri-allate	1.68–3.0	B	0	19,39,44,103
Trifluralin		B,D	0	19,115
Other organic compounds				
3-Chlorophenol	0.1–1000*	A,B	4	72,73
Dichloroaniline	0.1–1000*	A,B	NS	73
3,4-Dichlorophenol	0.1–1000*	A,B	4	72,73
Dimethyl phthalate	1064–3335*	B,P,U,V	NS	123
Fluorene	170–206*	B,P,U,V	NS	126
Hexoestrol	16.7	A,D,G	NS	130
p-Nitrophenol	40–56*	B,P,U,V	4	123,138
N-Nitrosodiphenylamine	5.0*	B,P,U,V	NS	123
Dioxin	0.0	C,D,J	2	117,135
Nitrobenzene	226–362*	B,P,U,V	NS	123
Phenol	188–450*	B,P,U,V	NS	123
2,3,4,5-Tetrachlorophenol	0.1–1000*	A,B	3	72,73
1,2,3-Trichlorobenzene	0.1–1000*	A,B	NS	72,73
1,2,4-Trichlorobenzene	127–251*	B,P,U,V	NS	123
2,4,5-Trichlorophenol	0.1–1000*	A,B	4	72,73
2,4,6-Trichlorophenol	–	B,P,U,V	4	123

[a]Not reported (NR), recommended rate (RR).
[b]*Lumbricus terrestris* Savigny (A), *Eisenia fetida* Savigny (B), *Lumbricus rubellus* Hoffmeister (C), *Aporrectodea caliginosa* Savigny (D), *Pheretima hupeiensis* Michaelsen (E), *Aporrectodea longa* Ude (F), *Allolobophora chlorotica* Savigny (G), *Helodrilus roseus* Beddard (H), *Eiseniella tetrahedra*

Savigny (I), *Aporrectodea rosea* Savigny (J), *Lampito mauritii* Kinberg (K), *Pheretima posthuma* Vaillant (L), *Pontoscolex corethrurus* Muller (M), *Dendrobaena rubida* Savigny (N), *Octolasion* spp. (O), *Diplocardia* spp. (P), *Eudrilus eugeniae* Kinberg (Q), *Pheretima californica* Kinberg (R), *Alma* sp. (S), *Pheretima divergens* Michaelsen (T), *Aporrectodea tuberculata* Eisen (U), *Perionyx excavatus* Michaelsen (V), species mixture or not reported (*).

[c]Relatively nontoxic (0), slightly toxic (1), moderately toxic (2), very toxic (3), extremely toxic (4), insufficient evidence to categorize (NS).

[d]1. Anton *et al.* 1990; 2. Atlavinyte 1975; 3. Atlavinyte 1981; 4. Atlavinyte *et al.* 1974; 5. Atlavinyte *et al.* 1977; 6. Atlavinyte *et al.* 1978; 7. Atlavinyte *et al.* 1980; 8. Atlavinyte *et al.* 1982; 9. Baker 1946; 10. Baker 1983; 11. Bauer 1964; 12. Benz and Altwegg 1975; 13. Bharathi and Subba Rao 1984; 14. Bigger and Decker 1966; 15. Black and Neely 1975a; 16. Black and Neely 1975b; 17. Blackshaw 1980; 18. Blankwaardt and van der Drift 1961; 19. Bouché 1974; 20. Bouché and Beugnot 1978; 21. Bouwman and Reinecke 1987; 22. Bracher and Bider 1982; 23. Brande van den and Heungens 1969; 24. Broadbent and Tomlin 1982; 25. Buahin and Edwards 1963; 26. Caseley and Eno 1966; 27. Cathey 1982; 28. Clements and Henderson 1977; 29. Clements *et al.* 1982; 30. Cook and Swait 1975; 31. Cook *et al.* 1980; 32. Davey 1963; 33. Dawson *et al.* 1938; 34. Dikshith and Gupta 1981; 35. Doane 1962; 36. Drandarevski *et al.* 1977; 37. Drift van der 1963; 38. Edwards 1965; 39. Edwards 1970; 40. Edwards 1983a; 41. Edwards 1984; 42. Edwards 1985; 43. Edwards and Arnold 1963; 44. Edwards and Arnold 1964; 45. Edwards and Arnold 1966; 46. Edwards and Brown 1982; 47. Edwards and Jeffs 1965; 48. Edwards and Lofty 1971; 49. Edwards and Lofty 1973; 50. Edwards and Lofty 1976; 51. Edwards and Stafford 1976; 52. Edwards and Stafford 1979; 53. Edwards and Thompson 1969; 54. Edwards and Thompson 1973; 55. Edwards *et al.* 1966; 56. Edwards *et al.* 1967a; 57. Edwards *et al.* 1968a; 58. Edwards *et al.* 1968b; 59. Edwards *et al.* 1971; 60. Edwards *et al.* 1972; 61. Edwards *et al.* 1974; 62. Escritt 1955; 63. Escritt and Arthur 1948; 64. Fayolle 1979; 65. Finlayson *et al.* 1975; 66. Fisher 1984; 67. Fleming and Hadley 1945; 68. Flickinger *et al.* 1980; 69. Fox 1964; 70. Fox 1974; 71. Galvyalis and Lugauskas 1978; 72. Gestel van and Ma 1988; 73. Gestel van and Ma 1990; 74. Ghabbour and Imam 1967; 75. Gilman and Vardanis 1974; 76. Goats 1983; 77. Goats and Edwards 1988; 78. Greenwood 1945; 79. Griffiths *et al.* 1967; 80. Grigoreva 1952; 81. Hans *et al.* 1990; 82. Haque and Ebing 1983; 83. Heimbach 1984; 84. Heungens 1968; 85. Heungens 1969a; 86. Hopkins and Kirk 1957; 87. Houpert *et al.* 1982; 88. Hoy 1955; 89. Hyche 1956; 90. Ilijin 1969; 91. Inglesfield 1984; 92. Kale and Krishnamoorthy 1979; 93. Kale and Krishnamoorthy 1982; 94. Karnak and Hamelink 1982; 95. Kelsey and Arlidge 1968; 96. Keogh and Whitehead 1975; 97. King and Dale 1977; 98. Kring 1969; 99. Lebrun *et al.* 1981; 100. Leemput van *et al.* 1989; 101. Leger and Millette 1977; 102. Legg 1968; 103. Lhoste 1975; 104. Lidgate 1966; 105. Lipa 1958; 106. Lofs-Holmin 1980; 107. Lofs-Holmin 1982b; 108. Lofs-Holmin 1982a; 109. Lofs-Holmin 1982b; 110. Luckman and Decker 1960; 111. Lyons *et al.* 1972; 112. Ma 1984; 113. Malone and Reichle 1973; 114. Martin 1976; 115. Martin 1982; 116. Martin and Wiggans 1959; 117. Martinucci *et al.* 1983; 118. McColl 1984; 119. Medts de 1981; 120. Milne and du Toit 1976; 121. Moeed 1975; 122. Morrison 1950; 123. Neuhauser *et al.* 1986; 124. Niklas and Kennel 1978; 125. Patel 1960; 126. Polivka 1951; 127. Potter *et al.* 1990; 128. Randall *et al.* 1973; 129. Raw 1959; 130. Raw 1960a; 131. Raw 1961; 132. Raw 1965; 133. Raw and Lofty 1962; 134. Raw and Lofty 1964; 135. Reinecke and Nash 1984; 136. Rhee van 1969c; 137. Roark and Dale 1979; 138. Roberts and Dorough 1984; 139. Ruppel and Laughlin 1977; 140. Ruppel *et al.* 1973; 141. Satchell 1955b; 142. Saunders and Forgie 1977; 143. Schread 1952; 144. Smirnoff and Heimpel 1961; 145. Stenerson 1979; 146. Stenerson *et al.* 1973; 147. Stringer and Lyons 1974; 148. Stringer and Lyons 1977; 149. Stringer and Wright 1973; 150. Stringer and Wright 1976; 151. Stringer and Wright 1980; 152. Takahashi and Sakai 1982; 153. Thompson 1971; 154. Thompson and Sans 1974; 155. Tomlin and Gore 1974; 156. Tomlin *et al.* 1981; 157. Walton 1928; 158. Way and Scopes 1965; 159. Way and Scopes 1968; 160. White 1980; 161. Wright 1977; 162. Wright and Stringer 1973.

[f]Includes topical application test.

[g]Includes forced feeding tests.

[h]Includes laboratory leaf disk/antifeedant tests.

(Adapted from Edwards and Bohlen, 1992.)

Appendix B
Glossary

Acanthodriline Having the male pores in segment 18 and the prostatic pores in segments 17 and 19.

Aclitellate Without a clitellum. Not necessarily confined to juvenile stages.

Amphimictic Sexual reproduction in earthworms involving two parents.

Anecic earthworms Earthworms that live in permanent vertical burrows and are characterized by medium to heavy dorsal pigmentation, large size, long generation time, and a diet of surface litter (cf. epigeic, endogeic).

Anthropochorous Transported by the agency of man (usually unintentional).

Astomate The condition in which the nephridia are closed, i.e. without a nephrostome.

Autotomy The process of self induced loss of segments – as when a worm is gripped by the tail.

Bioaccumulation The concentration of a given persistent compound or toxin by an organism so that the level in the body exceeds that in the organism's food-source or surrounding environment.

Bioindicator An organism or taxon used to monitor the biological status of the environment or ecosystem in which it lives.

Biomass The total mass of a population or group of organisms.

Brown bodies Rounded bodies in the coelomic fluid containing tissue debris such as setae and corpuscles, and foreign material.

Calciferous glands Specialized glands, usually situated in evaginations or folds in the esophageal wall, that open into the oesophagus and play a role in the calcium metabolism of earthworms.

Cast Earthworm feces.

Chaeta, chaetae Alternative for seta, setae.

Cladogram A branching diagram representing phylogenetic relationships among taxonomic groups of organisms with a common evolutionary origin.

Classical system The classification of the Oligochaeta originated by Michaelsen in 1900 and developed by Stephenson in 1930 in *The Oligochaeta*.

Clitellate Possessing a clitellum. Also used to describe the stage in a worm's growth during which it has a clitellum, when it is a more specific term than 'adult'.

Clitellum   A region of epidermal thickening, containing gland cells which secrete the cocoon material.

Copulatory chamber   An invagination of the body wall into the coelom which contains the male pore.

Copulatory pouches   An old term for spermathecae.

Copulatory setae   Setae in the same segment as, and near to, the spermathecae.

Denitrification   The biological reduction of nitrate or nitrite to gaseous nitrogen either as molecular nitrogen or oxides of nitrogen.

Diapause   As applied to earthworms, a non-active state during which the worm has an empty gut and stays tightly coiled in a mucus lined cell within the soil to protect itself against adverse environmental conditions. It may be optional or obligatory.

Drilosphere   The zone of soil affected by earthworm burrows, including the burrow lining and the narrow band of surrounding soil.

Endemic   Indigenous, native.

Endogeic earthworms   Earthworms that burrow through the upper mineral soil and are characterized by no or light pigmentation, medium size, intermediate generation time and a diet of mineral soil enriched in organic matter (cf. anecic, epigeic).

Epigeic earthworms   Earthworms that live on the soil surface and are characterized by dark pigmentation, small to medium size, short generation time, and a diet of surface litter (cf. anecic, endogeic).

Epilobous   The condition when the continuation of the prostomium q.v. backwards does not reach the first segmental groove.

Estivation   A summer diapause.

Genital tumescences   Raised areas of the epidermis from which genital setae grow, found in the Lumbricidae.

Geophagous   Having a diet consisting mainly of mineral soil.

Hemerophilic   Not adversely affected by human interference with the environment.

Hemerophobic   Adversely affected by human interference with the environment.

Heteromorphic   The condition when regeneration occurs so that a head regenerates instead of a tail, and vice versa.

Holandric   Testes restricted to segments 10 and 11, or a homeotic equivalent.

Hologynous   Ovaries restricted to segments 12 and 13, or a homeotic equivalent.

Homeotic   The state in which an organ or series of organs are in a segment or segments in which they are not normally found.

Humification   The breaking down of large particles of organic matter into complex amorphous colloids containing phenolic materials.

Humus   The dark-coloured, major fraction of organic matter formed during decomposition of organic residues, which is relatively resistant to decomposition and contains organic acids and other unknown constituents.

Immobilization   The conversion of an element from an inorganic form to an organic form by microorganisms or plants.

Intersegmental furrow   The actual boundary between two adjacent segments.

Intersegmental groove   The visible annular depression that indicates externally the position of the boundary between two adjacent segments.

Juvenile   The term used for earthworms from the time of emergence from the cocoon to when the first indications of maturity, such as genital tumescences, seminal grooves and genital pores appear.

Lumbricine setal arrangement   Having four pairs of setae per segment.

Macroic   Large. A substitute for meganephridial.

Male ducts   Male gonoducts. Sperm ducts q.v.

Male funnels   Funnel shaped internal end of sperm duct.

Male pores   Exterior openings of the sperm ducts.

Meganephridia   Large nephridia. A term now little used, the condition meganephridial now being encompassed by the term holonephric.

Megascolecoid   Referring to worms placed in the classical family Megascolecidae.

Meronephridia   Divided nephridia, individual tubules often present in large numbers. Can be either large – megameronephridia, or small – micromeronephridia.

Mesohumic   Endogeic earthworms that feed on both mineral and organic particles in the upper 10–15 cm of soil (cf. polyhumic, oligohumic).

Micronephridia   Small nephridia, usually numerous, often encompassed by the term meronephric.

Mineralization   The conversion of an element from an organic form to an inorganic form.

Mor   A type of forest soil in which there is little mixing of surface organic matter with mineral soil, with a sharp transition between the surface organic horizon and underlying mineral horizon.

Mull   A type of forest soil in which the surface horizon consists of organic matter and mineral soil thoroughly mixed together, with a gradual transition to the underlying horizon.

Mycorrhiza   The association between a fungus and the root system of a plant.

Oligohumic   Tropical endogeic earthworms that feed on soil of the deep horizons (30–40 cm deep) that is poor in organic matter (cf. mesohumic, polyhumic).

Oviducts   Female gonoducts.

Parthenogenesis   Reproduction in which the ova develops without being fertilized by a spermatozoa, hence it only involves one parent.

Peregrine   Earthworm species that are dispersed over a wide range of geographically distant localities; the agency of dispersal to new localities is usually transportation by humans.

Perichaetine setal arrangement   Many setae arranged in a ring right round each segment with only a break in the mid-dorsal and mid-ventral regions.

Peristomium   The most anterior segment of an earthworm surrounding the mouth, and which bears the prostomium.

Phylogeny, phylogenetic   Pertaining to past evolutionary development as opposed to individual development (autogeny).

Polyandric   Having testes in more than segments 10–11.

Polyhumic   Endogeic earthworms that feed on soil with a high organic matter content (cf. mesohumic, oligohumic).

Polymorphic   Pertaining to polymorphism. Polymorphs arising from parthenogenesis and geographical distribution are considered important in systematics.

Proandry, proandric   Testes restricted to segment 10 or a homeotic equivalent.

Progyny, progynous   Ovaries restricted to segment 12, or a homeotic equivalent.

Prolobous   The condition when there is no continuation of the prostomium backwards into the peristomium.

Prostate   Glands associated with the male ducts, usually tubular, opening beside the male pores in acanthodrilid, octochaetid and ocnerodrilid worms.

Prostomium   A lobe-like appendage attached to the front of the dorsal aspect of the peristomium.

Racemose   Applied to prostates when those organs are divided into many lobes (cf. a bunch of grapes).

Seminal grooves   Permanent longitudinal grooves on the ventral surface of an earthworm associated with the male and prostatic pores which form a passage for seminal and other fluids.

Seminal vesicles   Septal pockets in which spermatozoa mature.

Seta, setae   Stiff bristle-like structures protruding from follicles through the epidermis. Usually sigmoid in shape, except specialized setae. May be enlarged at the extreme portions of the body.

Somatic   Pertaining to any part of the body except the genital organs.

Sperm ducts   Ducts carrying sperm to the body's exterior.

Stomate   Pertaining to an 'open' nephridium, i.e. a nephridium with a funnel opening (usually) to the exterior.

Tanylobous   The condition when the continuation of the prostomium reaches backwards to the first segmental groove.

Taxon   Any unit in a system of classification.

Testis sac   Closed part of the coelom containing the testes and male funnels of a particular segment.

Vermicompost   Compost generated from the processing of organic waste materials by earthworms.

Vermicomposting   The composting of organic wastes by the action of earthworms.

Vermiculture   The intensive cultivation of earthworms for use as fish bait, feed supplement, or as a byproduct of vermicomposting (q.v.).

Vermistabilization   The stabilization or processing of sewage sludge or other organic wastes by vermicomposting.

Water-stable aggregate   A soil aggregate that is stable to the action of water, such as by wet-sieving.

Zygolobous   Condition when the prostomium as seen from above, is not differentiated from the peristomium by any superficial markings

Bibliography

Abbasi, S. A. and Soni, R. (1983) Stress-induced enhancement of reproduction in the earthworm *Octochaetus pattoni* exposed to chromium and merandy: implications in environmental management. *International Journal of Environmental Studies*, **22**, 43–7.

Abrahamsen, G. (1972) Ecological study of Lumbricidae (Oligochaeta) in Norwegian coniferous forest soils. *Pedobiologia*, **12**, 267–81.

Adu, J. K. and Oades, J. M. (1978) Utilization of organic materials in soil aggregates by bacteria and fungi. *Soil Biol. Biochem.*, **10**, 117–22.

Agarwal, G. W., Rao, K. S. K. and Negi, L. S. (1958) Influence of certain species of earthworms on the structure of some hill soils. *Curr. Sci.*, **27**, 213.

Agarwal, H. C., Yadav, D. V. and Pillai, M. K. K. (1978) Metabolism of ^{14}C-DDT in *Pheretima posthuma* and effect of pretreatment with DDT, Lindane, and Dieldrin. *Bull. Environ. Contam. Toxicol.*, **19**, 295–9.

Aichberger, R. von (1914a) Untersuchungen über die Ernährung des Regenwurmes. *Ztsch. Deutsch. Mikrob. Gesell.*, **58**, 69–72.

Aichberger, R. von (1914b) Studies on the nutrition of earthworms. *Kleinwelt*, **6**, 53–8, 69–72, 85–8.

Aisyazhnyuk, A. A. (1950). Use of 666 for the control of chafer grubs. *Agrobiologiya*, **5**, 141–2.

Aldag, R. and Graff, O. (1974) Einfluss der Regenwurmtatigkeit auf Proteingehalt und Proteinqualität junger Haferpflanzen. *Z. Landw. Forsch.*, **31**(11), 277–84.

Aldag, R. and Graff, O. (1975) N-Fraktionen in Regenwurmlösung und deren Ursprungsboden. *Pedobiologia*, **15**, 151–3.

Allee, W. C., Torvik, M. M., Lahr, J. P. and Hollister, P. L. (1930) Influence of soil reaction on earthworms. *Physiol. Zool.*, **3**, 164–200.

Allen, R. W. (1960) Relative susceptibility of various species of earthworms to the larvae of *Capillaria annulata*. *Proc. Helminthol. Soc. Washington*, **17**, 58–64.

Allevi, L., Citterio, B. and Ferrari, A. (1987) Vermicomposting of rabbit manure: modifications of microflora, in *Compost: Production, Quality and Use*, (eds M. De Bertoldi, M. P. Ferranti, P. L'Hermite and F. Zucconi) pp. 115–126.

Alonso, J. C., Alonso, J. A. and Carrascal, L. M. (1991) Habitat selection by foraging white storks, *Ciconia ciconia*, during the breeding season. *Can. J. Zool.*, **69**(7), 1957–62.

Altemuller, H. J. and M. Joschko (1992) Fluorescent staining of earthworm casts for thin section microscopy. *Soil Biol. Biochem.*, **24**, 1577–82.

Ammer, S. and Makeschin, F. (1994) Auswirkungen experimenteller saurer Beregnung und Kalkung auf die Regenwurmfauna (Lumbricidae, Oligochaeta) und die Humusform in einem Fichtenaltbestand (Höglwaldexperiment). *Forstw. Cbl.*, **113**, 70–85.

Anderson, C. W. (1993) The modulation of feeding behavior in response to prey type in the frog *Rana pipiens*. *J. Exp. Biol.*, **179**, 1–11.

Anderson, J. M., Ineson, P. and Huish, S. A. (1983) Nitrogen and cation mobilization by soil fauna feeding on leaf litter and soil organic matter from deciduous woodlands. *Soil Biol. Biochem.*, **15**, 463–7.

Anderson, J. M., Proctor, J. and Vallack, H. W. (1983) Ecological studies in four contrasting lowland rain forests in Gunung Mulu National Park, Sarawak. III Decomposition processes and nutrient loss from leaf litter. *J. Ecol.*, **71**, 503–27.

Anderson, N. C. (1983) Nitrogen turnover by earthworms in plots treated with farmyard manure and slurry, in *Earthworm Ecology: From Darwin to Vermiculture*, (ed. J. E. Satchell), Chapman & Hall, London, pp. 139–50.

Anstett, M. (1951) Sur l'activation macrobiologique des phénomènes d'humification. *C.R. Hebd. Seanc. Acad. Agric. France*, 230.

Anton, F., Laborda, E. and Laborda, P. (1990) Acute toxicity of the fungicide captan to the earthworm *Eisenia foetida* (Savigny). *Bull. Environ. Contam. Toxicol.*, **45**, 82–7.

Appelhof, M. (Ed.) (1981) *Workshop on the Role of Earthworms in the Stabilization of Organic Residues.* Vol. 1. *Proceedings*, Beech Leaf Press, Kalamazoo, Michigan.

Arbit, J. (1957) Diurnal cycles and learning in earthworms. *Am. Assoc. Adv. Sci.*, **126**, 654–5.

Arldt, T. (1908) Die Ausbreitung der terricolen Oligochaeten im Laufe der erdgeschichtlichen Entwicklung des Erdreliefs. *Zool. Jahrb. Syst.*, **26**.

Arldt, T. (1919) *Handbuch der Palaeogeographie*. Leipzig.

Armstrong, M. J. and Bragg, N. C. (1984) Soil physical parameters and earthworm populations associated with opencast coal working and land restoration. *Agric. Ecosyst. Environ.*, **11**(2), 131–43.

Arrhenius, O. (1921) Influence of soil reaction on earthworms. *Ecology*, **2**, 255–7.

Arthur, D. R. (1965) Form and function in the interpretation of feeding in lumbricid worms. *Viewpoints in Biology*, **4**, 204–51.

Ash, C. P. J. and Lee, D. L. (1980) Lead, cadmium, copper, and iron in earthworms from roadside sites. *Environ. Pollut. Ser. A.*, **22**, 59–67.

Aspock, H. and van der Laan, H. (1963) Ökologische Auswirkungen und Physiologische Besonderheiten des Pflanzenschutzmittels Sevin (1-Naphthyl-N-methylcarbamate). *Z. Angew. Zool.*, **50**, 343–80.

Atkin, L. and Proctor, J. (1988) Invertebrates in the litter and soil on Volcan Barva, Costa Rica. *J. Trop. Ecol.*, **4**, 307–10.

Atlavinyte, O. (1964) Distribution of earthworms (Lumbricidae) and larvae of insects in the eroded soil under cultivated crops. *Pedobiologia*, **4**, 245–50.

Atlavinyte, O. (1965) The effect of erosion on the population of earthworms in the soils under different crops. *Pedobiologia*, **5**, 178–88.

Atlavinyte, O. (1971) The activity of Lumbricidae, Acarina and Collembola in the straw humification process. *Pedobiologia*, **II**, 104–15.

Atlavinyte, O. (1974) Effect of earthworms on the biological productivity of barley. *Inst. Zool. Parasit. Acad. Sci. Lithuania*, **I**(65), 69–79.

Atlavinyte, O. (1975) *Ecology of Earthworms and their Effect on the Fertility of Soils in the Lithuanian SSR*, Mokslas Publishers, Vilnius.

Atlavinyte, O. (1981) The effect of pesticides on the abundance of mites and Collembola during the process of decomposition of organic substances in *Noveishie Dostizheniya Sel Skokhozyaistvennoi Entomologii Publ Po Materialam Ush Sezda Veo*, (ed. V.P. Semyanov). Vilnius, USSR, pp. 515–19.

Atlavinyte, O. and Daciulyte, J. (1969) The effect of earthworms on the accumulation of vitamin B$_{12}$ in soil. *Pedobiologia*, **9**, 165–70.

Atlavinyte, O. and Lugauskas, A. (1971) The effect of Lumbricidae on soil microorganisms. *Ann. Zool. Ecol. Anim. Special Publ.*, **4**, 73–80.

Atlavinyte, O. and Payarskaite, A. I. (1962) The effect of erosion on earthworms (Lumbricidae) during the growing season. *Zool. Zh.*, **41**, 1631–6.

Atlavinyte, O. and Pociene, C. (1973) The effect of earthworms and their activity on the amount of algae in the soil. *Pedobiologia*, **13**, 445–55.

Atlavinyte, O. and Vanagas, J. (1973) Mobility of nutritive substances in relation to earthworm numbers in the soil. *Pedobiologia*, **13**, 344–52.

Atlavinyte, O. and Vanagas, J. (1982) The effect of earthworms on the quality of barley and rye grain. *Pedobiologia*, **23**, 256–62.

Atlavinyte, O., Baydonaviciene, Z. and Budaviciene, I. (1968) The effect of Lumbricidae on the barley crops in various soils. *Pedobiologia*, **8**, 415–23.

Atlavinyte, O., Daciulyte, J. and Lugauskas, A. (1971) Correlations between the number of earthworms, microorganisms, and vitamin B$_{12}$ in soil fertilized with straw. *Liet. TSR Mokslu. Akad. Darb., Ser. B.*, **3**, 43–56.

Atlavinyte, O., Daciulyte, J. and Lugauskas, A. (1974) The effect of herbicides and insecticides on populations and activities of soil organisms. *Dinamika Mikrobiologicheskikh Protesessov v Pochve Obuslovlivayushchie Ee Faktory. Materialy Simpoziuma*, Kharkov, 1974. Akademiya Nauk Estonskoi SSR, Tallin, USSR, pp. 137–40.

Atlavinyte, O., Daciulyte, J. and Lugauskas, A. (1977) The effect of Lumbricidae on plant humification and soil organism biocenoses under application of pesticides. *Ecol. Bull.*, **25**, 222–8.

Atlavinyte, O. Daciulyte, J. and Lugauskas, A. (1978) The influence of TCA on soil organisms and the accumulation of vitamin B$_{12}$ in soil. *Microbiologicheskie Protesessov v Pochve I urozhainost Sel Skokhozyaistvennoi Kultur. Materialy K Respublikanskoi Konferentsii*, Vilnius, 1978, Akademiya Nauk Lithuanskoi, Vilnius, USSR, pp. 30–1.

Atlavinyte, O., Galvelis, A., Daciulyte, J. and Lugauskas, A. 91980) Accumulation of organophosphorus insecticides in earthworms and reaction of earthworms and micro-organisms to these substances. *Soil Biology as Related to Land Use Practices. Proceedings of the VII International Soil Zoology Colloquium* (ed. D.L. Dindal). Office of Pesticide and Toxic Substances, US Environmental Protection Agency, Washington, DC, pp. 13–24.

Atlavinyte, O., Galvelis, A., Daciulyte, J. and Lugauskas, A. (1982) Effects of enterobacteria on earthworm activity. *Pedobiologia*, **23**, 372–9.

Avel, M. (1959) Classe des Annélides Oligochaetes (Oligochaeta. Huxley, 1875). *Traite de Zool.*, **5**, 224–71.

Axelsson, B., Gardefers, D., Lohm, U. *et al.* (1971) Reliability of estimation of a standing crop of earthworms by hand sorting. *Pedobiologia*, **11**, 338–40.

Axelsson, B., Lohm, U., Lundkurst, H. *et al.* (1973) Effects of nitrogen fertilization on the abundance of soil fauna populations in a Scots pine stand. *Res. Notes Inst. Växtekologi Marklära*, **14**, 18.

Ayres, I. and Guerra, R. A. T. (1981) Water as a limiting factor in the distribution of earthworms (Annelida, Oligochaeta) in Central Amazonia. *Acta Amazon*, **11**, 77–86.

Bachelier, G. (1963) *La Vie Animale dans les Sols*, Orston, Paris.

Bahl, K. N. (1919) On a new type of nephridia found in Indian earthworms of the genus *Pheretima*. *Q. J. Micros. Sci.*, **64**.

Bahl, K. N. (1922) On the development of the 'enteronephric' type of nephridial system found in Indian earthworms of the genus *Pheretima*. *Q. J. Micros. Sci.*, **66**, 49–103.

Bahl, K. N. (1927) On the reproductive processes of earthworms: Pt I. The process of copulation and exchange of sperm in *Eutyphoeus waltoni*. *Q. J. Micros. Sci.*, **71**, 479–502.

Bahl, K. N. (1947) Excretion in the Oligochaeta. *Biol. Rev.*, **22**, 109–47.

Bahl, K. N. (1950) *The Indian Zoological Memoirs. I. Pheretima*, 4th edn, Lucknow Pub. House, Lucknow.

Baker, G. H. (1983) Distribution, abundance and species associations of earthworms (Lumbricidae) in a reclaimed peat soil in Ireland. *Holarctic Ecol.*, **6**, 74–80.

Baker, G. and Barrett, V. (1994) *Earthworm Identifier*, CSIRO, Australia.

Baker, G. H., Barrett, V. J., Grey-Gardner, R. and Buckerfield, J. C. (1992a) The life history and abundance of the introduced earthworms *Aporrectodea trapezoides* and *Aporrectodea caliginosa* in pasture soils in the Mount Lofty Range, South Australia. *Aust. J. Ecol.*, **17**(2), 177–88.

Baker, G., Buckerfield, J., Grey-Gardner, R. *et al.* (1992b) The abundance and diversity of earthworms in pasture soil in the Fleurieu Peninsula, South Australia. *Soil Biol. Biochem.*, **24**, 1389–95.

Baker, G. H., Barrett, V. J., Grey-Gardner, R. and Buckerfield, J. C. (1993a) Abundance and life history of native and introduced earthworms (Megascolediae and Lumbricidae) in pasture soils in the Mount Lofty Range, South Australia. *Trans. Roy. Soc. S. Aust.*, **117**(1-2), 47–53.

Baker, G. H., Barrett, V. J., Carter P. J. *et al.* (1993b) Seasonal changes in the abundance of earthworms (Lumbricidae and Acanthodrilidae) in soils used for cereal and lucerne production in South Australia. *Aust. J. Agric. Res.*, **44**(6), 1291–301.

Baker, G. M. (1982) Short-term effects of methiocarb formulations on pasture earthworms (Oligochaeta: Lumbricidae). *NZ J. Exp. Agric.*, **10**, 309–11.

Baker, W. L. (1946) D.D.T. and earthworm populations. *J. Econ. Ent.*, **39**, 404–5.

Bakhtin, P. U. and Polsky, M. N. (1950) The role of earthworms in structure formations of sod-podzolized soils. *Pochvovedenie*, 487–91.

Baldwin, F. M. (1917) Diurnal activity of the earthworm. *J. Anim. Behav.*, **7**, 187–90.

Ball, R. C. and Curry, L. L. (1956) Culture and agricultural importance of earthworms. *Mich. Stat. Univ. Agr. Exp. Stn Coop. Ext. Soc. Circ. Bull.*

Baltzer, R. (1956) Die Regenwürmer Westfalens. Eine tiergeographische, okologische und sinnesphysiologische Untersuchung. *Zool. Jahrb. Syst.*, **85**, 355–414.

Baluev, V. K. (1950) Earthworms of the basic soil types of the Iranov region. *Pochvovedenie*, 487–91.

Bano, K. and Dale, R. D. (1988) Reproductive potential and existence of endoge-
nous rhythm in reproduction of earthworm *Eudrilus eugeniae*. *Proc. Zool.
Soc.(Calcutta)*, **38**, 9–14.

Bano, K. and Kale, R. D. (1991) Earthworm fauna of southern Karnataka, India, in
Advances in Management and Conservation of Soil Fauna, (eds G. K. Veeresh, D.
Rajagopal and C. A. Viraktamath), Oxford and IBH, New Dehli, pp. 627–34.

Barker, R. J. (1958) Notes on some ecological effects of DDT sprayed on elms. *J.
Wildl. Manage.*, **22**, 269–74.

Barley, K. P. (1959a) The influence of earthworms on soil fertility. II. Consumption
of soil and organic matter by the earthworm *Allolobophora caliginosa. Aust. J.
Agric. Res.*, **10**, 179–85.

Barley, K. P. (1959b) Earthworms and soil fertility. IV. The influence of earthworms on the physical properties of a red-brown earth. *Aust. J. Agric. Res.*, **10**,
371–6.

Barley, K. P. (1961) The abundance of earthworms in agricultural land and their
possible significance in agriculture. *Adv. Agron.*, **13**, 249–68.

Barley, K. P. and Jennings, A. C. (1959) Earthworms and soil fertility. III. The influence of earthworms on the availability of nitrogen. *Aust. J. Agric. Res.*, **10**,
364–70.

Barley, K. P. and Kleinig, C. R. (1964) The occupation of newly-irrigated lands by
earthworms. *Aust. J. Sci.*, **26**, 290.

Barnes, B. T. and Ellis, F. B. (1979) The effects of different methods of cultivation
and direct drilling and of contrasting methods of straw dispersal on populations of earthworms. *J. Soil Sci.*, **30**, 669–79.

Barois, I. (1992) Mucus production and microbial activity in the gut of two species
of *Amynthas* (Megascolecidae) from cold and warm tropical climates. *Soil Biol.
Biochem.*, **24**, 1507–10.

Barois, I. and Lavelle, P. (1986) Changes in respiration rate and some physico-
chemical properties of a tropical soil during transit through *Pontoscolex corethrurus* (Glossoscolecidae, Oligochaeta). *Soil Biol. Biochem.*, **18**, 539–41.

Barois, I., Verdier, B., Kaiser, P. *et al.* (1987) Influence of the tropical earthworm
Pontoscolex corethrurus (Glossoscolecidae) on the fixation and mineralization of
nitrogen, in *On Earthworms*, (eds A. M. Bonvicini-Pagliai and P. Omodeo),
Mucchi Editore, Modena, pp. 151–8.

Barois, I., Villemin, G., Lavelle, P. and Toutain, F. (1993) Transformation of the soil
structure through *Pontoscolex corethrurus* (Oligochaeta) intestinal tract.
Geoderma, **56**, 57–66.

Barrett, T. J. (1949) *Harnessing the Earthworm*, Faber and Faber, London.

Basker, A., Macgregor, A. N. and Kirkman, J. H. (1993) Exchangeable potassium
and other cations in non-ingested soil and casts of two species of pasture earthworms. *Soil Biol. Biochem.*, **25**, 1673–7.

Basker, A., Kirkman, J. H. and Macgregor, A. N. (1994) Changes in potassium
availability and other soil properties due to soil ingestion by earthworms. *Biol.
Fertil. Soils*, **17**, 154–8.

Bas Lopez, S., Rivera, J. S., Lorenzzo, A. de C. and Sandez Canals, B. L. (1979) Data
on the feeding of *Salamandra salamandra* L., in Galicia, N.W. Spain. *Bol. Estac.
Cent. Ecol.*, **8**, 73–8.

Bassalik, K. (1913) On silicate decomposition by soil bacteria. *Z. Gärungs-physiol.*, **2**,
1–32.

Bather, E. A. (1920) *Pontoscolex latus*, a new worm from Lower Ludlow, Beds. *Ann. Mag. Nat. Hist.*, **9**, 5.

Bauchhenss, J. (1991) Earthworm taxocenosis in arable land subjected to various intensities of fertilization and pesticide use. *Bayer. Landwirtschaftl. Buch*, **68**, 335–54.

Bauer, K. (1964) Studien über Nebenwirkungen von Pflanzenschutzmitteln auf die Bodenfauna. *Mitt. Biol. Bund. Land. Forst. Berlin Dahlem.*, **112**.

Baweja, K. D. (1939) Studies of the soil fauna with special reference to the recolonisation of sterilised soil. *J. Anim. Ecol.*, **8**, 120–61.

Baylis, H. A. (1914) Preliminary account of *Aspidodrilus*, a remarkable epizoic oligochaete. *Ann. Mag. Nat. Hist.*

Baylis, H. A. (1915) A new African earthworm collected by Dr. C. Christy. *Ann. Mag. Nat. Hist.*

Beauge, A. (1912) Les vers de terre et la fertilité du sol. *J. Agric. Prat. Paris*, **23**, 506–7.

Becher, H. H. and Kainz, M. (1983) Effects of long-term manuring on soil structure. *Z. Acker-Pflanzenbau.*, **152**, 152–8.

Bejsovec, J. (1962) Rozsirovani Zarodu Helmintu Pasazi Zazivacim Traktem Adekvatnich Prenasecu. *Cs. Parcisitol.*, **9**, 95–109.

Bengston, S.-A., Nilsson, A., Nordström, S. and Rundgren, S. (1975) Habitat selection of lumbricids in Iceland. *Oikos*, **26**, 253–63.

Bengston, S.-A., Nilsson, A., Nordström, S. and Rundgren, 5. (1976) Effects of bird predation on lumbricid populations. *Oikos*, **27**, 9–12.

Bengtsson, G., Nordström, S. and Rundgren, S. (1983) Population density and tissue metal concentrations of lumbricids in forest soils near a brass mill. *Environ. Pollut. Ser. A*, **30**, 87–108.

Bengtsson, G., Gunnarsson, T. and Rundgren, S. (1986) Effects of metal pollution on the earthworm *Dendrobaena rubida* (Sav.) in acidified soils. *J. Water Air Soil Pollut.*, **28**, 361–83.

Benham, W. B. (1896) On *Kynotus cingulatus*, a new species of earthworm from Imerina in Madagascar. *Q. J. Micros. Sci.*

Benham, W. B. (1922) Oligochaeta of Macquarie Island. *Aust. Antarct. Exp. Sci. Rep., Zool. and Bot.*, **6**.

Benz, G. and Altwegg, A. (1975) Safety of *Bacillus thuringlensis* for earthworms. *J. Invert. Pathol.*, **26**, 125–6.

Berg, A. (1993) Food resources and foraging success of curlews *Numenius arquata* in different farmland habitats. *Ornis Fennica*, 70(1), 22–31.

Bernier, N. and Ponge, J. F. (1994) Humus form dynamics during the sylvogenetic cycle in a mountain spruce forest. *Soil Biol. Biochem.*, **26**, 183–220.

Beyer, L., Blume, H. P. and Irmler, U. (1991) The humus of 'parabraunerde' (Orthic Luvisol) under *Fagus sylvatica* L. and *Quercus robur* L. and its modification in 25 years. *Ann. Sci. Forest.*, **48**, 267–78.

Beyer, W. N. (1981) Metals and terrestrial earthworms (Annelida: Oligochaeta), in *Proceedings of the Workshop on the Role of Earthworms on the Stabilization of Organic Residues*, (ed. M. Appelhof), Beech Leaf Press, Kalamazoo, Michigan, pp. 137–50.

Beyer, W. N., Chaney, R. L. and Mulhern, B. M. (1982) Heavy metal concentrations in earthworms from soil amended with sewage sludge. *J. Environ. Qual.*, **11**, 381–5.

Beyer, W. N., Hensler, G. and Moore, J. (1987) Relation of pH and other soil variables to concentrations of Pb, Cu, Zn, Cd, and Se in earthworms. *Pedobiologia*, 30, 167–72.

Bhandari, G. S., Randhawa, N. S. and Maskin, M. S. (1967) On the polysaccharide content of earthworm casts. *Curr. Sci. (Bangalore)*, 36, 519–20.

Bharathi, C. and Subba Rao, B. V. S. S. R. (1984) Toxicity of phosphamidon to the common South Indian earthworm *Lampito mauritii*. *Bull. Environ. Contam. Toxicol.*, 32, 295–300.

Bharathi, C. and Subba Rao, B. V. S. S. R. (1984) Toxicity of phosphamidon to the common South Indian earthworm, *Lampito mauritii*. *Bull. Environ. Contam. Toxicol.*, 32, 295–300.

Bharucha-Reid, R. P. (1956) Latent learning in earthworms. *Science*, 123, 222.

Bhatnagar, T. (1975) Lombriciens et humification: Un aspect nouveau de l'incorporation microbienne d'azote induite par les vers de terre, in *Biodegradation et Humification*, (eds G. Kilbertus, O. Reisinger, A. Mourey and J. A. Cancela de Fonseca), Pierron, Sarreguemines, pp. 169–82.

Bigger, J. H. and Decker, G. C. (1966) Controlling root-feeding insects on corn. *Illinois Univ. Agr. Exp. Stn Bull.*, 716.

Billington, R. S. (1984) An economic assessment of the unit operations needed for a plant processing earthworm compost and earthworm protein. Divisional Note, DN 1252, National Institute of Agricultural Engineering, Silsoe, UK.

Binet, F. and Trehen, P. (1992) Experimental microcosm study of the role of *Lumbricus terrestris* (Oligochaeta:Lumbricidae) on nitrogen dynamics in cultivated soils. *Soil Biol. Biochem.*, 24, 1501–7.

Black, W. M. and Neely, D. (1975a) Effect of soil-injected benomyl on resident earthworm populations. *Pestic. Sci.*, 6, 543–5.

Black, W. M. and Neely, D. (1975b) Dutch elm disease control with soil injected benomyl; effect on resident earthworm population. *Proc. Am. Phytopath. Soc.*, 2, 82.

Blackshaw, R. P. (1980) The effect of benzimidazole fungicides on the ecology of soil fauna in winter wheat. PhD Thesis, University of Newcastle upon Tyne, England.

Blackshaw, R. P. (1990) Studies on *Artioposthia triangulata* (Dendy) (Tricladida: Terricola), a predator of earthworms. *Ann. Appl. Biol.*, 116, 169–76.

Blackshaw, R. P. (1991) Mortality of the earthworm *Eisenia fetida* (Savigny) presented to the terrestrial planarian *Artioposthia triangulata* (Dendy) (Tricladida: Terricola). *Ann. Appl. Biol.*, 118, 689–94.

Blair, J. M., Parmelee, R. W. and Lavelle, P. (1994). Influences of earthworms on biogeochemistry, in *Earthworm Ecology and Biogeography in North America*, (ed. P. F. Hendrix), Lewis Publishers, Chelsea, pp. 127–158.

Blair, J. M., Allen, M. F., Parmelee, R. W. *et al.* (1995) Changes in soil N pools in response to earthworm population manipulations under different agroecosystem treatments. *Soil Biol. Biochem.*, in press.

Blanchart, E. (1992) Restoration by earthworms (Megascolecidae) of the macroaggregate structure of a destructured savanna soil under field conditions. *Soil Biol. Biochem.*, 24, 1587–94.

Blanchart, E., Lavelle, P. and Spain, A. V. (1990) Effects of biomass and size of *Millsonia anomala* (Oligochaeta: Acanthodrilidae) on particle aggregation in a tropical soil in the presence of *Panicum maximum* (Gramineae). *Biol. Fertil. Soils*, 10, 113–20.

Blancke, E. and Giesecke, F. (1923) Mono- und Dimethyloharnstoffe in Ihrer Wirkung auf die Pflanzenproduktion und ihr Umsatz im Boden. *Z. Pflanz. Düng Bodenkunde,* **2**.

Blancke, E. and Giesecke, F. (1924) The effect of earthworms on the physical and biological properties of soil. *Z. Pflanz. Düng Bodenkunde,* 3(B), 198–210.

Blankwaardt, H. F. H. and van der Drift, J. (1961) Invloed van Grondontsmetting in Kassen op Regenwormen. *Meded. Dir. Tuinbouw.,* **24**, 490–6.

Blenkinsop, A. (1957) Some aspects of the problem of neotoata of open-cast sites. *Planning,* 4(3), 28–32.

Block, W. and Banage, W. B. (1968) Population density and biomass of earthworms in some Uganda soils. *Rev. Ecol. Biol. Sol,* 5, 515–21.

Blum, J. P. and Menzies, J. I. (1988) Notes on *Xenobatrachus* and *Xenorhina* (Amphibia: Microchylidae) from New Guinea with description of nine new species. *Alytes,* **7**, 125–63.

Boag, B., Neilson, R., Palmer, L. F. and Yeates, G. W. (1993) The New Zealand flatworm (*Artioposthia triangulata*): a potential alien predator of earthworms in northern Europe. *Proc. Brighton Pest Contr. Conf.,* 54, 397–402.

Bocock, K. L., Gilbert, O., Capstick, C. K. *et al.* (1960) Changes in leaf litter when placed on the surface of soils with contrasting humus types. I. Losses in dry weight of oak and ash leaf litter. *Soil Sci.,* **II**, 1–9.

Bodenheimer, F. S. (1935) Soil conditions which limit earthworm distribution. *Zoogeographica,* **2**, 572–8.

Boettcher, S. E. and Kalisz, P. J. (1991) Single-tree influence on earthworms in forest soils in eastern Kentucky (USA). *Soil Sci. Soc. Am. J.,* **55**, 862–5.

Bogh, P. S. (1992) Identification of earthworms: choice of method and distinction criteria. *Megadrilogica,* 4(10), 164–74.

Bohlen, P. J. and Edwards, C. A. (1995) Earthworm effects on soil N dynamics and respiration in microcosms receiving organic and inorganic nutrients. *Soil Biol. Biochem.,* **27**, 341–8.

Bohlen, P. J., Edwards, W. M. and Edwards, C. A. (1995a) Earthworm community structure and diversity in experimental agricultural watersheds in northeastern Ohio. *Plant and Soil,* 233–9.

Bohlen, P. J., Parmelee, R. W. and McCartney, D. A. (1995b) Carbon and nitrogen dynamics and microbial activity of crop residues in maize agroecosystems with modified earthworm populations. *Soil Biol. Biochem.,* in press.

Bohlen, P. J., Parmelee, R. W., Blair, J. M. *et al.* (1995c) Efficacy of methods for manipulating earthworm populations in large-scale field experiments in agroecosystems. *Soil Biol. Biochem.,* **27**, 993–9.

Bolton, P. J. and Phillipson, J. (1976) Burrowing, feeding, egestion and energy budgets of *Allolobophora rosea* (Savigny) (Lumbricidae). *Oecologia,* **23**, 225–45.

Bonvicini-Pagliai, A. M. and Omodeo, P. (eds) (1987) *On Earthworms.* Proc. 4th Int. Symp. Earthworm Ecol., Selected Symposia and Monographs 2, Mucchi Editore, Modena, Italy.

Bornebusch, C. H. (1930) The fauna of the forest soil. *Forstl. Forsogsv. Dan.,* **II**, 1–224.

Bornebusch, C. H. (1953) Laboratory experiments on the biology of worms. *Dansk Skovforen Tidsskr.,* **38**, 557–79.

Bosse, J. (1967) Restoration of biologically impoverished Weinberg soils, an example of earthworm colonization, in *Progress in Soil Biology* (eds O. Graff and J. E. Satchell). North Holland, Amsterdam, pp. 299–309.

Boström, U. and Lofs-Holmin, A. (1986) Growth of earthworms (*Allolobophora caliginosa*) fed on shoots and roots of barley, meadow fescue, and lucerne: studies in relation to particle size, protein, crude fiber and toxicity. *Pedobiologia*, **29**, 1–12.

Bouché, M. B. (1966) Sur un nouveau procédé d'obtention de la vacuité artificielle du tube digéstif des lumbricides. *Rev. Ecol. Biol. Sol*, **3**, 479–82.

Bouché, M. B. (1969) Comparison critique de methodes d'evaluation des populations de lumbricides. *Pedobiologia*, **9**, 26–34.

Bouché, M. B. (1971) Relations entre les structures spatiales et fonctionelles des écosystemes, illustrées par le rôle pédobiologique des vers de terre, in *La Vie dans les Sols, Aspects Nouveaux, Études Experimentales*, (ed. P. Pesson), Gauthier-Villars, Paris, pp. 187–209.

Bouché, M. B. (1972) *Lombriciens de France. Ecologie et Systématique*. Institut National de la Recherche Agronomique, Paris.

Bouché, M. B. (1974) Pesticides et lombriciens: problems methodologiques et economiques. *Phytiat. Phytopharm.*, **23**, 107–16.

Bouché, M. B. (1975a) Fonctions des lombriciens IV. Connections et utilizations des distosians pausee par les methodes de capture. In *Progress in Soil Zoology*, (ed. J. Vanek). *Proceedings of the 5th International Colloquium on Soil Zoology*, Prague. Junk, The Hague, pp. 571–82.

Bouché, M. B. (1975b) La reproductia de *Spermophorodrium albanianus* gen nov. (Lumbricidae) explique t'elle la functia des spermatophores. *Zool. Jahrb. Syst.*, **102**, 1–11.

Bouché, M. B. (1977) Stratégies lombriciennes, in *Soil Organisms as Components of Ecosystems*, (eds U. Lohm and T. Persson). *Ecol. Bull.* (Stockholm), **25**, 122–32.

Bouché, M. B. (1981) Contribution des lombriciens aux migrations d'éléments dans les sol tempérés. *Colloques Int. Centre Nat. Rech. Sci.*, no. 303, 145–53.

Bouché, M. B. (1982) Ecosystème prairial. 4.3 Un exemple d'activité animale: Le rôle des lombriciens. *Acta Oecol., Oecol. Gen.*, **3**, 127–54.

Bouché, M. B. (1988) Ecotoxicologie des lombriciens II. Surveillance de la contamination des milieux. *Acta. Oecol.*, **5**, 291–301.

Bouché, M. B. and Beugnot, M. (1972) Contribution a l'approche methodologique de l'étude des biocenoses et l'extraction des macroelements du sol par lavage-amisage. *Ann. Anim. Zool. Ecol.*, **4**, 537–44.

Bouché, M. B. and Beugnot, M. (1978) Action du chlorate de sodium sur le niveau des populations et l'activité biodegradatrice des lombriciens. *Phytiat. Phytopharm.*, **27**, 147–62.

Bouché, M. B. and Gardner, R. H. (1984) Earthworm functions. VIII. Population estimation techniques. *Rev. Ecol. Biol. Sol*, **21**(1), 37–63.

Bouché, M. B., Rafidison, Z. and Toutain, F. (1983) Etude de l'alimentation et du brassage pédo-intestinal du lombricienne *Nicodrilus velox* (Annelida, Lumbricidae) par l'analyse élémentaire. *Rev. Ecol. Biol. Sol*, **20**, 49–75.

Bouma, J. (1991) Influence of soil macroporosity on environmental quality. *Adv. Agron.*, **46**, 1–37.

Bouma, J., Belmans, C. F. M. and Dekker, L. W. (1982) Water infiltration and redistribution in a silt loam subsoil with vertical worm channels. *Soil Sci. Soc. Am. J.*, **46**, 917–21.

Bouwman, H. and Reinecke, A. J. (1987) Effects of carbofuran on the earthworm, *Eisenia fetida* using a defined medium. *Bull. Environ. Contam. Toxicol.*, **38**, 171–8.

Boyd, J. M. (1957a) The Lumbricidae of a dune-machair soil gradient in Tiree, Argyll. *Ann. Mag. Nat Hist.*, **12**, 274–82.

Boyd, J. M. (1957b) The ecological distribution of the Lumbricidae in the Hebrides. *Proc. R. Soc. Edinb.*, **66**, 311–38.

Boyd, J. M. (1958) The ecology of earthworms in cattle-grazed machair in Tiree, Argyll. *J. Anim. Ecol.*, **27**, 147–57.

Boykins, E. A. (1966) DDT residues in the food chains of birds. *Atlantic Nat.*, **21**, 18.

Boynton, D. and Compton, O. C. (1944) Normal seasonal changes of oxygen and carbon dioxide percentages in gas from the larger pores of three orchard subsoils. *Soil Sci.*, **57**, 107–17.

Bracher, G. A. and Bider, J. R. (1982) Changes in terrestrial animal activity of a forest community after an application of aminocarb. *Can. J. Zool.*, **60**, 1981–97.

Brande, J. van den and Heungens, A. (1969) Influence of repeated application of nematicides on the soil fauna in begonia culture. *Neth. J. Plant Pathol.*, **75**, 40–4.

Bray, J. R. and Gorham, E. (1964) Litter production in forests of the world. *Adv. Ecol. Res.*, **2**, 101–57.

Bretnall, G. H. (1927) Earthworms and spectral colours. *Science*, **66**, 427.

Bretscher, K. (1896) The Oligochaeta of Zürich. *Rev. Suisse Zool.*, **3**, 499–532.

Breza, M. (1959) Kebologichym viztahom daziloviek (Lumbricidae) abo medzihostitelov preuno helmintov z rodu *Metastrongilus*. I. Novy unimavy druh medzihostitelov. *Eisenia veneta* (Rosa) var. *hortensis* (Mich.). *Fol. veter. cas.*, **3**, 251–66.

Brinkhurst, R. O. (1992) Evolutionary relationships within the Clitellata. *Soil Biology and Biochemistry*, **24**(12), 1201–5.

Brinkhurst, R. O. (1994) Evolutionary relationships within the Clitellata: an update. *Megadrilogica*, **5**(10), 109–16.

Brinkhurst, R. O. and Jamieson, B. G. M. (1972) *Aquatic Oligochaeta of the World*, University of Toronto Press.

Briones, M. J. I., Mascato, R. and Mato, S. (1992) Relationships of earthworms with environmental factors studied by means of detrended canonical analysis. *Acta Oecol.*, **13**, 617–26.

Broadbent, A. B. and Tomlin, A. D. (1982) Comparison of two methods for assessing the effects of carbofuran on soil animal decomposers in cornfields. *Environ. Entomol.*, **11**, 1036–42.

Brown, B. R., Love, C. W. and Handley, W. R. C. (1963) Protein-fixing constituents of plants. *Rep. For. Res. London*, Part III, 90–3.

Brown, D. M. (1944) The cause of death in submerged worms. *J. Tenn. Acad. Sci.*, **19**(2), 147–9.

Brown, G. G. (1995) How do earthworms affect microfloral and faunal diversity? *Plant and Soil*, **170**, in press.

Bruel, W. E. van der (1964) Le sol, la pedofauna et les applications de pesticides. *Ann. Gembl.*, **70**, 81–101.

Brüsewitz, G. (1959) Untersuchungen über den Einfluss des Regenwurms auf Zahl und Leistungen von Mikrooganismen im Boden. *Arch. Microbiol*, **33**, 52–82.

Buahin, G. K. A., and Edwards, C. A. (1963). The side effects of toxic chemicals in the soil on arthropods and worms. *Rep. Rothamsted Exp. Stn for 1962*, pp. 156–7.

Buahin, G. K. A. and Edwards, C. A. (1964) The recolonisation of sterilised soil by invertebrates. *Rep. Rothamsted Exp. Stn for 1963*, pp. 149–50.

Buchanan, M. A., Russell, E. and Block, S. D. (1988) Chemical characterization and nitrogen mineralization potentials of vermicomposts derived from differing organic wastes, in *Earthworms in Environmental and Waste Management*, (eds C. A. Edwards and E. F. Neuhauser),   SPB Acad. Publ., The Netherlands, pp. 231–9.

Buckalew, D. W., Riley, R. K., Yoder, W. A. and Vail, W. J. (1982) Invertebrates as vectors of endomycorrhizal fungi and *Rhizobium* upon surface mine soils. *Proc. West Virginia Acad. Sci.*, **54**, 1.

Buckerfield, J. C. (1992) Earthworm populations in dryland cropping soils under conservation-tillage in South Australia. *Soil Biol. Biochem.*, **24**, 1667–72.

Buckerfield, J. C. and Doube, B. M. (1991) Effects of tillage, crop rotation and stubble management on earthworms. *S. Aust. Dept Agric. Tech. Pub.*, pp. 52–3.

Bull, C. T., Weller, C. M. and Tomashow, L. S. (1991) Relationship between root colonization and suppression of *Gaeumannomyces graminis* var *tritici* by *Pseudomonas flavescens*. *Phytopath.*, **81**, 954–9.

Buntley, C. J. and Papedick, R. I. (1960) Worm-worked soils of Eastern South Dakota, their morphology and classification. *Proc. Soil Sci. Soc. Am.*, **24**, 128–32.

Buse, A. (1990) Influence of earthworms on nitrogen fluxes and plant growth in cores taken from variously managed upland pastures. *Soil Biol. Biochem.*, **22**, 775–80.

Butt, K. (1991) The effects of temperature on the intensive production of *Lumbricus terrestris* (Oligochaeta: Lumbricidae). *Pedobiologia*, **35**, 257–64.

Butt, K. R. (1993). Utilization of solid paper mill sludge and spent brewery yeast as a feed for soil-dwelling earthworms. *Bioresource Technol.*, **44**, 105–7.

Butt, K. R., Roberts, R. D., Inskip, M. J. and Goodman, G. T. (1977) Mercury concentrations in soil, grass, earthworms, and small mammals near an industrial emission source. *Environ. Pollut.*, **12**, 135–40.

Byzova, J. B. (1965) Comparative rates of respiration in some earthworms. *Rev. Ecol. Biol. Sol*, **2**, 207–16.

Byzova, J. B. (1977) Haemoglobin content of *Allolobophora caliginosa* (Sav.) (Lumbricidae, Oligochaeta) during aestivation. *Doklady Biol. Sci.*, **236**, 763–5.

Cain, A. J. (1955) The taxonomic status of *Allolobophora iowana* Evans. *Ann. Mag Nat. Hist.*, **8**, 481–97.

Callahan, C. A. (1988) Earthworms as ecotoxicological assessment tools, in *Earthworms in Waste and Environmental Management*, (eds C. A. Edwards and E. F. Neuhauser), SPB Acad. Publ., The Hague, The Netherlands, pp. 295–301.

Callahan, C. A. and Linder, G. (1992) Assessment of contaminated soils using earthworm test procedures, in *Ecotoxicology of Earthworms*, (eds P. W. Greig-Smith, H. Becker, P. J. Edwards and F. Heimbach), Intercept, Andover, UK, pp. 197–208.

Carter, A., Heinonen, J. and de Vries, J. (1982) Earthworms and water movement. *Pedobiologia*, **23**, 395–7.

Carter, G. S. (1940) *A General Zoology of the Invertebrates*, 4th edn.

Caseley, J. C. and Eno, C. F. (1966) Survival and reproduction of two species of earthworm and a rotifer following herbicide treatments. *Proc. Soil Sci. Soc. Am.*, **30**, 346–50.

Casellato, S. (1987) On polyploidy in oligochaetes with particular reference to lumbricids, in *On Earthworms*, (eds A. M. Bonvicini-Pagliai and P. Omodeo), Mucchi Editore, Modena Italy, pp. 75–87.

Cathey, B. (1982) Comparative toxicities of five insecticides to the earthworm, *Lumbricus terrestris*. *Agric. Environ.*, **7**, 73–81.

Catling, P. M. and Freedman, B. (1980) Food and feeding behavior of sympatric snakes at Amherstburg, Ontario. *Can. Fld Nat.*, **94**, 28–33.

Causey, D. (1961) The earthworms of Arkansas, in *The Challenge of Earthworm Research*, (ed. R. Rodale), Soil and Health Foundation, PA, pp. 43–52.

Cernosvitov, L. (1928) Eine neue, an Regenwürmern schmarotzende Enchytreidenart. *Zool. Anz.*, **78**.

Cernosvitov, L. (1930a) Oligochaeten aus Turkestan. *Zool. Anz.*, **91** (I-4), 7–15.

Cernosvitov, L. (1930b) Prispevky k poznani fauny tatranskych Oligochaetu. *Vestniku Kral Ces. Spol. Nauk.*, **2**, 1–8.

Cernosvitov, L. (1931a) Revision des *Lumbricus submontanus* Vejdovsky, 1875. *Zool. Anz.*, **95**(1–2), 59–62.

Cernosvitov, L. (1931b) Zur Kenntnis der Oligochaeten fauna des Balkans. *Zool. Anaz.*, **95**(11–12), 312–27.

Cernosvitov, L. (1931c) Eine neue *Lumbricus*, Art aus der Umgebung von Prag. *Zool. Anz.*, **96**(7–8), 201–4.

Cernosvitov, L. and Evans, A. C. (1947) *Synopses of the British Fauna (6) Lumbricidae*, Linnaean Society, London.

Chadwick, L. C. and Bradley, J. (1948) An experimental study of the effects of earthworms on crop production. *Proc. Am. Soc. Hort. Sci.*, **51**, 552–62.

Chan, K. Y. and Heenan, D. P. (1993) Surface hydraulic properties of a red earth under continuous cropping with different management practices. *Aust. J. Soil Res.*, **31**, 13–24.

Chapman, G. (1950) On the movement of worms. *J. Exp. Biol.*, **27**, 29–39.

Chen, C. M. and Liu, C. L. (1963) Dynamics of the populations and communities of rice insect pests in the bank of Fung-Ting Lake region Hunar. *Acta Ent. Sin.*, **12**, 649–57.

Cheshire, M. V. and Griffiths, B. S. (1989) The influence of earthworms and crane-fly larvae on the decomposition of uniformly ^{14}C labelled plant materials in soil. *J. Soil Sci.*, **40**, 117–24.

Chio, H. and Sanborn, J. R. (1978) The metabolism of atrazine, chloramben and dicamba in earthworms (*Lumbricus terrestris*) from treated and untreated plots. *Weed Sci.*, **26**, 331–5.

Christensen, O. (1987) The effect of earthworms on nitrogen cycling in arable soil. *Proc. 9th International Colloquium on Soil Zoology*, Nauka, Moscow, pp. 106–18.

Christensen, O. (1988) The direct effect of earthworms on nitrogen turnover in cultivated soils. *Ecol. Bull.*, **39**, 41–4.

Clements, R. O. (1982) Some consequences of large and frequent pesticide applications to grassland, in *Proc. 3rd Aust. Conf. Grassl. Invert. Ecol.*, (ed. K. E. Lee), South Australia Government Printer, Adelaide, pp. 393–6.

Clements, R. O. and Henderson, I. F. (1977) Some consequences of prolonged absence of invertebrates from a perennial ryegrass sward. *Proc. 13th Int. Grassl. Congr.*, pp. 1261–3.

Clements, R. O., Henderson, I. F. and Bentley, B. R. (1982) The effects of pesticide application on upland permanent pasture. *Grass Forage Sci.*, **37**, 123–8.

Clements, R. O., Murray, P. J. and Sturdy, R. G. (1991) The impact of 20 years' absence of earthworms and three levels of nitrogen fertilizer on a grassland soil environment. *Agric. Ecosyst. Environ.*, **36**, 75–86.

Cockerell, T. D. A. (1924) Earthworms and the cluster fly. *Nature, Lond.*, **113**(2832), 193–4.

Cohen, S. and Lewis, H. B. (1949a) Nitrogenous metabolism of the earthworm *L. terrestris. Fedn Proc. Am. Soc. Exp. Biol.*, **8**, 191.

Cohen, S. and Lewis, H. B. (1949b) The nitrogen metabolism of the earthworm. *J. Biol. Chem.*, **180**, 79–92.

Coin, C. J. (1898) Beitrag zur Biologie von *Spiroptera turdi. Sitzsser. Deutsch. Net. Med. Ver. Bohmen, Prag.*

Coleman, D. C. and Sasson, A. (1978) Decomposer subsystem, in *Grasslands, Systems Analysis and Man*, (eds A. J. Breymeyer and G. M. van Dyne), Cambridge University Press, Cambridge, pp. 609–55.

Collier, J. (1978) Use of earthworms in sludge lagoons, in *Utilization of Soil Organisms in Sludge Management*, (ed. R. Hartenstein), Matl Tech. Inf. Services, PB286932, Springfield, VA, pp. 131–3.

Collins, M. M. (1980) The distribution of soil macrofauna on the west ridge of Gunung, Mount Mulu, Sarawak. *Oecologia*, **44**, 263–75.

Collins, P. T. (1992) Length-biomass relationships for terrestrial Gastropoda and Oligochaeta. *Am. Midl. Nat.*, **128**(2), 404–6.

Combault, A. (1909) Contribution a l'étude de la respiration et la circulation des Lombriciens. *J. Anat. Paris*, **45**, 358–9.

Contreras, E. (1980) Studies on the intestinal actinomycete flora of *Eisenia lucens* (Annelida, Oligochaeta). *Pedobiologia*, **20**, 411–16.

Cook, M. E. and Swait, A. A. J. (1975) Effects of some fungicide treatments on earthworm populations and leaf removal in apple orchards. *J. Hortic. Sci.*, **50**, 495–9.

Cook, A. G., Critchley, B. R., Critchley, U. *et al.* (1980) Effects of cultivation and DDT on earthworm activity in a forest soil in the sub-humid tropics. *J. Appl. Ecol.*, **17**, 21–9.

Cooke, A. (1983) The effects of fungi on food selection by *Lumbricus terrestris* L., in *Earthworm Ecology*, (ed. J. E. Satchell), Chapman & Hall, London, pp. 365–73.

Cooke, A. and Luxton, M. (1980) Effect of microbes on food selection by *Lumbricus terrestris. Rev. Ecol. Biol. Sol*, **17**, 365–70.

Cooper, E. L. and Roch, P. (1984) Earthworm leukocyte interactions during early states of graft rejection. *J. Exp. Zool.*, **232**(1), 67–72.

Cooper, E. L. and Roch, P. (1986) Second-set allograft responses in the earthworm *L. terrestris.* Kinetics and characteristics. *Transplantation*, **41**(4), 514–20.

Corrall, G. (1978) Some effect of *Lumbricus terrestris* L. on seed movement, seedling emergence, establishment and growth. M.Sc. Dissertation, University of Wales.

Cortez, J. and Bouché, M. B. (1992) Do earthworms eat living roots? *Soil Biol. Biochem.*, **9**, 913–15.

Cortez, J. and Hameed, R. (1988) Effets de la maturation des litières de ray-gras (*Lolium perenne* L.) dans le sol sur leur consommation et leur assimilation par *Lumbricus terrestris* L. *Rev. Ecol. Biol Sol*, **25**, 397–412.

Cortez, J., Hameed, R. and Bouché, M. B. (1989) C and N transfer in soil with or without earthworms fed with ^{14}C- and ^{15}N-labelled wheat straw. *Soil Biol. Biochem.*, **21**, 491–7.

Cotton, D. C. F. and Curry, J. P. (1980a) The effects of cattle and pig slurry fertilizers on earthworms (Oligochaeta, Lumbricidae) in grassland managed for silage production. *Pedobiologia*, **20**, 181–8.

Cotton, D. C. F. and Curry, J. P. (1980b) The response of earthworm populations (Oligochaeta, Lumbricidae) to high applications of pig slurry. *Pedobiologia*, **19**, 425–38.

Cragg, J. B. (1961) Some aspects of the ecology of moorland animals. *J. Anim. Ecol.*, **30**, 205–54.

Cramp, S. and Conder, P. J. (1965) *5th Rept. Br. Trust Ornith. and Roy. Soc. Prot. Birds on Toxic Chemicals.*

Crompton, E. (1953) Grow the soil to grow the grass. Some pedological aspects of marginal land improvement. *J. Minist. Agric. Fish.*, **50**(7), 301–8.

Crossley, D. A., Reichle, D. E. and Edwards, C. A. (1971) Intake and turnover of radioactive cesium by earthworms (Lumbricidae). *Pedobiologia*, **11**, 71–6.

Cuendet, G. (1977) Etude du comportement alimentaire des Mouettes rieuses (*Larus ridibundus*) et de leur influence sur les populations de vers de terre. *Ornithol. Beob.*, **2**, 87–8.

Cuendet, G. (1984) Les peuplements lombriciens des pelouses alpines du Munt Las Schera (Parc national suisse). *Rev. Suisse Zool.*, **91**(1), 217–28.

Cuendet, G. (1985) Repartition des lombriciens (Oligochaeta) dans la Basse Engadine, le Parc National et le Val Muestair (Grisons, Suisse). *Rev. Suisse. Zool.*, **92**(1), 145–63.

Curry, J. P. (1976) Some effects of animal manures on earthworms in grassland. *Pedobiologica*, **16**, 425–38.

Curry, J. P. (1987) The invertebrate fauna of grassland and its influence on productivity. III. Effects on soil fertility and plant growth. *Grass Forage Sci.*, **42**, 325–41.

Curry, J. P. (1988) The ecology of earthworms in reclaimed soils and their influence on soil fertility, in *Earthworms in Waste and Environmental Management*, (eds C. A. Edwards and E. F. Neuhauser), SPB Acad. Publ., The Hague, The Netherlands, pp. 251-61.

Curry, J. P. (1992) The role of earthworms in straw decomposition and nitrogen turnover in arable land in Ireland. *Soil Biol. Biochem.*, **24**, 1409–12.

Curry, J. P. and Bolger, T. (1984) Growth, reproduction and litter and soil consumption by *Lumbricus terrestris* L. in reclaimed peat. *Soil Biol. Biochem.*, **16**, 253–7.

Curry, J. P. and Boyle, K. E. (1987) Growth rates, establishment, and effects on herbage yield of introduced earthworms in grassland on reclaimed cutover peat. *Biol. Fertil. Soils*, **3**, 95–8.

Curry, J. P. and Byrne, D. (1992) The role of earthworms in straw decomposition and nitrogen turnover in arable land in Ireland. *Soil Biol. Biochem.*, **24**, 1409–12.

Curry, J. P. and Cotton D. C. F. (1983) Earthworms and land reclamation, in *Earthworm Ecology*, (ed. J. E. Satchell), Chapman & Hall, London, pp. 215–18.

Czerwinski, Z., Jakubczyk, H. and Nowak, E. (1974) Analysis of a sheep pasture ecosystem in the Pieniny Mountains (The Carpathians). XII. The effect of earthworms on the pasture soil. *Ekol. Pol.*, **22**, 635–50.

Daniel, O. (1991) Leaf-litter consumption and assimilation by juveniles of *Lumbricus terrestris* L. (Oligochaeta, Lumbricidae) under different environmental conditions. *Biol. Fertil. Soils*, **12**, 202–8.

Daniel, O. and Anderson, J. M. (1992) Microbial biomass and activity in contrasting soil material after passage through the gut of the earthworm *Lumbricus rubellus* Hoffmeister. *Soil Biol. Biochem.*, **24**, 465–70.

Daniel, O., Jager, P., Cuendet, G. and Bieri, M. (1992) Sampling of *Lumbricus terrestris* (Oligochaeta, Lumbricidae). *Pedobiologia*, **36**(4), 213–20.

Darwin, C. (1881) *The Formation of Vegetable Mould through the Action of Worms, with Observations of their Habits*, Murray, London.

Dash, H. K., Beura, B. N. and Dash, M. C. (1986) Gut load, transit time, gut microflora and turnover of soil, plant and fungal material by some tropical earthworms. *Pedobiologia*, **29**, 13–20.

Dash, M. C. and Patra, U. C. (1977) Density, biomass and energy budget of a tropical earthworm population from a grassland site in Orissa, India. *Rev. Ecol. Biol. Sol*, **14**, 461–71.

Dash, M. C. and Senapati, B.K. (1982) Environmental regulation of Oligochaeta reproduction in tropical pastures. *Pedobiologia*, **23**, 270–1.

Dash, M. C., Patra, U. C. and Thambi, A. V. (1974) Comparison of primary production of plant material and secondary production of oligochaetes in a tropical grassland of Orissa, India. *Trop. Ecol.*, **15**, 16–21.

Dash, M. C., Mishra, P. C. and Behera, N. (1979) Fungal feeding by a tropical earthworm. *Tropical Ecol.*, **20**, 9–12.

Dash, M. C., Binapani, S., Behera, N. and Dei, C. (1984) Gut load and turnover of soil, plant and fungal material by *Drawida calebi*, a tropical earthworm. *Rev. Ecol. Biol. Sol*, **21**, 387–93.

Datta, L. G. (1962) Learning in the earthworm *Lumbricus terrestris. Am. J. Psych.*, **75**, 531–53.

Daughberger, P. (1988) Temperature and moisture preference of three earthworm species (Oligochaeta: Lumbricidae). *Pedobiologia*, **32**, 57–64.

Davey, S. P. (1963) Effects of chemicals on earthworms: a review of the literature. *Spec. Sci. Rep. Wild. 74. USDA Fish and Wildlife Service.*

Davies, H. (1960) A revised list of the lumbricids found in the Bristol district. *Proc. Bristol. Nat. Soc.*, **30**, 51–4.

Davis, B. N. K. (1966) Soil animals as vectors of organochlorine insecticides for ground feeding birds. *J. Appl. Ecol. (Suppl).*, **3**, 133–9.

Davis, B. N. K. (1968) The soil macrofauna and organochlorine residues at twelve agricultural sites near Huntingdon. *Ann. Appl. Biol.*, **61**, 29–45.

Davis, B. N. K. (1971) Laboratory studies on the uptake of dieldrin and DDT by earthworms. *Soil Biol. Biochem.*, **3**, 221–33.

Davis, B. N. K. and French, M. C. (1969) The accumulation and loss of organochlorine insecticide residues by beetles, worms and slugs in sprayed fields. *Soil Biol. Biochem.*, **1**, 45–55.

Davis, B. N. K. and Harrison, R. B. (1966) Organochlorine insecticide residues in soil invertebrates. *Nature, Lond.*, **211**, 1424–5.

Davis, C. A. and Vohs, P. A. (1993) Availability of earthworms and scarab beetles to sandhill cranes in native grasslands along the Platte River. *Prairie Nat.*, **25**, 199–212.

Dawson, A. B. (1920) The intermuscular nerve cells of the earthworm. *J. Comp. Neurol.*, **32**, 155–71.

Dawson, R. B., Boyns, B. M. and Shorrock, R. W. (1938) The use of derris in the control of earthworms. *J. Sports Turf Res. Inst.*, **5**, 249–57.

Dawson, R. C. (1947) Earthworm microbiology and the formation of water-stable aggregates. *Soil Sci.*, **69**, 175–84.

Dawson, R. C. (1948) Earthworm microbiology and the formation of water-stable soil aggregates. *Proc. Soil Sci. Soc. Am.*, **12**, 512–16.

Day, G. M. (1950) The influence of earthworms on soil microorganisms. *Soil Sci.*, **69**, 175–84.

Dean-Ross, D. (1983) Methods for the assessment of the toxicity of environmental chemicals to earthworms. *Regul. Toxicol. Pharmacol.*, **3**, 48–59.

Denneman, W. D. (1991) A comparison of the diet composition of two *Sorex araneus* populations under different heavy metal stress. *Acta Theriologica*, 35(1–2), 25–38.

Derby, J. M., Houssiau, M., Lemasson-Florenville, M. *et al.* (1982) Impact de populations lombriciennes introduites sur le pH et sur la dynamique de l'azote dans un sol traite avec du lisier de porcs. *Pedobiologia*, 23, 157–71.

Devigne, J. and Jevniaux, C. (1961) Sur l'origine des chitinases intestinales des lombrics. *Arch. Int. Physiol. Biochim.*, 68(5), 833–4.

Dhawan, C. L., Sharma, R. L., Singh, A. and Handa, B. K. (1955) Preliminary investigations on the reclamation of saline soils by earthworms. *Proc. Nat. Inst. Sci. India*, 24, 631–6.

Dhennin, L., Heim de Balsac, H. and Verge, J. (1963) Recherches sur le role eventual de *Lumbricus terrestris* dans la transmission experimentale du virus de las fievre aphteuse. *Bull. Acad. Vet. Fr.*, 36, 153–5.

Dickschen, F. and Topp, W. (1987) Feeding activities and assimilation efficiencies of *Lumbricus rubellus* (Lumbricidae) on a plant-only diet. *Pedobiologia*, 30, 31–7.

Dietz, S. and Bottner, P. (1981) Etude par autoradiographie de l'endouissement d'une litière marquée au ^{14}C en milieu herbacé, in *Migration Organo-minerales dans les Sols Tempérés CNRS, Coll. Int.* CNRS 303, Paris, pp. 125–32.

Dikshith, T. S. S. and Gupta, S. K. (1981) Carbaryl induced biochemical changes in the earthworm *Pheretima posthuma*. *Ind. J. Biochem. Biophys.*, 18, 154.

Dindal, D. L., Schwert, D. P., Moreau, J.-P. and Theoret, L. (1977) Earthworm communities and soil nutrient levels as affected by municipal wastewater irrigation, in *Soil Organisms as Components of Ecosystems*, (eds U. Lohm and T. Persson). *Ecol. Bull. (Stockholm)*, 25, 284–90.

Dkhar, M. C. and Mishra, R. R. (1986) Microflora in earthworm casts. *J. Soil Biol. Ecol.*, 6, 24–31.

Doane, C. C. (1962) Effects of certain insecticides on earthworms. *J. Econ. Ent.*, 55, 416–18.

Dobson, R. M. (1956) *Eophila oculata* at Verulamium: a Roman earth-worm population. *Nature, Lond.*, 177, 796–7.

Dobson, R. M. and Lofty, J. R. (1956) Rehabilitation of marginal grassland. *Rep. Rothamsted Exp. Stn for 1955.*

Dobson, R. M. and Lofty, J. R. (1965) Observations of the effect of BHC on the soil fauna of arable land. *Congr. Int. Sci. Sol, Paris*, 3, 203–5.

Doeksen, J. (1950) An electrical method of sampling soil for earthworms. *Trans. 4th Int. Congr. Soil Sci.*, pp. 129–31.

Doeksen, J. (1964a) Notes on the activity of earthworms. I. The influence of *Rhododendron* and *Pinus* on earthworms. *Jaarb. IBS*, 177–80.

Doeksen, J. (1964b) Notes on the activity of earthworms. 3. The conditioning effect of earthworms on the surrounding soil. *Jaarb. IBS*, 187–91.

Doeksen, J. (1967) Notes on the activity of earthworms. V. Some causes of mass migration. *Meded. Inst. Biol. Scheik. Onderz. LandbGewass.*, 353, 199–221.

Doeksen, J. (1968) Notes on the activity of earthworms. VI. Periodicity in the oxygen consumption and the uptake of feed. *Meded. Inst. Biol. Scheik. Onderz.LandbGewass.*, 354, 123–8.

Doeksen, J. and Couperus, H. (1926) An estimation of the growth of earthworms. *Meded. Inst. Biol. Scheik. Onderz. LandbGewass.*, 195, 173–5.

Doeksen, J. and Couperus, H. (1968) Met vastellen van groei bij regenwormen. *Jaarb. IBS*, 173–5.

Doeksen, J. and van der Drift, J. (1963) *Soil Organisms. Proceedings of the Colloquium on Soil Fauna Soil Microflora and their Relationships.* Oosterbeek, North Holland, Amsterdam.

Doeksen, J. and van Wingerden, C. G. (1964) Notes on the activity of earthworms. 2. Observations on diapause in the earthworm *A. caliginosa. Jaarb. IBS*, 181–6.

Doerell, E. C. (1950) How do earthworms react to the application of minerals? *Deutsche Landwirtschaft. Presse*, 4(1), 9.

Domsche, K. M. and H. J. Banse (1972) Mykologische Untersuchungen An Regenwurm Exkrementen. *Soil Biol. Biochem.*, 4, 31–8.

Dotterweich, H. (1933) The function of storage of calcium by animals as a buffer reserve in the regulation of reaction. The calciferous glands of earthworms. *Pflügers Arch. gew. Physiol.*, 232, 263–86.

Doube, B. M., Ryder, M. H., Davoren, C. W. and Stephens, P. M. (1994a). Enhanced root nodulation of subterranean clover *Trifolium subterraneum* by *Rhizobium trifolii* in the presence of the earthworm *Aporrectodea trapezoides. Biol. Fertil. Soils*, 6, 237–51.

Doube, B. M., Stevens, P. M., Davoren, C. W. and Ryder, M. H. (1994b) Earthworms and the introduction and management of beneficial soil microorganisms, in *Soil Biota: Management in Sustainable Farming Systems*, (eds C. E. Pankhurst, B. M. Doube, U.V.S.R. Gupta and P. R. Grace), CSIRO, Melbourne, pp. 32–41.

Doube, B. M., Stephens, D. M., Davoren, C. W. and Ryder, M. H. (1994c) Interactions between earthworms, beneficial soil microorganisms and root pathogens. *Appl. Soil Ecol.*, 1(1), 3–10.

Doube, B. M., Williams, P. M. and Willmott, P. (1995) The effect of two species of earthworm (*Aporrectodea trapezoides* and *A. rosea*) on growth of wheat, barley, and faber beans in three soil types in the greenhouse. *Soil Biol. Biochem.*, in press.

Dowdy, W. W. (1944) Influence of temperature on vertical migration. *Ecology (Brooklyn)*, 25, 449–60.

Drandarevski, C. A., Eichler, D. and Domsch, K. H. (1977) Behaviour of trifluralin in soil and its influence on microbiological soil processes. *Z. Pflanzenkrank. Pflanzenschutz.*, 84, 18–30.

Dreidax, L. (1931) Investigations on the importance of earthworms for plant growth. *Arch. Pflanzenbau*, 7, 413–67.

Drewes, C. D. and Callahan, C. A. (1988) Electrophysiological detection of sublethal neurotoxic effects in intact earthworms, in *Earthworms in Waste and Environmental Management*, (eds C. A. Edwards and E. F. Neuhauser), SPB Acad. Publ., The Hague, The Netherlands, pp. 355–66.

Drewes, C. D., Zoran, M. J. and Callahan, C. A. (1987) Sublethal neurotoxic effects of the fungicide benomyl on earthworms (*Eisenia fetida*). *Pestic. Sci.*, 19, 197–208.

Drift, J. van der (1963) The influence of biocides on the soil fauna. *Neth. J. Pl. Path.*, 69, 188–99.

Dunger, W. (1969) Fragen der naturlichen und experimentellen Besiedlung kulturfeindlichenboden durch Lumbriciden. *Pedobiologia*, 9, 146–51.

Dunger, W. (1991) Primary succession of humiphagous soil animals on coal mined areas. *Zool. Jbch. Abt. Fuer System Oekol Geog Tiere*, 118(3–4), 423–47.

Dustman, E. H. and Stickel, L. F. (1966) Pesticide residues in the ecosystem. Pesticides and their effects on soils and water. *Am. Soc. Agron. Spec. Publ.*, 8, 109–21.

Dutt, A. K. (1948) Earthworms and soil aggregation. *J. Am. Soc. Agron.*, **40**, 407.

Dzangaliev, A. D. and Belousova, N. K. (1969) Earthworm populations in irrigated orchards under various soil treatments. *Pedobiologia*, **9**, 103–5.

Easton, E. G. (1983) A guide to the valid names of Lumbricidae (Oligochaeta), in *Earthworm Ecology: from Darwin to Vermiculture*, (ed. J.E. Satchell), Chapman & Hall, London, pp. 475–85.

Easton, T. H. Jr (1942) Earthworms of the North-eastern United States. *J. Wash. Acad. Sci.*, **32**(8), 24–9.

Easton, T. H. Jr and Chandler, R. F. (1942) The fauna of forest-humus layers in New York. *Mem. 247. Cornell Agr. Exp. Stn.*

Eberhardt, A. I. (1954) *Sarcophaga carnaria* als obligatorischer Regenwurmparasit. *Naturwissenschaften*, **41**(18), 436.

Ebing, K. W. and Haque, A. (1979) Summarizing report of previous studies concerning earthworms to test the ecological effects of organic chemicals on soil organisms. 3rd Meeting OECD Expert Group C 'Degradation/Accumulation', Tokyo.

Edwards, C. A. (1965) Effects of pesticide residues on soil invertebrates and plants. *Proc. 5th Symp. Br. Ecol. Soc.*, Blackwell, Oxford, pp. 23–61.

Edwards, C. A. (1969a) Insecticide residues in soils. *Residue Rev.*, **13**, 83.

Edwards, C. A. (1969b) Effects of gamma irradiation on populations of soil invertebrates. *Proc. 2nd Symp. Radioecol.* Ann Arbor, MI, pp. 68–77.

Edwards, C. A. (1970a) Persistent pesticides in the environment. *Critical Reviews in Environmental Control*, Chem. Rubber Co., Cleveland, pp. 7–67.

Edwards, C. A. (1970b) Effects of herbicides on the soil fauna. *Proc. 10th Weed Control Conf. 1970*, **3**, 105–262.

Edwards, C. A. (1975) Effects of direct drilling on the soil fauna. *Outlook on Agriculture*, **8**, 243–4.

Edwards, C. A. (1982) Production of earthworm protein for animal feed from potato waste, in *Upgrading Waste for Feed and Food*, (eds. D. A. Ledward, A. J. Taylor, and R. A. Lawrie), Butterworths, London.

Edwards, C. A. (1983a) Earthworms organic wastes and food. *Span. Shell Chem. Co.*, **26**(3), 106–8.

Edwards, C. A. (1983b) *Development of a standardized laboratory method for assessing the toxicity of chemical substances to earthworms*. Commission of the European Communities, Brussels, Luxembourg, EUR 8714 EN.

Edwards, C. A. (1984) *Report on the second stage in development of a standardized laboratory method for assessing the toxicity of chemical substances to earthworms*. Commission of the European Communities, Brussels, Luxembourg, EUR 9360 EN.

Edwards, C. A. (1985) Production of feed protein from animal wastes by earthworms. *Phil. Trans. Roy. Soc. London*, **310**, 299–307.

Edwards, C. A. (1988) Breakdown of animal, vegetable, and industrial organic wastes by earthworms. *Agric. Ecosyst. Environ.*, **24**, 21–31.

Edwards, C. A. (1989a) The importance of integration in lower input agricultural systems. *Agric. Ecosyst. Environ.*, **25**, 25-37.

Edwards, C. A. (1989b) Impact of herbicides on soil ecosystems. *Crit. Rev. Plant Sci.*, **8**, 221–57.

Edwards, C. A. (1991) The assessment of populations of soil-inhabiting invertebrates. *Agric. Ecosyst. Environ.*, 34, 145–76.

Edwards, C. A, and Arnold, M. K. (1963) The side-effects of toxic chemicals in the soil on arthropods and worms. *Rep. Rothamsted Exp. Sta.*, Part 1, 1962, p. 156.

Edwards, C. A. and Arnold, M. K. (1964) The side effects of toxic chemicals in the soil on arthropods and earthworms. *Rep. Rothamsted Exp. Sta.*, Part 1, 1963, p. 147.

Edwards, C. A. and Arnold, M. (1966) Effects of insecticides on soil fauna. *Rep. Rothamsted Exp. Stn for 1965*, p. 19–56.

Edwards, C. A. and Bohlen, P. J. (1992) The effects of toxic chemical on earthworms. *Rev. Environ. Contam. Toxicol.*, **125**, 23–99.

Edwards, P. J. and Brown, S. M. (1982) Use of grassland plots to study the effect of pesticides on earthworms. *Pedobiologia*, **24**, 145–50.

Edwards, C. A. and Burrows, I. (1988) The potential of earthworm composts as plant growth media, in *Earthworms in Environmental and Waste Management*, (eds C. A. Edwards and E. F. Neuhauser), SPB Acad. Publ., The Netherlands, pp. 211–20.

Edwards, C. A. and Dennis, E. B. (1960) Some effects of aldrin and DDT on the soil fauna of arable land. *Nature, Lond.*, **188**(4572), 767.

Edwards, C. A. and Fisher, S. W. (1991) The use of cholinesterase measurements in assessing the impacts of pesticides on terrestrial and aquatic invertebrates, in *Cholinesterase Inhibiting Insecticides: Impacts on Wildlife and the Environment*, (ed. P. Mineau), Elsevier, The Hague, The Netherlands, pp. 255–76.

Edwards, C. A. and Fletcher, K. E. (1988) Interactions between earthworms and microorganisms in organic-matter breakdown. *Agric. Ecosyst. Environ.*, **24**, 235–47.

Edwards, C. A. and Heath, G. W. (1963) The role of soil animals in breakdown of leaf material, in *Soil Organisms*, (eds J. Doeksen and J. van der Drift). North Holland, Amsterdam, pp. 76–80.

Edwards, C. A. and Jeffs, K. A. (1965) The persistence of some insecticides in soil and their effects on soil animals. *Proc. 12th Int. Congr. Entomol.*, pp. 559–60.

Edwards, C. A. and Lofty, J. R. (1971) Nematicides and the soil fauna. *Proc. 6th Brit. Insectic. Fungic. Conf.*, **1**, 158–66.

Edwards, C. A. and Lofty, J. R. (1969a) Effects of cultivations on earthworm populations. *Rep. Rothamsted Exp. Stn for 1968*, pp. 247–8.

Edwards, C. A. and Lofty, J. R. (1969b) The influence of agricultural practice on soil micro-arthropod populations, in *The Soil Ecosystem*, (ed. J. G. Sheals), Systematics Association publication No. 8, pp. 237–47.

Edwards, C. A. and Lofty, J. R. (1972a) *Biology of Earthworms*, Chapman & Hall, London.

Edwards, C. A. and Lofty, J. R. (1972b) Effects of pesticides on soil invertebrates. *Rep. Rothamsted Exp. Stn for 1971*, pp. 210–12.

Edwards, C. A. and Lofty, J. R. (1973) Pesticides and earthworms. *Rep. Rothamsted Exp. Stn for 1972*, P. I, pp. 211–12.

Edwards, C. A. and Lofty, J. R. (1975a) The invertebrate fauna of the Park Grass plots. *Rep. Rothamsted Exp. Stn for 1974*, Pt. 2, pp. 133–54.

Edwards, C. A. and Lofty, J. R. (1975b) The influence of cultivation on soil animal populations, in *Progress in Soil Zoology*, (ed. J. Vanek), Academia Publishing House, Prague, pp. 399–408.

Edwards, C. A. and Lofty, J. R. (1976) The influence of invertebrates on root crops grown with minimal or zero cultivation. *Proc. 6th Int. Coll. of the ISSS Soil Zoology Committee.*

Edwards, C. A. and Lofty, J. R. (1977) *Biology of Earthworms*, 2nd edn, Chapman & Hall, London.

Edwards, C. A. and Lofty, J. R. (1978) The influence of arthropods and earthworms upon root growth of direct drilled cereals. *J. Appl. Ecol.*, 15, 789–95.

Edwards, C. A. and Lofty, J. R. (1979) The effects of straw residues and their disposal on the soil fauna, in *Straw Decay and its Effect on Dispersal and Utilization*, (ed. E. Willey), New York, pp. 37–44.

Edwards, C. A. and Lofty, J. R. (1980) Effects of earthworm inoculation upon the root growth of direct drilled cereals. *J. Appl. Ecol.*, 17, 533–43.

Edwards, C. A. and Lofty, J. R. (1982). The effects of direct drilling and minimal cultivation on earthworm populations. *J. Appl. Ecol.*, 19, 723–34.

Edwards, C. A. and Neuhauser, E. F. (1988) *Earthworms in Waste and Environmental Management*, SPB Acad. Publ., The Hague, The Netherlands.

Edwards, C. A. and Niederer, A. (1988) The production and processing of earthworm protein, in *Earthworms in Waste and Environmental Management*, (eds C. A. Edwards and E. F. Neuhauser), SPB Acad. Publ., The Hague, The Netherlands.

Edwards, C. A. and Stafford, C. J. (1976) Effects of a herbicide on the soil fauna. *Rep. Rothamsted Exp. Sta.*, 1975, Part 1, p. 129.

Edwards, C. A. and Stafford, C. J. (1979) Interactions between herbicides and the soil fauna. *Ann. App. Biol.*, 91, 132–7.

Edwards, C. A. and Thompson, A. R. (1969) Insecticides and the soil fauna. *Rep. Rothamsted Exp. Sta.*, 1968, pp. 216–17.

Edwards, C. A. and Thompson, A. R. (1973) Pesticides and the soil fauna. *Residue Rev.*, 45, 1–79.

Edwards, C. A., Arnold, M. K. and Thompson, A. R. (1966) Effects of insecticides on soil fauna. *Rep. Rothamsted Exp. Sta.*, 1965, p. 186.

Edwards, C. A., Dennis, E. B. and Empson, D. W. (1967a) Pesticides and the soil fauna. I. Effects of Aldrin and DDT in an arable field. *Ann. Appl. Biol.*, 59(3), 11–22.

Edwards, C. A., Thompson, A. R. and Beynon, K. I. (1967b) Some effects of chlorfenvinphos, an organophosphorus insecticide, on populations of soil animals. *Rev. Ecol. Biol. Sol*, 5(2), 199–214.

Edwards, C. A., Thompson, A. R. and Beynon, K. I. (1968a) Some effects of chlorfenvinfos, an organophosphorus insecticide, on populations of soil animals. *Rev. Ecol. Biol. Sol.*, 5, 199–224.

Edwards, C. A., Thompson, A. R. and Lofty, J. R. (1968b) Changes in soil invertebrate populations caused by some organophosphorus insecticides. *Proc. 4th Brit. Insectic. Fungic. Conf.*, pp. 48–55.

Edwards, C. A., Reichle, D. E. and Crossley, D. A. Jr (1970a) The role of soil invertebrates in turnover of organic matter and Nutrients, in *Ecological Studies, Analysis and Synthesis*, Springer-Verlag, Berlin, pp. 147–72.

Edwards, C. A., Whiting, A. E. and Heath, G. W. (1970b) A mechanized washing method for separation of invertebrates from soil. *Pedobiologia*, 10(5), 141–8.

Edwards, C. A., Lofty, J. R. and Stafford, C. J. (1971) Pesticides and the soil fauna. *Rep. Rothamsted Exp. Stn for 1970*, Pt 1, p. 194.

Edwards, C. A., Lofty, J. R. and Stafford, C. J. (1972a) Insecticides and total soil fauna. *Rep. Rothamsted Exp. Stn for 1971*, pp. 210–11.

Edwards, C. A., Lofty, J. R. and Stafford C. J. (1972b) Pesticides and earthworms. *Rep. Rothamsted Exp. Sta.*, 1971, Part 1, pp. 211–12.

Edwards, C. A., Lofty, J. R. and Stafford, C. J. (1974) Soil fauna: pesticides and earthworms. *Rep. Rothamsted Exp. Sta.*, 1973, Part 1, p. 204.

Edwards, C. A., Burrows, I., Fletcher, K. E. and James, B. A. (1985) The use of earthworms for composting food wastes, in *Composting of Agricultural and Other Wastes*, (ed. J.K.R. Gasser), Elsevier, Amsterdam, pp. 229–42.

Edwards, W. M., Norton, L. D. and Redmond, C. E. (1988) Characterizing macropores that affect infiltration into nontilled soils. *Soil Sci. Soc. Am. J.*, **43**, 851–6.

Edwards, W. M., Shipitalo, M. J., Owens, L. B. and Norton, L. D. (1990) Effect of *Lumbricus terrestris* L. burrows on hydrology of continuous no-till corn fields. *Geoderma*, **46**, 73–84.

Edwards, W. M., Shipitalo, M. J., Dick, W. A. and Owens, L. B. (1992) Rainfall intensity affects transport of water and chemicals through macropores in no-till soils. *Soil Sci. Soc. Am. J.*, **56**, 52–8.

Edwards, C. A., Bohlen, P. J., Linden, D. R. and Subler, S. (1995) Earthworms in agroecosystems, in *Earthworm Ecology and Biogeography in North America*, (ed. P. F. Hendrix), Lewis Publishers, Boca Raton, FL, pp. 185–213.

Ehlers, W. (1975) Observations on the earthworm channels and infiltration on tilled and untilled loess soil. *Soil Sci.*, **119**, 242–9.

Eijsackers, H. (1982) Soil fauna and soil microflora as possible indicators of soil pollution. *Environ. Mgt. Assess.*, **3**, 317–19.

Eitminaviciute, I., Bagdanaviciene, Z., Budaviciene, I. *et al.* (1971) Untersuchungen der Beziehungen zwischen Gruppen von Wirbellosenlebewesen und Mikrooorganismen sowie der B-Vitamingruppe in unterschiedlichen Boden, in *IV Coll. Pedobiol.*, (ed. J. d'Aguilar), Institut National des Recherches Agriculturelles Publ. 71–7, Paris, pp. 93–7.

Ela, S. D., Gupta, S. C. and Rawls, W. J. (1992) Macropore and surface seal interactions affecting water infiltration into soil. *Soil Sci. Soc. Am. J.*, **56**, 714–21.

El-Duweini, A. K. (1965) Studies on the anatomy of *Pheretima californica*. *Bull. Zool. Soc. Egypt*, **20**, 11–30.

El-Duweini, A. K. and Ghabbour, S. I. (1965a) Population density and biomass of earthworms in different types of Egyptian soils. *J. Appl. Ecol.*, **2**, 271–87.

El-Duweini, A. K. and Ghabbour, S. I. (1965b) Temperature relations of three Egyptian oligochaete species. *Oikos*, **16**, 9–15.

El-Duweini, A. K. and Ghabbour, S. I. (1968) Nephridial systems and water balance of three Oligochaeta genera. *Oikos*, **19**, 61–70.

El-Duweini, A. K. and Ghabbour, S. I. (1971) Nitrogen contribution by live earthworms to the soil. *Ann. Zool. Ecol. Anim. Special Publ.*, **4**, 495–501.

Ellenby, C. (1945) Influence of earthworms on larval emergence in the potato root eelworm, *Heterodera rostochiensis* Wollenweber. *Ann. Appl. Biol.*, **31**(4), 332–9.

Elliot, P. W., Knight, D. and Anderson, J. M. (1990) Denitrification in earthworm casts and soil from pasture under different fertilizer and drainage regimes. *Soil Biol. Biochem.*, **22**, 601–5.

Ernst, D. (1995) *The Farmer's Earthworm Handbook: Managing your Underground Money-Makers*, Lessiter Publications, Brookfield, Wisconsin.

Escherich, K. (1911) *Termitenleben aus Ceylon*. Jena.

Escritt, J. R. (1955) Calcium arsenate for earthworm control. *J. Sports Turf Res. Inst.*, **9**(31), 28–34.

Escritt, J. R. and Arthur, J. H. (1948) Earthworm control – a resumé of methods available. *J. Bd. Greenkeep. Res.*, **7**(23), 49.

Evans, A. C. (1946) Distribution of numbers of segments in earthworms and its significance. *Nature, Lond.,* **158**, 98.

Evans, A. C. (1947a) Some earthworms from Iowa, including a description of a new species. *Ann. Mag Nat. Hist.,* **11**(14), 514.

Evans, A. C. (1947b) Method of studying the burrowing activity of earthworms. *Ann. Mag. Nat. Hist.,* **11**(14), 643–50.

Evans, A. C. (1948a) Some effects of earthworms on soil structure. *Ann. Appl. Biol.,* **35**, 1–13.

Evans, A. C. (1948b) Relations of worms to soil fertility. *Discovery, Norwich,* **9**(3), 83–6.

Evans, A. C. (1948c) Identity of earthworms stored by moles. *Proc. Zool. Soc. Lond.,* **118**, 1356–9.

Evans, A. C. and Guild, W.J. Mc. L. (1947a) Some notes on reproduction in British earthworms. *Ann. Mag. Nat. Hist.,* 654.

Evans, A. C. and Guild, W. J. Mc. L. (1947b) Cocoons of some British Lumbricidae. *Ann. Mag. Nat. Hist.,* 714–19.

Evans, A. C. and Guild, W. J. Mc. L. (1947c) Studies on the relationships between earthworms and soil fertility. I. Biological studies in the field. *Ann. Appl. Biol.,* **34**, 307–30.

Evans, A. C. and Guild, W. J. Mc. L. (1948a) Studies on the relationships between earthworms and soil fertility. IV. On the life cycles of some British Lumbricidae. *Ann. Appl. Biol.,* **35**, 471–84.

Evans, A. C. and Guild, W. J. Mc. L. (1948b) Studies on the relationships between earthworms and soil fertility. V. Field populations. *Ann. Appl. Biol.,* **35**, 485–93.

Fayolle, L. (1979) Consequences of the impact of pollutants on earthworms. III: Laboratory tests. *Doc. Pedozool.,* **1**, 34–65.

Feldkamp, J. (1924) Untersuchungen über die Geschlechtsmerkmale und die Begattung der Regenwürmer. *Zool. Jb (Anat.),* **46**, 609–32.

Fender, W. M. (1992) Oligochaeta: Megascolecidae, in *Soil Biology,* (ed. D. L. Dindal), Wiley and Sons, New York, pp. 357–86.

Fenton, G. R. (1947) Ecological note on worms in forest soil. *J. Anim. Ecol.,* **16**, 76–93.

Ferreira, M. E. and Cruz, M. C. P. D. (1992) Effects of a compost from municipal wastes digested by earthworms on the dry matter production of maize and on soil properties. *Cientifica (Jaboticabal),* **20**(1), 217–27.

Ferrière, G. (1980) Fonctions des lombriciens. VII. Une méthode d'analyse de la matière organique végétale ingérée. *Pedobiologia,* **20**, 263–73.

Ferrière, G. and Bouché, M. B. (1985) Première mesure écophysiologique d'un débit d'azote de *Nicodrilus longus* (Ude) (Lumbricidae Oligochaeta) dans la prairie de Citeaux. *CR Acad. Sci.,* **301**, 789–94.

Fieldson, R. S., Billington, R. S. and Audsley, E. (1985) A study of the economic feasibility of on-farm vermiculture with centralized processing of worked waste to convert animal wastes to horticultural composts. Divisional Note, DN. 1265, National Institute of Agricultural Engineering. Silsoe, U K.

Finck, A. (1952) Ökologische und Bodenkundliche Studien über die Leistungen der Regenwürmer für die Bodenfruchtbarkeit. *Z. PflErnähr. Düng.,* **58**, 120–45.

Finlayson, D. G., Campbell, C. J. and Roberts, H. A. (1975) Herbicides and insecticides: their compatibility and effects on weeds, insects and earthworms in the minicauliflower crop. *Ann. App. Biol.,* **79**, 95–108.

Fisher, C. (1988) The nutritional value of earthworm meal for poultry, in *Earthworms in Waste and Environmental Management*, (eds C. A. Edwards and E. F. Neuhauser), SPB Acad. Publ., The Netherlands, pp. 181–92.

Fisher, S. W. (1984) A comparison of standardised methods for measuring the biological activity of pesticides to the earthworm *L. terrestris*. *Ecotoxicol. Environ. Safety*, **8**, 564–71.

Flack, F. and Hartenstein, R. (1984) Growth of the earthworm *Eisenia foetida* on microorganisms and cellulose. *Soil Biol. Biochem.*, **16**, 491–5.

Fleckenstein, J. and Graff, O. (1982) Schwermetallaufnahme aus Mullkompost durch den Regenwurm *Eisenia fetida* (Savigny 1826). *Landbauforsch. Voelkenrodw.*, **32**, 198–202.

Fleming, W. E. and Hadley, C. H. (1945) DDT ineffective for control of an exotic earthworm. *J. Econ. Ent.*, **38**, 411.

Fleming, W. E. and Hanley, I. M. (1950) A large scale test with DDT to control the Japanese beetle. *J. Econ. Ent.*, **43**, 586–90.

Flickinger, E. L., King, K. A., Stout, W. F. and Mohn, M. M. (1980) Wildlife hazards from Furadan 3G applications to rice in Texas, U.S.A. *J. Wildl. Man.*, **44**, 190–7.

Ford, J. (1935) Soil communities in Central Europe. *J. Anim. Ecol.*, **6**, 197–8.

Fosgate, O. T. and Babb, M. R. (1972) Biodegradation of animal waste by *Lumbricus terrestris*. *J. Dairy Sci.*, **55**, 870–2.

Fox, C. J. S. (1964) The effects of five herbicides on the numbers of certain invertebrate animals in grassland soils. *Can. J. Pl. Sci.*, **44**, 405–9.

Fox, C. J. S. (1974) Effect of a carbamate and three organophosphorus insecticides on the numbers of wireworms, earthworms, springtails and mites in grassland soil. *Phytoprot.*, **55**, 103–5.

Fragoso, C. and Lavelle, P. (1987) The earthworm community of a tropical rain forest, in *On Earthworms*, (ed. A. M. Bonvicini-Pagliai and P. Omodeo), Mucchi Editore, Modena, Italy, pp. 281-95.

Fragoso, C. and Lavelle, P. (1992) Earthworm communities of tropical rain forests. *Soil Biol. Biochem.*, **24**, 1397–408.

Fragoso, C., Kanyonyo, J. K., Lavelle, P. and Moreno, A. (1992) A survey of communities and selected species of earthworms for their potential use in low-input tropical agricultural systems. CCF Project TS2-0292-F(E)B Report, pp. 7–34.

Franz, H. and Leitenberger, L. (1948) Biological-chemical investigations into the formation of humus through soil animals. *Ost. zool. Z.*, **1**,498–518.

Fraser, C. H. T. (1958) Maze learning in earthworms. Unpublished MS Thesis.

Friend, H. (1923) *British Earthworms and How to Identify Them*, Epworth Press, London.

Gacheva, E. S. H. (1990) Study of metastrongylosis in pigs in Georgia (USSR). *Soobshceniya Akad. Nauk Gruzinskoi SSR*, **140**(3), 609–12.

Gaddie, R. E. Sr and Douglas, D. E. (1975) *Earthworms for Ecology and Profit. I. Scientific Earthworm Farming*, Bookworm Publishing, California.

Galvyalis, A. G. and Lugauskas, A. (1978) Effect of chlorophos, carbophos, simazine and sodium trichloroacetate on earthworms and microscopic fungi. *Liet. TSR Mokslu. Akad. Darb. Ser C*, **2**, 17–26.

Gange, A. (1993) Translocation of mycorrhizal fungi by earthworms during early succession. *Soil Biol. Biochem.*, **25**, 1021–6.

Gansen, P. S. van (1956) Les cellules chloragogenes des Lombriciens. *Bull. Biol. Fr. Belg.*, **90**, 335–56.

Gansen, P. S. van (1957) Histophysiologie du tube digestif d' *Eisenia foetida* (Sav.) region buccale, pharynx et glandes pharyngiennes. *Bull. Biol. Fr. Belg.*, **91**, 225–39.

Gansen, P. S. van (1958a) Physiologie des cellules chloragogenes d'un lombricien. *Enzymologia*, **20**, 98–108.

Gansen, P. S. van (1958b) *Physiologie des cellules chloragogenes d'un lombricien* Eisenia foetida *Savigny*. Pub. Imp. Med. Sci. Bruxelles.

Gansen, P. van (1963) Structure and functions of the digestive canal of the earthworm *Eisenia foetida* Savigny. *Annales de la Société Royal Zoologique de Belgique*, **93**, 1–120.

Garcia, T. A. and Jaime, B. (1990) The use of earthworm meal (*Eudrilus eugeniae*) in the food of *Penaeus schmitti* postlarvae. *Rev. Investig. Marinas*, **11**(2), 147-56.

Gardiner, M. S. (1972) *Biology of the Invertebrates*, McGraw Hill, New York.

Gardner, M. R. (1953) The preparation of latex casts of soil cavities for the study of the tunneling habits of animals. *Science, N.Y.*, **118**, 380–1.

Gast, J. (1937) Contrast between the soil profiles developed under pines and hardwood. *J. For.*, **35**, 11–16.

Gates, G. E. (1929) The earthworm fauna of the United States. *Science, N.Y.*, **70**, 266–7.

Gates, G. E. (1949) Miscellanea Megadrilogica. *Am. Nat.*, **83**, 139–52.

Gates, G. E. (1954) On regenerative capacity of earthworms of the family Lumbricidae. *Am. Midl. Nat.*, **50**(2), 414–19.

Gates, G. E. (1959) On a taxonomic puzzle and the classification of the earthworms. *Bull. Mus. Comp. Zool. Harv.*, **121**, 229–61.

Gates, G. E. (1960) On natural regeneration by earthworms of the megascolecid genus *Perionyx* Perrier, 1872. *Wasmann J. Biol.*, **18**, 291–6.

Gates, G. E. (1961) Ecology of some earthworms with special reference to seasonal activity. *Am. Midl. Nat.*, **66**, 61–86.

Gates, G. E. (1962) An exotic earthworm now domiciled in Louisiana. *Proc. Louisiana Acad. Sci.*, **25**, 7–15.

Gates, G. E. (1963) Miscellanea Megadrilogica. VII. Greenhouse earthworms. *Proc. Biol. Soc. Wash.*, **76**, 9–18.

Gates, G. E. (1966) Requiem for Megadrile Utopias. A contribution toward the understanding of the earthworm fauna of North America. *Proc. Biol. Soc. Wash.*, **79**, 239–54.

Gates, G. E. (1967) On the earthworm fauna of the great American Desert and adjacent areas. *Great Basin Nat.*, **27**, 142–76.

Gates, G. E. (1968) Contributions to a revision of the Lumbricidae. III. *Eisenia hortensis* (Michaelson) (1890). *Brevicora*, **300**, 1–12.

Gates, G. E. (1969a) On the earthworms of the Ascension and San Juan Islands. *Brevicora*, **323**, 4.

Gates, G. E. (1969b) On two American genera of the family Lumbricidae. *J. Nat. Hist.*, **9**, 305–7.

Gates, G. E. (1970) Miscellanea Megadrilogica. VIII. *Megadrilogica*, 1(2), 1–14.

Gates, G. E. (1971) On reversion to former ancestral conditions in megadrile-oligochaetes. *Evolution*, **25**(1), 245–8.

Gates, G. E. (1972a) Burmese Earthworms. An introduction to the systematics and biology of megadrile oligochaetes with special reference to Southeast Asia. *Trans. Am. Phil. Soc., NS*, **62**(7), 1–326.

Gates, G. E. (1972b) Contributions to the North American earthworms (Annelida: Oligochaeta). No. 3. Toward a revision of the earthworm family Lumbricidae. IV. The *trapezoides* species group. *Bull. Tall Timbers Res. Stn*, **12**, 1–146.

Gates, G. E. (1974) On Oligochaeta gonads. *Megadrilogica*, **I**(9), 1–4.

Gates, G. E. (1976) More on oligochaete distribution in North America. *Megadrilogica*, **2**, 1–8.

Gavrilov, K. (1939) Sur la reproduction de *Eiseniella tetraedra* (Sav) f. *typica*. *Acta. Zool. Stockholm*, **20**, 439–64.

Gavrilov, K. (1960) La sexualidad y la reproduction de los Oligochaetos. *Acta Trab. I. Congr. Sudan Zool.*, **2**, 145–55.

Gavrilov, K. (1963) Earthworms, producers of biologically active substances. *Zh. Obshch. Biol.*, **24**, 149–54.

Gavrilov, K. (1967) Dates complementaries sobre *Eukerria subandina* (Rosa, 1895) (Oligochaeta, Ocnerodrilidae). *Acta Zool. Lill.*, **22**, 255–306.

Genov, T. (1963) Detection of the cysticercoid *Parieterotaenia paradoxa* (Rudolphi, 1802) (Dilepididae Fuhrmann, 1907) in *Allolobophora caliginosa* (Sav.) f. *trapezoides* (A. Dug.) (Lumbricidae). *Zool. Zh.*, **42**, 1578–9.

Geoghegan, M. J. and Brain, R. C. (1948) Aggregate formation in soil. I. Influence of some bacterial polysaccharides on the binding of soil particles. *Biochem. J.*, **43**, 5–13.

Gerard, B. M. (1960) The biology of certain British earthworms in relation to environmental conditions. Ph.D. thesis, London.

Gerard, B. M. (1963) The activities of some species of Lumbricidae in pasture land, in *Soil Organisms*, (ed. J. Doeksen and J. van der Drift), North Holland, Amsterdam, pp. 49–54.

Gerard, B. M. (1964) *Synopses of the British Fauna.* (6) *Lumbricidae.* Linnaean Society, London.

Gerard, B. M. (1967) Factors affecting earthworms in pastures. *J. Anim. Ecol.*, **36**, 235–52.

Gerard, B. M. and Hay, R. K. M. (1979) The effects on earthworms of ploughing cultivation, direct drilling and nitrogen cultivation in a barley monoculture system. *J. Agric. Sci.*, **93**, 147–55.

Gersch, M. (1954) Effect of carcinogenic hydrocarbons on the skin of earthworms. *Naturwissenschaft*, **41**, 337.

Gest, H. and Favinger, J. L. (1992) Enrichment of purple photosynthetic bacteria from earthworms. *FEMS Microbiol. Letters*, **91**, 265–9.

Gestel, van C. A. M. and Ma, W. C. (1988) Toxicity and bioaccumulation of chlorophenols in earthworms in relation to bioavailability in soil. *Ecotox. Environ. Safety*, **15**, 289–97.

Gestel, van C. A. M., Dis, W. A., van Breemen, E. M. and Sparenburg, P. M. (1989) Development of a standardized reproduction toxicity test with the earthworm species *Eisenia foetida andrei* using copper, pentachlorophenol and 2,4 dichloraniline. *Ecotoxicol. Environ. Safety*, **18**, 305–12.

Gestel, van C. A. M. and Ma, W. C. (1990) An approach to quantitative structure–activity relationships (QSARs) in earthworm toxicity studies. *Chemosphere.* **8**, 1023–33.

Ghabbour, S. I. (1966) Earthworms in agriculture: a modern evaluation. *Rev. Ecol. Biol. Soc.*, **111**(2), 259–71.

Ghabbour, S. I. (1975) Ecology of water relations in Oligochaeta. I. Survival in various relative humidities. *Bull. Zool. Soc. Egypt*, **27**, 1–10.

Ghabbour, S. I. and Imam, M. (1967) The effect of five herbicides on three oligochaete species. *Rev. Ecol. Biol. Sol.*, **4**, pp. 119–22.

Ghabbour, S. I. and Shakir, S. H. (1982) Population density and biomass of earthworms in agro-ecosystems of the Mariut coastal desert region. *Pedobiologia*, **23**, 189–98.

Ghilarov, M. S. (1956a) Significance of the soil fauna studies for the soil diagnostics. *6th Congr. Sci. Sol. Paris*, **3**, 130–44.

Ghilarov, M. S. (1956b) Soil fauna investigation as a method in soil diagnostics. *Bull. Lab. Zool. 'Filipo Silvestri' Portici*, **33**, 574–85.

Ghilarov, M. S. (1963) On the interrelations between soil dwelling invertebrates and soil microorganisms, in *Soil Organisms*, (eds J. Doeksen and J. van der Drift), North Holland, Amsterdam, pp. 255–9.

Ghilarov, M. S. (1965) Zoological methods in soil diagnostics, in *Nauka*, Moscow, p. 278.

Ghilarov, M. S. and Byzova, J. B. (1961) Vlijanie Chimiceskich Obrabotok Lesa Na Pocuennuja Faunu. *Iesn. Ch-Vo.*, **10**, 589.

Ghilarov, M. S. and Mamajev, B. M. (1963) Soil-inhabiting insects in irrigated regions of Uzbekistan. *Rast Vreidilelei Bod.*, **8**, 21–2.

Ghilarov, M. S. and Mamajev, B. M. (1966) Über die Ansiedlung von Regenwürmern in den artesisch bewässerten Oasen der Würste KystKum. *Pedobiologia*, **6**, 197–218.

Gish, C. D. (1970) Organochlorine insecticide residues in soils and soil invertebrates from agricultural land. *Pest. Mon. J.*, **3**(4), 241–52.

Gish, C. D. and Christensen, R. E. (1973) Cadmium, nickel, lead and zinc in earthworms from roadside soil. *Environ. Sci. Technol.*, **7**, 1060–2.

Gjelstrup, P. and Hendriksen, N. B. (1991) *Histiostoma murchiei* Hughes and Jackson, in *The Acari: Reproduction, Development, and Life-History Strategies*, (eds R. Schuster and P. W. Murphy), Chapman & Hall, London, pp. 441–5.

Gilman, A. P. and Vardanis, A. (1974) Carbofuran. Comparative toxicity and metabolism in the worms *L. terrestris* and *E. foetida*. *J. Agric. Food Chem.*, **22**, 625–8.

Goats, G. C. (1983) A comparison of field and laboratory methods for testing toxicity to earthworms. *Proc. 10th Int. Cong. Plant Prot.*, **2**, 713.

Goats, G. C. and Edwards, C. A. (1988) The prediction of field toxicity to earthworms by laboratory methods, in *Earthworms in Waste and Environmental Management*, (eds C.A. Edwards and E. F. Neuhauser). SPB Acad. Publ., The Hague, The Netherlands, pp. 283–94.

Goffart, H. (1949) Die Wirkung neuer Insektiziden Mittel auf Regenwürmer, in *Anx. f. Schädlingskunde*, **22**(72), 4.

Gotwald, W. W. (1974) Foraging behavior of *Anomma* driver ants in Ghana cocoa farms (Hymenoptera: Formicidae). *Bull. Inst. Fondam. Afr. Noire Ser. A. Sci. Nat.*, **36**, 705–13.

Graff, O. (1953a) Investigations in soil zoology with special reference to the terricole Oligochaeta. *Z. PflErnähr. Düng.*, **61**, 72–7.

Graff, O. (1953b) Die Regenwürmer Deutschlands. *Schrift. Forsch. Land. Braunschweig-Volk*, **7**, 81.

Graff, O. (1967) Translocation of nutrients into the subsoil through earthworm activity. *Landw. Forsch.*, **20**, 117–27.

Graff, O. (1969) Regenwurmtätigkeit in Ackerböden unter verschiedenem Bedeckungsmaterial, gemessen an der Lösungsablage. *Pedobiologia*, **9**, 120–8.

Graff, O. (1971) Stikstoff, Phosphor und Kalium in der Regenwurmlösung auf der Wiesenversuchsfläche des Sollingprojektes. *Ann. Zool. Ecol. Anim. Special Publ.*, **4**, 503–12.

Graff, O. (1974) Gewinnung von Biomasse aus Abfallstoffen durch Kultur des Kompostregenwurms *Eisenia foetida* (Savigny 1826). *Landbauforsch Volkenrode*, **2**, 137–42.

Graff, O. (1982) Vergleich der Regenwurmarten *Eisenia foetida* und *Eudrilus eugeniae* hinsichtlich ihrer Eignung zur Proteingewinnung aus Abfallstoffen. *Pedobiologia*, **23**, 277–82.

Graff, O. (1983). Darwin on earthworms – the contemporary background and what the critics thought, in *Earthworm Ecology, From Darwin to Vermiculture*, (ed. J. E. Satchell), Chapman & Hall, London, pp. 5–18.

Graff, O. and Makeschin, F. (1980) Beeinflussung des Ertrags von Weidelgras (*Lolium multiflorum*) durch Ausscheidungen von Regenwürmern dreier verschiedener Arten. *Pedobiologia*, **20**, 176–80.

Graff, O. and Satchell, J. E. (eds) (1967) *Progress in Soil Biology*, North Holland, Amsterdam.

Graham, R. C. and Wood, H. B. (1991) Morphological development and clay redistribution in lysimeter soils under chaparral and pine. *Soil Sci. Soc. Am. J.*, **55**, 1638–46.

Grant, J. D. (1983) The activities of earthworms and the take of seeds, in *Earthworm Ecology, From Darwin to Vermiculture*, (ed. J. E. Satchell), Chapman & Hall, London, pp. 107–22.

Grant, W. C. (1955a) Studies on moisture relationships in earthworms. *Ecology*, **36**(3), 400–7.

Grant, W. C. (1955b) Temperature relationships in the megascolecid earthworm, *Pheretima hupeiensis*. *Ecology*, **36**(3), 412–17.

Grant, W. C. (1956) An ecological study of the peregrine earthworm, *Pheretima hupeiensis* in the Eastern United States. *Ecology*, **37**(4), 648–58.

Granval, P. and Muys, B. (1992) Management of forest soils and earthworms to improve woodcock (*Scolopax* sp.). *Gib. Faune Sauv.*, **9**, 243–55.

Grappelli, A., Tomati, U. and Palma, G. (1985) Plant consisting of modular elements for degrading organic wastes by means of earthworms. Patent entry at Consiglio Nazionale dell Ricerche, Rome (Italy).

Gras, H. (1984) The tail flattening reflex in *Lumbricus*: Reconstituting after tail amputation and modification in segmental nerve roots. *J. Neurobiol.*, **15**(4), 249–61.

Grassi, B. and Rovelli, G. (1892) Recherche embriologiche sui Cestodi. *Att. Asc. Catania*, **4**, 15–108.

Gray, J. and Lissmann, H. W. (1938) Studies on animal locomotion. VII. Locomotory reflexes in the earthworm. *J. Exp. Biol.*, **15**(4), 506–17.

Green, E. and Penton, S. (1981) Full scale vermicomposting at the Lufkin water pollution control plant, in *Proc. Workshop on the Role of Earthworms in the Stabilization of Organic Residues*, (ed. M. Appelhof), Beach Leaf Press, Kalamazoo, Michigan, Vol. 1, pp. 229–31.

Greenwood, D. E. (1945) Wireworm investigations. *Conn. Agric. Exp. Stn Bull.*, **488**, 344–7.

Gregory, P. T. and Nelson, K. J. (1991) Predation on fish and intersite variation in the diet of common garter snakes, *Thamnophis sirtalis*, on Vancouver Island (Canada). *Can. J. Zool.*, **69**, 988–94.

Greig-Smith, P. W., Becker, H., Edwards, P. J. and Heimbach, F. (1992) *Ecotoxicology of Earthworms*, Intercept, Andover, UK.

Griffiths, D. C., Raw, F. and Lofty, J. R. (1967) The effects on soil fauna of insecticides tested against wireworms (*Agriotes* spp.) in wheat. *Ann. Appl. Biol.*, **60**, 479–90.

Griffiths, M. (1978) *The Biology of the Monotremes*. Academic Press, New York.

Grigor'eva, T. G. (1952) The action of BHC introduced into the soil on the soil fauna. *Dokl. vseoyuz. Akad. selkhoz Nauk. Lenina.*, **17**, 16–20 (summary *Rev. Appl. Ent.* (A), **41**, 336).

Grove, A. J. and Newell, G. E. (1962) *Animal Biology*, University Tutorial Press, London.

Gruia, L. (1969) Répartition quantitative des lombricides dans les sols de Roumanie. *Pedobiologia*, **9**, 99–102.

Guerrero, R. D. (1983) The culture and use of *Perionyx excavatus* as a protein resource in the Philippines, in *Earthworm Ecology from Darwin to Vermiculture*, (ed. J. E. Satchell), Chapman & Hall, London, pp. 309–13.

Guild, W. J. Mc. L. (1948) Effect of soil type on populations. *Ann. Appl. Biol.*, **35**(2), 181–92.

Guild, W. J. Mc. L. (1951a) Earthworms in Agriculture. *Scot. Agric.*, **30**(4), 220–3.

Guild, W. J. Mc. L. (1951b) The distribution and population density of earthworms (Lumbricidae) in Scottish pasture fields. *J. Anim. Ecol.*, **20**(I), 88–97.

Guild, W. J. Mc. L. (1952a) Variation in earthworm numbers within field populations. *J. Anim. Ecol.*, **21**(2), 169.

Guild, W. J. Mc. L. (1952b) The Lumbricidae in upland areas. II. Population variation on hill pasture. *Ann. Mag. Nat. Hist.*, **12**(5), 286–92.

Guild, W. J. Mc. L. (1955) Earthworms and soil structure, in *Soil Zoology*, (ed. D. K. Mc. E. Kevan), Butterworths, London, pp. 83–98.

Gunn, A. (1992) The use of mustard to estimate earthworm populations. *Pedobiologia*, **36**(2), 65–7.

Gunn, A. and Sadd, J. W. (1994) The effect of ivermectin on the survival, behavior and cocoon production of the earthworm, *Eisenia fetida*. *Pedobiologia*, **38**, 327–37.

Gunthart, E. (1947) Die Bekämpfung der Engerlinge mit Hexachlorocyclohexan Präparation. *Mitt. Schweiz Ent. Ges*, **20**, 409–50.

Gurianova, O. Z. (1940) Effect of earthworms and of organic fertilisers on structure formation in chernozem soils. *Pedology*, **4**, 99–108.

Haimi, J. and Enbrok, M. (1992) Effects of endogeic earthworms on soil processes and plant growth in coniferous forest soil. *Biol. Fertil. Soils*, **13**, 6–10.

Haimi, J. and Huhta, V. (1990) Effects of earthworms on decomposition processes in raw humus forest soil: a microcosm study. *Biol. Fertil. Soils*, **10**, 178–83.

Haimi, J., Huhta, V. and Boucelham, M. (1992) Growth increase of birch seedling under the influence of earthworms – a laboratory study. *Soil Biol. Biochem.*, **24**, 1525–8.

Hallett, L., Viljoen, S. A. and Reinecke, A. J. (1992) Moisture requirements in the life cycle of *Perionyx excavatus* (Oligochaeta). *Soil Biol. Biochem.*, **24**, 1333–40.

Hamblyn, C. J. and Dingwall, A. R. (1945) Earthworms. *N. Z. J. Agric.*, **71**, 55–8.

Hameed, R., Bouché, M. B. and Cortez, J. (1994a) Etudes *in situ* des transferts d'azote d'origine lombricienne (*Lumbricus terrestris* L.) vers les plantes. *Soil Biol. Biochem.*, **26**, 495–501.

Hameed, R., Cortez, J. and Bouché, M. B. (1994b) Biostimulation de la croissance de *Lolium perenne* L. par l'azote excrete par *Lumbricus terrestris* L.– mesure au laboratoire de ce debit. *Soil Biol. Biochem.*, **26**, 483–93.

Hamilton, W. E. and Dindal, D. L. (1989) Influence of earthworms and leaf litter on edaphic variables in sewage sludge-treated soil microcosms. *Biol. Fertil. Soils*, **7**, 128–33.

Hamilton, W. E. and Vimmerstedt, J. P. (1980) Earthworms on spoil banks, in *Soil Biology as Related to Land Use Practices*, (ed. D. L. Dindal), EPA, Washington, pp. 409–17.

Hamoui, V. (1991) Life-cycle and growth of *Pontoscolex corethrurus* in the laboratory. *Rev. d'Ecol. Biol. Sol*, **28**(4), 469–78.

Hampson, M. C. and Coombes, J. W. (1989) Pathogenesis of *Synchytrium endobioticum*. VIII. Earthworms as vectors of wart disease of potato. *Plant and Soil*, **116**(2), 147–50.

Hand, P. and Hayes, W. A. (1988) The vermicomposting of cow slurry, in *Earthworms in Waste and Environmental Management*, (eds C. A. Edwards and E. F. Neuhauser), SPB Acad. Publ., The Hague, pp.49–64.

Handreck, K. A. (1986) Vermicomposting as components of potting media. *Biocycle*, October 1986.

Hanel, E. (1904) Ein Beitrag zur 'Psychologie' der Regenwurmer. *Z. Allg. Physiol.*, **4**, 244–58.

Hans, R. K., Gupta, R. C. and Beg, M. U. (1990) Toxicity assessment of four insecticides to the earthworm *Pheretima posthuma*. *Bull. Environ. Contam. Toxicol.*, **45**, 358–64.

Hanschko, D. (1958) The earthworm faunas of different plant communities. *Forsch. Beratung.*, **3**, 16–29.

Haque, A. and Ebing, W. (1980) Uptake of the herbicide ^{14}C-monolinuron by earthworms and metabolism in soil and earthworms. *Jahresb. Biolg. Bundesans Land u Fortswirtschaft, Berlin u Baunschweig*, 1979, p. 99.

Haque, A. and Ebing, W. (1983) Toxicity determination of pesticides to earthworms in the soil substrate. *Z. Pflanzenkrank Pflanzenschutz*, **90**, 395–408.

Haque, A., Schuphan, I. and Ebing. W. (1982) Bioavailability of conjugated and soil-bound (^{14}C) hydroxymonolinuron-β-D-glucoside residues to earthworms and rye grass. *Pestic. Sci.*, **13**, 219–28.

Harinikumar, K. M. and Bagyaraj, D. J. (1994) Potential of earthworms, ants, millipedes, and termites for dissemination of vesicular-arbuscular mycorrhizal fungi in soil. *Biol. Fertil. Soils*, **18**, 115–18.

Harinikumar, K. M., Bagyaraj, D. J. and Kale, R. D. (1991) Vesicular arbuscular mycorrhizal propagules in earthworm casts, in *Advances in Management and Conservation of Soil Fauna*, (eds G. K. Veeresh, D. Rajagopal and C. A. Viraktamath), Oxford and IBH, New Delhi, pp. 605–10.

Harman, W. J. (1965) Life history studies of the earthworm *Sparganophilus eiseni* in Louisiana. *Southwest Nat.*, **10**, 22–4.

Harmsen, G. and van Schreven, D. (1955) Mineralisation of organic nitrogen in soil. *Adv. Agron.*, **7**, 299–398.

Hartenstein, R. (1978) The most important problem in sludge management as seen by a biologist, in *Utilization of Soil Organisms in Sludge Management*, (ed. R. Hartenstein), Natl. Tech. Inf. Serv., PB286932, Springfield, Virginia, pp. 2–8.

Hartenstein, R. (1982) Soil macroinvertebrates, aldehyde oxidase, catalase, cellulase and peroxidase. *Soil Biol. Biochem.*, **15**, 51–4.

Hartenstein, R., Neuhauser, E. F. and Kaplan, D. L. (1979) Reproductive potential of the earthworm *Eisenia foetida. Oecologia*, **43**, 329–40.

Hartenstein, R., Neuhauser, E. F. and Collier, J. (1980) Accumulation of heavy metals in the earthworm *Eisenia foetida. J. Environ. Qual.*, **9**, 23–6.

Harwood, M. and Sabine, J. R. (1978) The nutritive value of worm meal, in *Proc. 1st Australasian Poultry Stockfeed Conv.*, Sydney, pp. 164–71.

Hasenbein, G. (1951) A pregnancy test on earthworms. *Arch. Gynakol.*, **181**, 5–28.

Haswell, W. A. and Hill, J. P. (1894) A proliferating cystic parasite of the earthworms. *Proc. Linn. Soc. N.S. Wales*, **8**(2), 365–76.

Hazelhoff, L., van Hoof, P., Imeson, A. C. and Kwaad, F. J. P. M. (1981) The exposure of forest soil to erosion by earthworms. *Earth Surf. Proc. Landforms*, **6**, 235–50.

Heath, G. W. (1962) The influence of key management on earthworm populations. *J. Br. Grassld Soc.*, **17**(4), 237–44.

Heath, G. W. (1965) The part played by animals in soil formation, in *Experimental Pedology*, (eds E. G. Hallsworth and D. V. Crawford), Butterworths, London, pp. 236–43.

Heath, G. W. and King, H. G. C. (1964) The palatability of litter to soil fauna. *Proc. VIII Int. Congr. Soil Sci. Bucharest*, pp. 979–86.

Heath, G. W., Arnold, M. K. and Edwards, C. A. (1966) Studies in leaf litter breakdown. I. Breakdown rates among leaves of different species. *Pedobiologia*, **6**, 1–12.

Heck, L. von. (1920) Über die Bildung einer Assoziation beim Regenwurm auf Grund von Dressurversuchen. *Lotos Naturwiss. Z.*, **68**, 168–89.

Heimbach, F. (1984) Correlations between three methods for determining the toxicity of chemicals to earthworms. *Pestic. Sci.*, **15**, 605–11.

Heimbach, F. and Edwards, P. J. (1983) The toxicity of 2-chloroacetamide and benomyl to earthworms under various test conditions in an artificial soil test. *Pestic. Sci.*, **14**, 635–6.

Heimburger, H. V. (1924) Reactions of earthworms to temperature and atmospheric humidity. *Ecology*, **5**, 276–83.

Helmke, P. A., Robarge, W. P., Korotev, R. L. and Schomberg, P. J. (1979) Effects of soil-applied sewage sludge on concentrations of elements in earthworms. *J. Environ. Qual.*, **8**, 322–7.

Hendriksen, N. B. (1990) Leaf litter selection by detritivore and geophagous earthworms. *Biol. Fertil. Soils*, **10**, 17–21.

Hendriksen, N. B. (1991) Gut load and food-retention time in the earthworms *Lumbricus festivus* and *L. castaneus*: a field study. *Biol.Fertil. Soils*, **11**, 170–3.

Hendrix, P. L. (ed.) (1995) *Earthworm Ecology and Biogeography in North America*, Lewis Publ., Boca Raton, FL.

Hendrix, P. F., Crossley, D. A. Jr, Coleman, D. C. *et al.* (1987) Carbon dynamics in soil microbes and fauna in conventional and no-tillage agroecosystems. *INTECOL Bull.*, **15**, 590–63.

Hendrix, P. J., Mueller, B. R., Bruce, R. R. *et al.* (1992) Abundance and distribution of earthworms in relation to landscape factors on the Georgia Piedmont, U.S.A. *Soil Biol. Biochem.*, **24**, 1357–61.

Hensen, V. (1877) Die Tätigkeit des Regenwurms (*L. terrestris*) für die Fruchtbarkeit dess Erdbodens. *Z. wiss. Zool.*, **28**, 354–64.

Herlant-Meewis, H. (1956) Croissance et reproduction du Lombricien, *Eisenia foetida* (Sav.). *Ann. Sci. Nat. Zool. Biol. Anim.*, **18**, 185–98.

Hess, W. N. (1924) Reactions to light in the earthworm, *Lumbricus terrestris*. *J. Morph.*, **39**, 515–42.

Hess, W. N. (1925a) Nervous system of the earthworm, *Lumbricus terrestris*. *J. Morph.*, **40**, 235–60.

Hess, W. N. (1925b) Photoreceptors of *Lumbricus terrestris*, with special reference to their distribution. *J. Morph.*, **41**, 235–60.

Heungens, A. (1966) Bestrijding van Regenwormen in Sparregrond en in vitro. *Med. Rijksfak. Landb. Sch. Gent*, **31**, 329–42.

Heungens, A. (1969a) L'influence de la fumure et des pesticides aldrine, carbaryl et DBCP sur la faune du sol dans la culture des azalees. *Rev. Ecol. Biol. Sci.*, **6**(2), 131–45.

Heungens, A. (1969b). The physical decomposition of pine litter by earthworms. *Pl. Soil*, **31**(1), 22–30.

Hilton, J. W. (1983) Potential of dried worm meal as a replacement for fish meal in a trout diet formulation. *Aquaculture*, **32**, 277–83.

Hirst, J. M., Storey, I. F., Ward, W. C. and Wilcox, H. G. (1955) The origin of apple scab epidemics in the Wisbech area in 1953 and 1954. *Pl. Path.*, **4**, 91.

Hobmaier, A. and Hobmaier, M. (1929) Die Entwicklung der Larve des Lungenwurmes *Metastrongylus elongatus (Strongylus paradoscus)* des Schweines und ihr Invasionsweg. *Munch. Tierärzt. Wschr.*, **80**, 365–9.

Hoeksema, K. J. and Jongerius, A. (1959) On the influence of earthworms on the soil structure in mulched orchards. *Proc. Int. Symp. Soil Struct. Ghent 1958*, pp. 188-94.

Hoeksema, K. J., Jongerius, A. and van der Meer, K. (1956) On the influence of earthworms on the soil structure in mulched orchards. *Boor en Spade*, **8**, 183–201.

Hofer, H. (1988) Variation in resource presence, utilization and reproductive success within a population of European badgers (*Meles meles*). *Mamm. Rev.*, **18**, 25–36.

Hoffman, J. A. and Purdy, L. H. (1964) Germination of dwarf bunt (*Tilletia controversa*) teliospores after ingestion by earthworms. *Phytopath.*, **54**, 878–9.

Hogben, L. and Kirk, R. L. (1944) Body temperature of worms in moist and dry air. *Proc. Roy. Soc. Lond.*, **132B**(868), 239–52.

Hogg, T. W. (1895) Immunity of some low forms of life from lead poisoning. *Chem. News*, **71**, 223–4.

Holmstrup, M. (1994) Physiology of cold hardiness in cocoons of five earthworm taxa (Lumbricidae: Oligochaeta). *J. Comp. Physiol. B*, **164**, 222–8.

Holmstrup, M., Hansen, B. T., Nielsin, A. and Østergaard, I. K. (1990) Frost tolerance of lumbricid earthworm cocoons. *Pedobiologia*, **34**, 361–6.

Holmstrup, M. Østergaard, I. K., Nielsen, A. and Hansen, B. T. (1991) The relationship between temperature and cocoon incubation time for some lumbricid earthworm species. *Pedobiologia*, **35**(3), 179–84.

Hoogerkamp, M., Rogaar, H. and Eijsackers, H. J. P. (1983) Effect of earthworms on grassland on recently reclaimed polder soils in the Netherlands, in *Earthworm Ecology, from Darwin to Vermiculture*, (ed. J. E. Satchell), Chapman & Hall, London, pp. 85–105.

Hook, R. I. van (1974) Cadmium, lead and zinc distributions between earthworms and soils: potentials for biological accumulation. *Bull. Environ. Contam. Toxicol.*, **12**, 509–12.

Hopkins, A. R. and Kirk, V. M. (1957) Effects of several insecticides on the English red worm. *J. Econ. Ent.*, **50**(5), 699–700.

Hopp, H. (1946) Earthworm fight erosion too. *Soil Conserv.*, **11**, 252–4.

Hopp, H. (1947) The ecology of earthworms in cropland. *Soil. Sci. Soc. Am. Proc.*, **12**, 503–7.

Hopp, H. (1973) *What Every Gardener should know about Earthworms*, Garden Way Publ., Vermont.

Hopp, H. and Hopkins, H. T. (1946a) Earthworms as a factor in the formation of water-stable aggregates. *J. Soil Water Conserv.*, **I**, 11-13.

Hopp, H. and Hopkins, H. T. (1946b) The effect of cropping systems on the winter populations of earthworms. *J. Soil Water Conserv.*, **I** (1), 85–8, 98.

Hopp, H. and Slater, C. S. (1948) Influence of earthworms on soil productivity. *Soil Sci.*, **66**, 421–8.

Hopp., H. and Slater, C. S. (1949) The effect of earthworms on the productivity of agricultural soil. *J. Agric. Res.*, **78**, 325–39.

Hossein, E. A., Ravasz, K., Zicsi, A. *et al.* (1991) Über das Vorkommen und die Bedeutung von nocardioform Actinomyceten im Darm von Regenwürmern, in *Advances in Management and Conservation of Soil Fauna*, (eds G. K. Veeresh, D. Rajagopal and C. A. Viraktamath), Oxford and IBH, New Delhi, pp. 585–90.

Houpert, G., Jenot, M. and Lardier, P. A. (1982) La sensibilité accrue d'*Eisenia fetida* (Lumbricidae) aux insecticides carbamates en presence d'atrazine. *Bull. Ec. Natl. Super Agron. Ind. Aliment.*, **24**, 3–9.

Howell, C. D. (1939) The response to light in the earthworm *Pheretima agrestis* Goto and Hatai with special reference to the function of the nervous system. *J. Exp. Zool.*, **81**, 231–59.

Howell, D. N. (1974) The worm turns: An investigation of experimentation in the learning abilities of earthworms. *Megadrilogica*, **I**, 1–6.

Hoy, H. M. (1955) Toxicity of some hydrocarbon insecticides to earthworms. *N. Z. J.Sci. Technol (A)*, **37**(4), 367–72.

Hubl, H. (1953) Die inkrorischen Zellelemente im Gehirn der Lumbriciden. *Arch. EntwMech. Org.*, **146**, 421–32.

Hubl, H. (1956) Über die Beziehungen der Neurosekretion zum Regenerations Geschehen bei Lumbriciden nebst Beschreibung eines neuartigen neurosekretorischen Zelltyps in Unterschlundganglion. *Arch. EntwMech Org.*, **149**, 73–87.

Huff, M. L. and Hurson, B. R. (1977) Soil faunal populations, in *Landscape Reclamation Practice*, (ed. B. Hackett), IPC Science and Technology Press, Guildford, pp. 125–47.

Hughes, M. K., Lepp, N. W. and Phipps, D. A. (1980) Aerial heavy metal pollution and terrestrial ecosystems. *Adv. Ecol. Res.*, **11**, 218–37.

Huhta, V. and Karpinnen, E. (1967) Effect of silvicultural practices upon arthropod, annelid and nematode populations in a coniferous forest soil. *Ann. Zool. Fenn.*, **4**, 87–143.

Huhta, V. (1979) Effects of liming and deciduous litter on earthworm (Lumbricidae) populations of a spruce forest, with an inoculata experiment on *Allolobophora caliginosa*. *Pedobiologia*, **19**, 340–5.

Huhta, V. and Haimi, J. (1988) Reproduction and biomass of *Eisenia foetida* in domestic waste, in *Earthworms in Waste and Environmental Management*, (ed. C. A. Edwards and E. F. Neuhauser), SPB Acad. Publ., The Netherlands, pp. 65–70.

Huhta, V., Matlu, N. and Valpas, A. (1969) Further notes on the effect of silvicultural practices on the fauna of coniferous forest soil. *Ann. Zool. Fennici,* **6,** 327–34.

Huhta, V., Hyvönen, R., Koskenniemi, A., Vilkamaa, P., Kaasalainen, P. and Sulander, M. (1986) Response of soil fauna to fertilization and manipulation of pH in coniferous forests. *Acta. For. Fenn.,* **195,** 1–30.

Hunt, L. B. (1965) Kinetics of pesticide poisoning in Dutch Elm Disease control. *U.S. Fish Wildl. Serv. Circ.,* **226,** 12–13.

Hunt, L. B. and Sacho, R. J. (1969) Response of robins to DDT and methoxychlor. *J. Wildl. Manage,* **33,** 267–72.

Huss, M. J. (1989) Dispersal of cellular slime molds by two soil invertebrates. *Mycologia,* **81,** 677–82.

Hutchinson, S. A. and Kamel, M. (1956) The effect of earthworms on the dispersal of soil fungi. *J. Soil Sci.,* **7**(2), 213–18.

Hyche, L. L. (1956) Control of mites infesting earthworm beds. *J. Econ. Ent.,* **49,** 409–10.

Hyman, L. H. (1940) Aspects of regeneration in annelids. *Am. Nat.,* **74,** 513–27.

Ibanez, I. A., Herrera, C. A., Belazques, L. A. and Hebel, P. (1993) Nutritional and toxicological evaluation on rats of earthworm (*Eisenia foetida*) meal as a protein source for animal feed. *Anim. Feed Sci. Tech.,* **42**(1–2), 165–72.

Ilijin, A. M. (1969) The toxic effect of herbicides upon ants and earthworms. *Zool. Zh.,* **48,** pp. 141–3.

Ingelsfield, C. (1984) Toxicity of the pyrethroid insecticides cypermethrin and WL85871 to the earthworm, *Eisenia foetida* (Savigny). *Bull. Environ. Contam. Toxicol.,* **33,** 568–70.

Inoue, T. and Kondo, K. (1962) Susceptibility of *Branchiura sowerbyi, Limrodrilus ocialis* and *L. willeyi* to several agricultural chemicals. *Botyu-bagaku (Japan),* **27,** 97–9.

Ireland, M. P. (1975) The effect of the earthworm *Dendrobaena rubida* on the solubility of lead, zinc, and calcium in heavy metal contaminated soil in Wales. *J. Soil Sci.,* **26,** 313–18.

Ireland, M. P. (1979) Metal accumulation by the earthworms *Lumbricus rubellus, Dendrobaena veneta,* and *Eiseniella tetraedra* living in heavy metal polluted sites. *Environ. Pollut.,* **13,** 201–6.

Ireland, M. P. (1983) Heavy metal uptake and tissue distribution in earthworms, in *Earthworm Ecology from Darwin to Vermiculture,* (ed. J. E. Satchell), Chapman & Hall, London, pp. 247–65.

Ireland, M. P. and Wooton, R. J. (1976) Variations in the lead, zinc and calcium content of *Dendrobaena rubida* (Oligochaeta) in a base metal mining area. *Environ. Pollut.,* **10,** 201–8.

Iwahara, S. and Fujita, O. (1965) Effect of intertrial interval and removal of the suprapharyngeal ganglion upon spontaneous alternation in the earthworm *Pheretima communissima. Jap. Psych. Res.,* **7,** 1–14.

Jacks, G. V. (1963) The biological nature of soil productivity. *Soils and Fert.,* **26**(3), 147–50.

Jacob, A. and Wiegland, K. (1952) Transformations of the mineral nitrogen of fertilisers in the soil. *Z. PflErnähr. Düng.,* **59,** 48–60.

Jaggi, W. and Högger, C. (1993) Eine Regenwurmart erschwert die Bewirtschaftung von Wiesen im Toggenburg. *Landwirtschaft Schweiz Band,* **6,** 169–76.

James, S. W. (1991) Soil, nitrogen, phosphorus, and organic matter processing by earthworms in tallgrass prairie. *Ecology*, **72**, 2101–9.

James, S. W. and T. R. Seastedt (1986) Nitrogen mineralization by native and introduced earthworms: effects on big bluestem growth. *Ecology*, **67**, 1094–7.

Jamieson, B. G. M. (1971a) A review of the megascolecid earthworm genera (Oligochaeta) of Australia, pt. I. Reclassification and checklist of megascolecid genera of the World. 1 *Proc. Roy. Soc. Queensland*, **82**(b), 75–86.

Jamieson, B. G. M. (1971b) A review of the megascolecid earthworm genera (Oligochaeta) of Australia, part II. The subfamilies Ocnerodrilinae and Acanthodrilinae *Proc. Roy. Soc. Queensland*, **82**(8), 95–108.

Jamieson, B. G. M. (1971c) A review of the megascolecid earthworm genera (Oligochaeta) of Australia III, the subfamily Megascolecinae. *Mem. Queensland Mus.*, **16**, 69–102.

Jamieson, B. G. M. (1978) Phylogenetic and phenetic systematics of the opisthoporous Oligochaeta (Annelida: Clitellata). *Evol. Theory*, **3**, 195–233.

Jamieson, B. G. M. (1988) On the phylogeny and higher classification of the Oligochaeta. *Cladistics*, **4**(4), 367–401.

Janda, V. and Gavrilov, K. (1939) Untersuchungen über die Vermehrungsfähigkeit von Individuen einiger Oligochaeten-Arten, die schon vor Erreichung der Geschlechtstreife isolviert wurden. *Vestnik Cesposlov. Zool. Spolecuosti v Praze*, **6–7**, 254–9.

Jeanson, C. (1964) Micromorphology and experimental soil zoology: contribution to the study, by means of giant-sized thin section, of earthworm-produced artificial soil structure, in *Soil Micromorphology*, (ed. A. Jongerius), Proc. 2nd Int. Work meet. soil micromorphology, Arnhem, The Netherlands, pp. 47–55.

Jeanson-Luusinang, C. (1961) Sur une methode d'étude du comportement de la fauna du sol et de sa contribution to pédogenese. *CR Acad. Sci.*, **253**, 2571–3.

Jeanson-Luusinang, C. (1963) Action des Lombricides sur la microflore totale, in *Soil Organisms*, (eds J. Doeksen and J. van der Drift), North Holland, Amsterdam, pp. 260–5.

Jefferson, P. (1955) Studies on the earthworms of turf. A. The earthworms of experimental turf plots. *J. Sports Turf Res. Inst.*, **9**(31), 6–27.

Jefferson, P. (1956) Studies on the earthworms of turf. B. Earthworms and soil. *J. Sports Turf Res. Inst.*, **9**(31), 6–27.

Jensen, J. (1993) Applications of vermiculture technology for managing organic waste resources. *Proc. Ninth Int. Conf. Solid Waste Manage.*, 14–17 November, 1993.

Jin-you, X., Xian-Kuan, Z., Zhi-ren, P. *et al.* (1982a) Experimental research on the substitution of earthworm for fish meal in feeding broilers. *J. South China Norm. Coll.*, **1**, 88-94.

Jin-you, X., Xi-cong, H. and Wen-xi, L. (1982b) An observation on the results of using earthworms as supplementary food for suckling pigs. *J. South China Norm. Coll.*, **1**, 1–8.

Joachim, A. W. R. and Panditesekera, D. G. (1948) Soil fertility studies. IV. Investigations on crumb structure on stability of local soils. *Trop. Agric.*, **104**, 119–39.

Johnson, M. L. (1942) The respiratory function of the haemoglobin of the earthworm. *J. Exp. Biol.*, **18**(3), 266–77.

Johnstone-Wallace, D. B. (1937) The influence of wild white clover on the seasonal production and chemical composition of pasture herbage and upon soil temperatures, soil moistures and erosion control. *4th Int. Grassl. Congr. Rep.*, pp. 188–96.

Jolly, J. M., Lappin-Scott, H. M., Anderson, J. M. and Clegg, C. D. (1993) Scanning electron microscopy of two earthworms: *Lumbricus terrestris* and *Octolasion cyaneum*. *Microb. Ecol.*, **26**, 235–45.

Joschko, M., Graff, O., Mueller, P. C. *et al.* (1991) A non-destructive method for the morphological assessment of earthworm burrow systems in three dimensions by X-ray computed tomography. *Biol. Fertil. Soils*, **11**, 88–92.

Joschko, M., Söchtig, W. and Larink, O. (1992) Functional relationships between earthworm burrows and soil water movement in column experiments. *Soil Biol. Biochem.*, **24**, 1545–7.

Joshi, N. V. and Kelkar, B. V. (1952) The role of earthworms in soil fertility. *Indian J. Agric. Sci.*, **22**, 189–96.

Joyner, J. W. and Harmon, N. P. (1961) Burrows and oscillative behavior therein of *Lumbricus terrestris*. *Proc. Indiana Acad. Sci.*, **71**, 378–84.

Judas, M. (1988) Washing-sieving extraction of earthworms from a broad-leaved litter. *Pedobiologia*, **31**(5–6), 421–4.

Judas, M. (1989) Predator-pressure on earthworms: field experiments in a beechwood. *Pedobiologia*, **33**(5), 339–54.

Judas, M. (1992) Gut content analysis of earthworms (Lumbricidae) in a beechwood. *Soil Biol. Biochem.*, **12**, 1413–17.

Julin, E. (1949) Se sueska daggmaskarterna. *Arkiv. F. Zool.*, **42A**, **17**, 1–58.

Kahsnitz, H. G. (1922) Investigations on the influence of earthworms on soil and plant. *Bot. Arch.*, **I**, 315–51.

Kale, R. D. and Bano, K. (1988) Earthworm cultivation and culturing techniques for production of Vee Comp. 83. E UAS. *Mysore J. Agric. Sci.*, **2**, 339–44.

Kale, R. D. and Bano, K. (1991) Time and space relative population growth of *Eudrilus eugeniae*, in *Advances in Management and Conservation of Soil Fauna*, (eds G. K. Veeresh, D. Rajagopal and C. A. Virakamath), Oxford and IBH, New Dehli, pp. 657–64.

Kale, R. D. and Krishnamoorthy, R. V. (1979) Pesticidal effects of Sevin (1-naphthyl-n-methyl carbamate) on the survivability and abundance of earthworm *Pontoscolex corethrurus*. *Proc. Indian Acad.Sci. Sect. B.*, **88**, 391–6.

Kale, R. D. and Krishnamoorthy, R. V. (1982) Residual effect of Sevin on the acetylcholinesterase activity of the nervous system of the earthworm *Pontoscolex corethrurus*. *Curr. Sci.*, **51**, pp. 885–6.

Kale, R. D., Bano, K. and Krishnamoorthy, R. V. (1982) Potential of *Perionyx excavatus* for utilizing organic wastes. *Pedobiologia*, **23**, 419–25.

Kalmus, H. (1955) On the colour forms of *Allolobophora chlorotica* Sav. *Ann. Mag. Nat. Hist.*, **12**(8), 795.

Kaplan, D. L. (1978) The biochemistry of sludge decomposition, sludge stabilization and humification, in *Utilization of Soil Organisms in Sludge Management*, (ed. R. Hartenstein), Matl Tech. Inf. Services, PB286932, Springfield, Virginia, pp. 78–86.

Kaplan, D. L., Hartenstein, R., Neuhauser, E. F. and Malecki, M. R. (1980) Physicochemical requirements in the environment of the earthworm *Eisenia foetida*. *Soil Biol. Biochem.*, **12**, 347–52.

Karmanova, E. M. (1959) Biology of the nematode *Hystrichis tricolor* Dujardin, 1845, and some data on epizootics in ducks: A histochemical study. *J. Univ. Bombay*, **30B**, 113–25.

Karmanova, E. M. (1963) Interpretation of the developmental cycle in *Dioctophyme renale. Med. Parazitol. Paraziter, Boleznii*, **32**, 331–4.

Karnak, R. E. and Hamelink, J. L. (1982) A standardized method for determining the acute toxicity of chemicals to earthworms. *Ecotoxicol. Environ. Safety*, **6**, 216–22.

Kaushal, B. R., Bisht, S. B. S. and Kalia, S. (1994) Effect of diet on cast production by the megascolecid earthworm *Amynthas alexandri* in laboratory culture. *Biol. Fertil. Soils*, **17**, 14–17.

Keilin, D. (1915) Recherches sur les larves de Dipteres cyclorrhaphes. *Bull. scient. Fr. Belg.*, **47**, 15–198.

Keilin, D. (1925) Parasitic autotomy of the host as a mode of liberation of coelomic parasites from the body of the earthworm. *Parasitology*, **17**, 70–2.

Kelsey, J. M. and Arlidge, G. Z. (1968) Effects of Isobenzan on soil fauna and soil structure. *N. Z. J. Agric. Res.*, **11**, 245–60.

Keogh, R. G. (1979) Lumbricid earthworm activities and nutrient cycling in pasture ecosystems, in *Proc. 2ⁿᵈ Aust. Conf. Grassland Invert. Ecol.*, (eds T. K. Crosby and R. P. Pottinger), Government Printer, Wellington, pp. 49–51.

Keogh, R. G. and Christensen, M. J. (1976) Influence of passage through *Lumbricus rubellus* Hoffmeister earthworms on viability of *Pithomyces chartarum* (Berk. & Curt.). *N. Z. J. Agric. Res.*, **19**, 255–6.

Keogh, R. G. and Whitehead, P. H. (1975) Observations on some effects of pasture spraying with benomyl and carbendazim on earthworm activity and litter removal from pasture. *N.Z. J. Exp. Agric.*, **3**, 103–4.

Ketterings, Q. M., Blair, J. M. and Marinissen, J. Y. C. (1995) The effects of earthworm activity on soil aggregate stability and carbon and nitrogen storage in a legume cover crop-nitrogen-based ecosystem, in press.

Kevan, D. K. Mc. E. (ed.) (1955) *Soil Zoology*, Butterworths, London, pp. 23–8, 452–88.

Khambata, S. R. and Bhatt, J. V. (1957) A contribution to the study of the intestinal microflora of Indian earthworms. *Arch. Mikrobiol.*, **28**, 69–80.

King, H. G. C. and Heath, G. W. (1967) The chemical analysis of small samples of leaf material and the relationship between the disappearance and composition of leaves. *Pedobiologia*, **7**, 192–7.

King, J. W. and Dale, J. L. (1977) Reduction of earthworm activity by fungicides applied to putting green turf. *Ark. Farm. Res.*, **26**, 12.

Kirberger, C. (1953) Untersuchungen über die Temperaturabhagigkeit von Lebensprozessen bei verschiedenen Wirbellosen. *Z. vergl. Physiol.*, **35**, 175–98.

Kitazawa, Y. (1971) Biological regionality of the soil fauna and its function in forest ecosystem types, in *Productivity of Forest Ecosystems*, (ed. P. Duvigneau), Proc. Brussels Symp. Ecol. and Conserv., vol. 4, pp. 485–98.

Kleinig, C. R. (1966) Mats of unincorporated organic matter under irrigated pasture. *Aust. J. Agric. Res.*, **17**, 327–33.

Knight, D., Elliot, P. W., Anderson, J. M. and Scholefield, P. (1992) The role of earthworms in managed, permanent pastures in Devon, England. *Soil Biol. Biochem.*, **24**, 1511–17.

Knollenberg, R. W., Merritt, R. W. and Lawson, D. L. (1985) Consumption of leaf litter by *Lumbricus terrestris* (Oligochaeta) in a Michigan woodland floodplain *Am. Midl. Nat.*, **113**, 1–6.

Knop, J. (1926) Bakterien und Bacteroiden bei oligochäten. *Z. Morph. Ökol. Tiere*, **6**.

Knuutinen, J., Palm, H. and Hakala, H. (1990) Polychlorinated phenols and their metabolism in soil and earthworms of sawmill environment. *Chemosphere*, **20**, 609–23.

Kobatake, M. (1954) The antibacterial substances extracted from lower animals. I. The earthworm. *Kekkabu (Tuberculosis)*, **29**, 60–3.

Kobayashi, S. (1937). On the breeding habit of the earthworms without male pores. 1. Isolating experiments in *Pheretima hilgendorfi [Michaelsen]. Sci. Rep. Tohoku Univ.*, **II**, 473–85.

Kollmannsperger, F. (1934) The Oligochaeta of the Bellinchen Region. Inaugural dissertation. Dilligen (Saargebiet).

Kollmannsperger, F. (1952) Über die Bedeutung des Regenwurmes fur die Fruchtbarkeit des Bodens. *Decheniana*, **105/106**, 165.

Kollmannsperger, F. (1955) Über Rhythmen bei Lumbriciden. *Decheniana*, **180**, 81–92.

Kollmannsperger, F. (1956) Lumbricidae of humid and arid regions and their effect on soil fertility. *VI Congr. Int. Sci Sol. Rapp. C.*, pp. 293–7.

Korschelt, E. (1914) Über Transplantationsversuche, Ruhezustände und Lebensdauer der Lumbriciden. *Zool. Anz.*, **43**, 537–55.

Korschgen, L. J. (1970) Soil–food-chain–pesticide wildlife relationships in aldrin-treated fields. *J. Wildl. Manage.*, **34**, 186–99.

Kozlovskaya, L. S. and Zaguralskaya, L. M. (1966) Relationships between earthworms and microbes in W. Siberia. *Pedobiologia*, **6**, 244–57.

Kozlovskaya, L. S. and Zhdannikova, E. N. (1961) Joint action of earthworms and microflora in forest soils. *Dokl. Akad. Nauk. SSSR.*, **139**, 470–3.

Kreis, B., Edwards, P., Cuendet, G. and Tarradellas, J. (1987) The dynamics of PCBs between earthworm populations and agricultural soils. *Pedobiologia*, **30**, 379–88.

Kretzschmar, A. (1978) Quantification écologique des galeries de lombriciens: techniques et premières estimations. *Pedobiologia*, **18**, 31–8.

Kretzschmar, A. (1987) Soil partitioning effect of an earthworm burrow system. *Biol. Fertil. Soils*, **3**, 121–5.

Kretzschmar, A. (1990) 3D images of natural and experimental earthworm burrow systems. *Rev. Ecol. Biol. Sol*, **27**(4), 407–14.

Kretzschmar, A. (1992) ISEE 4, Fourth International Symposium on Earthworm Ecology. *Soil Biol. Biochem.*, **24**(12), 1193–774.

Kretzschmar, A. and Bruchou, C. (1991) Weight response to the soil water potential of the earthworm *Aporrectodea longa. Biol. Fertil. Soils*, **12**, 209–12.

Kring, J. B. (1969) Mortality of the earthworm *Lumbricus terrestris* L. following soil applications of insecticides to a tobacco field. *J. Econ. Ent.*, **62**(4), 963.

Krishnamoorthy, R. V. (1990) Mineralization of phosphorus by faecal phosphatases of some earthworms of Indian tropics. *Proc. Indian Acad. Sci. (Anim. Sci.)* **99**, 509–18.

Krishnamoorthy, R. V. and Vajranabhaiah, S. N. (1986) Biological activity of earthworm casts: an assessment of plant growth promoter levels in the casts. *Proc. Ind. Acad. Sci. (Anim. Sci.)*, **95**, 341–51.

Krištůfek, V., Pizl, V. and Szabo, I. M. (1990) Composition of the intestinal streptomycete community of earthworms (Lumbricidae), in *Microbiology in Poecilotherms*, (ed. R. Lessel), Elsevier, Amsterdam, pp. 137–40.

Krištůfek, V., Ravasz, K. and Pizl, V. (1992) Changes in density of bacteria and microfungi during gut transit in *Lumbricus rubellus* and *Aporrectodea caliginosa* (Oligochaeta: Lumbricidae). *Soil Biol. Biochem.*, **24**, 1499–500.

Krištůfek, V., Ravasz, K. and Pizl, V. (1993) Actinomycete communities in earthworm guts and surrounding soil. *Pedobiologia*, **37**, 379–84.

Krivanek, J. O. (1956) Habit formation in the earthworm, *Lumbricus terrestris*. *Physiol. Zool.*, **29**, 241–50.

Krivolutsky, D. A., Tichomirova, A. L. and Turcaninova, V. A. (1972) Strukturänderungen des Tierbesatzes (Hand und Bodenwirbellose) unter dem Einfluss der Kontamination des Bodens mit Sr⁹⁰. *Pedobiologia*, **12**, 374–80.

Krivolutsky, D., Turcaninova, V. and Mikhaltsova, Z. (1982) Earthworms as bioindicators of radioactive soil pollution. *Terrest. Oligochaeta*, 263–5.

Krüger, F. (1952) Über die Beziehung des Sauerstoffverbauchs zum Gewicht bei *Eisenia foetida* Sav. *Z. vergl. Physiol.*, **34**, 1–5.

Kruglikov, B. A., Shtefan, M. K., Tarasenko, T. Y. A. and Melnik, R. I. (1987) Persistence of hoof-and-mouth disease virus in bodies of earthworms. *Veterinariya*, **1**, 26–7.

Krupka, R. M. (1974) On the anticholinesterase activity of benomyl. *Pest. Sci.*, **5**, 211-16.

Kruse, E. A. and Barrett, G. W. (1985) Effects of municipal sludge and fertilizer on heavy metal accumulation in earthworms. *Environ. Poll. (Ser. A)*, **38**, 235–44.

Kruuk, H. and Parish, T. (1985) Food, availability and weight of badgers (*Meles meles*) in relation to agricultural changes. *J. App. Ecol.*, **22**(3), 705-15.

Kubiena, W. L. (1953) *Bestimmungsbuch und Systematik der Böden Europas*, Stuttgart.

Kubiena, W. L. (1955) Animal activity in soils as a decisive factor in establishment of humus forms, in *Soil Zoology*, (ed. D. K. Mc. E. Kevan), Butterworths, London, pp. 73–82.

Kühle, J. C. (1983) Adaptation of earthworm populations to different soil treatments in an apple orchard, in *New Trends in Soil Biology*, (ed. Ph. Lebrun), Proc. 8th Intl Coll. Soil Zool., Louvain-la-Neuve, 1982, pp. 487–501.

Kuhnelt, W. (1961) *Soil Biology*, Faber and Faber, London.

Kuperman, R. G. (1993) Relationships between acid deposition, soil invertebrate, communities, microbial activity and litter decomposition in oak–hickory forests. PhD Thesis, Ohio State University.

Kurcheva, G. F. (1960) The role of invertebrates in the decomposition of the oak leaf litter. *Pochvovedenie*, (4), 1623.

Ladell, W. R. S. (1936) A new apparatus for separating insects and other arthropods from the soil. *Ann. Appl. Biol.*, **23**, 862–79.

Lakhani, K. H. and Satchell, J. E. (1970) Production by *Lumbricus terrestris* (L.). *J. Anim. Ecol.*, **39**, 473–92.

Lal, R. (1974) No-tillage effects on soil properties and maize (*Zea mays* L.) production in Western Nigeria. *Pl. Soil*, **40**, 321–31.

Lan van der, H. and Aspöck, H. (1962) Zur Wirkung von Sevin auf Regenwürmer. *Anz. Schädlingsk.*, **35**, 180–2.

Langmaid, K. K. (1964) Some effects of earthworm invasion in virgin podsols. *Can. J. Soil Sci.*, **44**, 34–7.

Lauer, A. R. (1929) Orientation in the earthworm. *Ohio. J. Sci.*, **29**,179.

Lavelle, P. (1971) Étude préliminaire de la nutrition d'un ver de terre Africain. *Millsonia anomala* (Acanthodrilidae, Oligochaetes). *Ann. Zool. Ecol. Anim. Special Publ.*, **4**, 131–46.

Lavelle, P. (1974) Les vers de terre de la savanne de Lamto, in *Analyse d'un Ecosysteme Tropical Humide: La Savanne de Lamto (Cote d'Ivoire)*. *Bull. de Liaison des Chercheurs de Lamto, No. Spec.*, **5**, 133–6.

Lavelle, P. (1975) Consommation annuelle d'une population naturelle de vers de terre (*Millsonia anomala* Omodes, Acanthodrilidae: Oligochaetes) dans la savanne de Lamto (Cote d'Ivoire), in *Progress in Soil Zoology*, (ed. J. Vanek), Academia Publishing House, Prague, pp. 299–304.

Lavelle, P. (1978) Les vers de terre de la savanne de Lamto (Côte d'Ivoire). Peuplements, populations et fonctions de l'écosystème. *Publ. Lab. Zool. E.N.S.*, **12**, 1–301.

Lavelle, P. (1979) Relations entre types ecologiques et profils demographiques chez les vers de terre de la savanne de Lamto (Cote d'Ivoire). *Rev. Ecol. Biol. Sol*, **16**, 85–101.

Lavelle, P. (1981) Strategies de reproductia chez les vers de terre. *Acta Oecologia Gen.*, **2**, 11–133.

Lavelle, P. (1983) The structure of earthworm communities, in *Earthworm Ecology, from Darwin to Vermiculture*, (ed. J. E. Satchell), Chapman & Hall, London, pp. 449–66.

Lavelle, P. (1988) Earthworms and the soil system. *Biol. Fertil. Soil*, **6**, 237–51.

Lavelle, P. (1992) Conservation of soil fertility in low-input agricultural systems of the humid tropics by manipulating earthworm communities (macrofauna project). European Economic Community Project No. TS2-0292-F (EDB).

Lavelle, P. and Martin, A. (1992) Small-scale and large-scale effects of endogeic earthworms on soil organic matter dynamics in soil of the humid tropics. *Soil Biol. Biochem.*, **24**, 1491–8.

Lavelle, P. and Pashanasi, B. M. (1989) Soil macrofauna and land management in Peruvian Amazonia. *Pedobiologia*, **33**, 283–91.

Lavelle, P., Douhalei, N. and Sow, B. (1974) Influence de l'humidité du sol sur la consommation et la croissance de *Millsonia anomala* (Oligochaetes-Acanthrodilidae) dans la Savanne de Lamto. *Ann. Univ. d'Abidjan*, 7(1), 305–14.

Lavelle, P., Sow, B. and Schaefer, R. (1980) The geophagous earthworms community in the Lamto savanna (Ivory Coast): Niche partitioning and utilization of soil nutritive resources, in *Soil Biology as Related to Land Use Practices*, (ed. D. L. Dindal), Proc. 7th Intl Soil Zool. Coll., Syracuse, 1979, Environmental Protection Agency, Washington DC, pp. 653–72.

Lavelle, P., Rangel, P. and Kanyonyo, J. (1983) Intestinal mucus production by two species of tropical earthworms *M. lamtoniana* and *P. corethrurus*, in *New Trends in Soil Biology*, (ed. P. Lebrun), Dieu Brichart Press, Louvain le Neuve, Belgium, pp. 405–10.

Lavelle, P., Barois, I., Martin, A. *et al.* (1989) Management of earthworm populations in agroecosystems: a possible way to maintain soil quality, in *Ecology of Arable Land*, (ed. M. Clarkolm and L. Bergswain), Kluiven, Stockholm, pp. 109–22.

Lavelle, P., Melendez, G., Pashanashi, B. and Schaefer, R. (1992) Nitrogen mineralization and reorganization in casts of the geophagous tropical earthworm *Pontoscolex corethrurus* (Glossoscolecidae). *Biol. Fertil. Soils*, **14**, 49–53.

Lavelle, P., Blanchart, E., Martin, A. *et al.* (1993) Impact of soil fauna on the properties of soils in the humid tropics, in *Myths and Science of Soils in the Tropics*,(eds P. A. Sanchez and R. Lal), SSSA Spec. Publ. 29, Madison, WI, pp. 157–85.

Laverack, M. S. (1960a) Tactile and chemical perception in earthworms. I. Responses to touch, sodium chloride, quinine and sugars. *Comp. Biochem. Physiol.*, **I**, 155–63.

Laverack, M. S. (1960b) The identity of the porphyrin pigments of the integument of earthworms. *Comp. Biochem. Physiol.*, **I**(4), 259–66.

Laverack, M. S. (1961a) Tactile and chemical perception in earthworms. II. Responses to acid pH solutions. *Comp. Biochem. Physiol.*, **2**(1), 22–34.

Laverack, M. S. (1961b) The effect of temperature changes on the spontaneous nervous activity of the isolated nerve cord of *Lumbricus terrestris*. *Comp. Biochem. Physiol.*, **3**(2), 136–40.

Laverack, M. S. (1963) *The Physiology of Earthworms*, Pergamon Press, London.

Lawrence, R. D. and Millar, H. R. (1945) Protein content of earthworms. *Nature, Lond.*, **155**(39), 517.

Leaky, R. J. G. and Proctor, J. (1987) Invertebrates in the litter and soil at a range of altitudes on Gunug Silam. *J. Trop. Ecol.*, **3**, 119–29.

Lebrun, P., De Medts, A. and Wauthy, G. (1981) Comparative ecotoxicology and bioactivity of three carbamate insecticides on an experimental population of the earthworm *Lumbricus herculeus*. *Pedobiologia*, **21**, 225–35.

Lee, K. E. (1951) Role of earthworms in New Zealand soil. *Tuatara*, **4**(1), 22–7.

Lee, K. E. (1958) Biological studies of some grassland tussock soils. X. Earthworms. *N. Z. J. Agric. Res.*, **1**, 998–1002.

Lee, K. E. (1959) A key for the identification of New Zealand earthworms. *Tuatara*, **8**(I), 13–60.

Lee, K. E. (1981) Earthworms (Annelida: Oligochaeta) of Vanua Tu (New Hebrides Islands). *Aust. J. Zool.*, **29**, 535–72.

Lee, K. E. (1983) The influence of earthworms and termites on soil nitrogen cycling, in *New Trends in Soil Biology*, (eds Ph. Lebrun, H. M. André, A. de Medts, *et al.*). Proceedings of the VIII International Colloquium of Soil Zoology, Louvain-la-Neuve, Belgium, pp. 35–48.

Lee, K. E. (1985) *Earthworms: Their Ecology and Relationships with Soils and Land Use*, Academic Press, Sydney.

Lee, K. E. (1987) Peregrine species of earthworms, in *On Earthworms*, (eds A. M. Bonvicini-Pagliai and D. Omodeo), Selected Symposia and Monographs 2, Mucchi Editore, Modena, Italy, pp. 281–95.

Lee, K. and R. C. Foster (1991) Soil fauna and soil structure. *Aust. J. Soil Res.*, **29**, 745–76

Leemput, L. van, Swysen, E., Woestenborghs, R. *et al.*(1989) On the terrestrial toxicity of the fungicide imazalil to the earthworm species *Eisenia foetida*. *Ecotoxicol. Environ. Safety*, **18**, 313–20.

Leger, R. G. and Millette, G. J. F. (1977) The resistance of earthworms *Lumbricus terrestris* and *Allolobophora turgida* to Captan 50 w.p. *Rev. Can. Biol.*, **36**, 351–3.

Legg, D. C. (1968) Comparison of various worm-killing chemicals. *J. Sports Turf Res. Inst.*, **44**, 47–8.

Lensi, R., Domenach, A. M. and Abbadie, L. (1992) Field study of nitrification and denitrification in a wet savanna of West Africa (Lamto, Côte d'Ivoire). *Plant and Soil*, **147**, 107–13.

Lesser, E. J. (1910) Chemische Prozesse bei Regenwürmern. 3. Über anoxybiotische Zersetzung des Glykogens. *Z. Biol.*, 50.

Lhoste, J. (1975) Preliminary investigations into the action of beet herbicides on the environment. *3e Réunion Internationale sur le desherbage Selectif en Cultures de Betteraves*, Paris, 1975, pp. 483–93.

Lidgate, H. J. (1966) Earthworm control with chlordane. *J. Sports Turf Res. Inst.*, **42**, 5–8.

Ligthart, T. N., Peek, G. J. W. C. and Taber, E. J. (1993) A method for the three-dimensional mapping of earthworm burrow systems. *Geoderma*, **57**, 129–41.

Lindquist, B. (1941) Investigations on the significance of some Scandinavian earthworms in decomposition of leaf litter and the structure of mull soil. *SvenskSkogs v Fören Tidskr.*, **39**(3), 179–242.

Lipa, J. J. (1958) Effect on earthworm and Diptera populations of BHC dust applied to soil. *Nature, Lond.*, **181**, 863.

Ljungström, P. O. (1964) Ekologin hos daggmaskar i Stockholmstrakten. *Fält. Biologen.*, **2**, 8–12.

Ljungström, P. O. (1969) On the earthworm genus *Udeina* in South Africa. *Zool. Anz.*, **182**, 370–9.

Ljungström, P. O. (1970) Introduction to the study of earthworm taxonomy. *Pedobiologia*, **10**, 265–85.

Ljungström, P. O. (1972a) Introduced earthworms of South Africa. On their taxonomy, distribution, history of introduction and on the extermination of endemic earthworms. *Zool. Jahrb. Syst.*, **99**, 1–18.

Ljungström, P. O. (1972b) Taxonomical and ecological notes on the earthworm genus *Udeina* and a requiem for the South African acanthodrilines. *Pedobiologia*, **12**, 100–10.

Ljungström, P. O. and Emiliani, F. (1971) Contribucion al conocimiento de la ecologiz y distribution geografica de las lombrices de tierra (Oligoquetos) de la Prov. de Santa Fe (Argentina). *Idia (Buenos Aires)*, August 1971, 19–32.

Ljungström, P. O. and Reinecke, A. J. (1969) Ecology and natural history of the microchaettid earthworms of South Africa. 4. Studies on influence of earthworms upon the soil and the parasitological question. *Pedobiologia*, **9**(1–2), 152–7.

Ljungström, P.-O., Orellana, J. A. de, and Priano, L. J. J. (1973) Influence of some edaphic factors on earthworm distribution in Sante Fe Province, Argentina. *Pedobiologia*, **13**, 236–47

Loehr, R. C., Martin, J. H., Neuhauser, E. F. and Malecki, M. R. (1984) *Waste Management Using Earthworms – Engineering and Scientific Relationships*, Final Report, Project ISP-8016764. National Science Foundation.

Loehr, R. C., Martin, J. H. and Neuhauser, E. F. (1988) Stabilization of liquid municipal sludge using earthworms, in *Earthworms in Waste and Environmental Management* (ed. C. A. Edwards and E. F. Neuhauser), SPB Acad. Publ., The Hague, The Netherlands, pp. 303–13.

Lofaldli, L., Kalas, J. A. and Fiske, P. (1992) Habitat selection of great snipe (*Gallinago media*) during breeding. *Ibis*, **134**, 35–43.

Lofs-Holmin, A. (1980) Measuring growth of earthworms as a method of testing sublethal toxicity of pesticides. *Swed. J. Agric. Res.*, **10**, 25–33.

Lofs-Holmin, A. (1982a) Influence on routine pesticide spraying on earthworms (Lumbricidae) in field experiments with winter wheat. *Swed. J. Agric. Res.*, **12**, 121–3.

Lofs-Holmin, A. (1982b) Measuring cocoon production of the earthworm *Allolobophora caliginosa* (Sav.) as a method of testing sublethal toxicity of pesticides; an experiment with benomyl. *Swed. J. Agric. Res.*, **12**, 117–19.

Lofs-Holmin, A. (1983) Influence of agricultural practices on earthworms (Lumbricidae). *Acta Agric. Scand.*, **33**, 225–334.

Lofs-Holmin, A. and Bostrom, U. (1988) The use of earthworms and other soil animals in pesticide testing, in *Earthworms in Waste and Environmental Management*, SPB Acad. Publ., The Hague, The Netherlands, pp. 303–13.

Lofty, J. R. (1972) The effects of gamma radiation on earthworms. *Pedobiologia*, personal communication.

Long, W. H., Anderson, H. L. and Isa, A. L. (1967) Sugarcane growth responses to chlordane and microarthropods, and effects of chlordane on soil fauna. *J. Econ. Ent.*, **60**, 623–9.

Lopez-Hernandez, D., Lavelle, P., Fardeau, J. C. and Niño, M. (1993) Phosphorous transformation in two P-sorption contrasting tropical soils during transit through *Pontoscolex corethrurus* (Glossoscolecidae: Oligochaeta). *Soil Biol. Biochem.*, **25**, 789–92.

Loquet, M., Bhatnagar, T., Bouché, M. B. and Rouelle, J. (1977) Essai d'estimation de l'influence écologique des lombrices sur les microorganismes. *Pedobiologia*, **17**, 400–17.

Low, A. J. (1955) Improvements in the structural state of soils under leys. *J. Soil Sci.*, **6**, 179–99.

Low, A. J. (1972) The effect of cultivation on the structure and other physical characteristics of grassland and arable soils (1945–1970). *J. Soil Sci.*, 363–80.

Luckman, W. H. and Decker, G. C. (1960) A 5-year report of observations in the Japanese beetle control area at Sheldon, Illinois. *J. Econ. Ent.*, **53**, 821–7.

Luff, M. L. and Hutson, B. R. (1977) Soil faunal populations, in *Landscape Reclamation Practice*, (ed. B. Hackett), IPC Science and Technology Press, Guildford, UK, pp. 125–47.

Lukose, J. (1960) A note on an association between two adult earthworms. *Curr. Sci.*, **29**, 106–7.

Lund, E. E., Wehr, E. E. and Ellis, D. J. (1963) Role of earthworms in transmission of *Heterakis* and *Histomonas* to turkeys and chickens. *J. Parasit.*, **49**(5), 50.

Lunt, H. A. and Jacobson, G. M. (1944) The chemical composition of earthworm casts. *Soil Sci.*, **58**, 367.

Lyons, C., Milsom, N., Morgan, N. G. and Stringer, A. (1972) The effects of repeated applications of the grass suppressant maleic hydrazide on an orchard sward and on the soil fauna. *Proc. 11th Brit. Weed Contr. Conf.*, 1972, pp. 356–9.

Ma, W. (1984) Sublethal toxic effects of copper on growth, reproduction and litter breakdown activity in the earthworm *Lumbricus rubellus*, with observations on the influence of temperature and soil pH. *Environ. Pollut.*, **33**, 207–19.

Ma, W., Edelman, T., van Beersum I. and Jans, T. (1983) Uptake of cadmium, zinc, lead and copper by earthworms near a zinc-smelting complex: influence of soil pH and organic matter. *Bull. Environ. Contam. Toxicol.*, **30**, 424–7.

McCalla, T. M. (1953) Microbiology studies of stubble mulching. *Nebr. Agric. Exp.Stn Bull.*, p. 417.

McColl, H. P. (1984) Nematicides and field populations of enchytraeids and earthworms. *Soil Biol. Biochem.*, **16**, 139–43.

McColl, H. P. and Latour, M. L. de (1978) Interactions between the earthworm *Allolobophora caliginosa* and ryegrass (*Lolium perenne*) in subsoil after topsoil stripping, in *Proc. 3rd Australasian Conf. Grassland Ecol.*, (ed. K. E. Lee), pp. 321–30.

McColl, H. P., Hart, P. B. S. and Cook, F. J. (1982) Influence of earthworms on some soil chemical and physical properties, and the growth of ryegrass on a soil after topsoil stripping – a pot experiment. *N. Z. J. Agric. Res.*, **25**, 229–43.

McCredie, T. A., Parker, C. A. and Abbott, I. (1992) Population dynamics of the earthworm *Aporrectodea trapezoides* in a Western Australian pasture. *Biol. Fertil. Soils*, **12**(4), 285–9.

MacDonald, D. W. (1980) The red fox, *Vulpes vulpes*, as a predator upon earthworms, *Lumbricus terrestris*. *Z. Tierpsychol.*, **52**, 171–200.

MacDonald, D. W. (1983) Predation on earthworms by terrestrial vertebrates, in *Earthworm Ecology from Darwin to Vermiculture*, (ed. J. E. Satchell), Chapman & Hall, London, pp. 393–414.

McIlveen, W. D. and Cole, H. Jr (1976) Spore dispersal of Endogonadaceae by worms, ants, wasps, and birds. *Can. J. Bot.*, **54**, 1486–9.

McInroy, D. (1971) Evaluation of the earthworm *Eisenia foetida* as food for man and domestic animals. *Feedstuffs*, **43**, 37–47.

MacKay, A. D. and Kladivko, E. J. (1985) Earthworms and the rate of breakdown of soybean and maize residues in soil. *Soil Biol. Biochem.*, **17**, 851–7.

MacKay, A. D., Syers, J. K., Springett, J. A. and Gregg, P. E. H. (1982). Plant availability of phosphorus in superphosphate and a phosphate rock as influenced by earthworms. *Soil Biol. Biochem.*, **14**, 281–7.

McLeod, J. H. (1954) Note on a staphylinid (Coleoptera) predator of earthworms. *Can. Ent.*, **86**, 236.

McRill, M. (1974) The ingestion of weed seed by earthworms. *Proceedings of the 12th Weed Control Conference*, **2**, 519–24.

Madge, D. S. (1966) How leaf litter disappears. *New Scientist*, **32**, 113–15.

Madge, D. S. (1969) Field and laboratory studies on the activities of two species of tropical earthworms. *Pedobiologia*, **9**, 188–214.

Madill, J., Coates, K. A., Wetzel, M. J. and Gelder, S. R. (1992) Common and scientific names of Aphanoneuran and Clitellate annelids of the United States of America and Canada. *Soil Biol. Biochem.*, **24**(12), 1259–62.

Madsen, E. L. and Alexander, M. (1982) Transport of *Rhizobium* and *Pseudomonas* through soil. *Soil Sci. Soc. Am. J.*, **46**, 557–60.

Magalhaes, P. S. (1892) Notes d'helminthologie brésilienne. *Bull. Soc. Zool. France*, **17**, 145–6.

Maldague, M. and Couture, G. (1971) Utilization de litieres radioactives par *Lumbricus terrestris*. *Ann. Zool. Ecol. Anim. Special Publ.*, **4**, 147–52.

Malecki, M. R., Neuhauser, E. F. and Loehr, R. C. (1982) The effect of metals on the growth and reproduction of *Eisenia foetida* (Oligochaeta, Lumbricidae). *Pedobiologia*, **24**, 129–37.

Malone, C. R. and Reichle, D. E. (1973) Chemical manipulation of soil biota in a fescue meadow. *Soil Biol. Biochem.*, **5**, 629–39.

Mamytov, A. (1953) The effect of earthworms on the water stability of mountain-valley serozem soils. *Pochvovedenie*, **8**, 58–60.

Mangold, O. (1951) Experiments in analysis of the chemical senses of earthworms. I. Methods and procedure for leaves of plants. *Zool. Jb. (Physiol.)*, **62**, 441–512.

Mangold, O. (1953) Experimente zur Analyse des chemischen Sinns des Regenwurms. 2. Versuche mit Chinin, Säuren und Süsstoffen. *Zool. Jb. Abt. Allem. Zool. Physiol. Tiere.*, **63**, 501–57.

Mansell, G. P., Syers, J. K. and Gregg, P. E. H. (1981) Plant availability of phosphorus in dead herbage ingested by surface-casting earthworms. *Soil Biol. Biochem.*, **13**, 163–7.

Marapao, B. P. (1959) The effect of nervous tissue extracts on neurosecretion in the earthworm *Lumbricus terrestris*. *Catholic U. Amer. Biol. Stud.*, **55**, 1–34.

Máriaglieti, K. (1979) On the community structure of the gut-microbiota of *Eisenia lucens* (Annelida, Oligochaeta). *Pedobiologia*, **19**, 213–20.

Marinissen, J. Y. C. and Dexter, A. R. (1990) Mechanisms of stabilization of earthworm casts and artificial casts. *Biol. Fertil. Soils*, **9**, 163–7.

Marinissen, J. Y. C. and van den Bosch, F. (1992) Colonization of new habitats by earthworms. *Oecologica*, **91**(3), 371–6.

Marquerie, J. M. and Simmons, J. W. (1988) A method to assess potential bioavailability of contaminants, in *Earthworms in Waste and Environmental Management*, (ed. C.A. Edwards and E. F. Neuhauser), SPB Acad. Publ., The Hague, The Netherlands, pp. 367–76.

Marshall, V. G. (1971) Effects of soil arthropods and earthworms on the growth of Black Spruce. *Ann. Zool. Ecol. Anim. Special Publ.*, **4**, 109–18.

Marshall, V. G. (1977) *Effects of manures and fertilizers on soil fauna: a review.* Commonwealth Bureau of Soils, Harpenden.

Martin, A. (1991) Short- and long-term effects of the endogeic earthworm *Millsonia anomala* (Omodeo) (Megascolecidae, Oligochaeta) of tropical savannas on soil organic matter. *Biol. Fertil. Soils*, **11**, 234–8.

Martin, A., Balesdent, J. and Mariotti, A. (1991) Earthworm diet related to soil organic matter dynamics through ^{13}C measurements. *Oecologia*, **91**, 23–9.

Martin, A., Mariotti, A., Balesdent, J. and Lavelle, P. (1992) Soil organic matter assimilation by a geophagous tropical earthworm based on carbon-13-measurements. *Ecology*, **73**, 118–28.

Martin, A. W. (1957) Recent advances in knowledge of invertebrate renal function, in *Invertebrate Physiology*, (ed. B. T. Scheer), University of Oregon Pub., pp. 247–76.

Martin, L. W. and Wiggans, S. C. (1959) The tolerance of earthworms to certain insecticides, herbicides, and fertilizers. *Okla. Agric. Exp. Stn Process Ser.* 344.

Martin, M. J. and Coughtrey, P. J. (1975) Preliminary investigation of the levels of cadmium in a contaminated environment. *Chemosphere*, **4**, 155–60.

Martin, N. A. (1976) Effect of four insecticides on the pasture ecosystem. V. Earthworms (Oligochaeta: Lumbricidae) and Arthropoda extracted by wet sieving and salt flotation. *N. Z. J. Agric. Res.*, **19**, 111–15.

Martin, N. A. (1977) Guide to the lumbricid earthworms of lumbricid pastures. *N. Z. J. Exp. Agric.*, **5**, 301–9.

Martin, N. A. (1982) The effects of herbicides used on asparagus on the growth rate of the earthworm *Allolobophora caliginosa*. *Proc. 35th NZ Weed and Pest Control Conf.*, pp. 328–31.

Martinucci, G. B., Crespi, P., Omodeo, P., *et al.* (1983) Earthworms and TCDD (2,3, 7,8-tetrachlorodibenzo-p-dioxin) in Seveso, in *Earthworm Ecology from Darwin to Vermiculture* (ed. J. E. Satchell), Chapman & Hall, London, pp. 275–83.

Mather, J. G. and Christensen, O. (1988) Surface movements of earthworms in agricultural land. *Pedobiologia*, **32**, 399–405.

Mather, J. G. and Christensen, O. (1992) Surface migration of earthworms in grassland. *Pedobiologia*, **36**(1), 51–7.

Mazaud, D. and Bouché, M. B. (1980) Introductions sur population et migrations of lombriciens marques, in *Soil Biology as Related to Land Use Practices*, (ed. D. L. Dindal). Proc. 7th Intl Soil Zool Coll., Syracuse, 1979, EPA, Washington DC, pp. 687–701.

Mba, C. (1987) Vermicomposting and biological N-fixation, in *Proc. 9th Int. Symp. on Soil Biol. and Conserv. of the Biosphere*, (ed. J. Szegi), Akadémiai Kiadó, Budapest, pp. 547–52.

Medts, A. de. (1981) Effects de residus de pesticides sur les lombriciens en terre de culture. *Pedobiologia.*, 21, 439–45.

Meggitt, F. J. (1914) On the anatomy of a fowl tapeworm, *Amoebotaenia sphenoides v. Linstow*. *Parasitology*, 7, 262–77.

Meijer, J. (1972) An isolated earthworm population in the recently reclaimed Lauwerseepolder. *Pedobiologia*, 12, 409–11.

Meinhardt, U. (1976) Dauerhafte Markierung von Regenwürmern durch ihre Lebendfärbung. *Nachrichtenbl. Dtsch. Pflanzenschutdienstes (Braunschw.)*, 28, 84–6.

Mekada, H., Hayashi, N., Yokota, H. and Okomura, J. (1979) Performance of growing and laying chickens fed diets containing earthworms. *J. Poult. Sci.*, 16, 293–7.

Mellanby, K. (1961) Earthworms and the soil. *Countryside*, 14(4), 1.

Mendes, E. G. and Almeida, A. M. (1962) The respiratory metabolism of tropical earthworms. III. The influence of oxygen tension and temperature. *Bol. Fac. filos. cience. e. letras Univ. S. Paulo Zool.*, 24, 43–65.

Merker, E. and Braunig, G. (1927) Die Empfindlichkeit feuchthäutiger Tiere im Lichte. 3. Die Atemnot feuchthäutiger Tiere in Licht der Quarzquecksiblerlampe. *Zool. Jb. Abt. Allgem. Zool. Physiol. Tiere*, 43, 275–338.

Meyer, L. (1943) Experimental study of macrobiological effects on humus and soil formation. *Bodenk. u. PflErnähr.*, 29(74), 119–40.

Michaelsen, W. (1903) *Die geographische Verbreitung der Oligochaeten*. Berlin.

Michaelsen, W. (1910) Die Oligochätenfauna der vorderindischceylonischen Region. *Abh. Naturw. Hamburg.*, 19.

Michaelsen, W. (1919) Über die Beziehungen der Hirudineen zu den Oligochäten. *Mitt. naturh. Mus. Hamburg*, 36.

Michaelsen, W. (1921) Zur Stammesgeschichte und Systematik der Oligochäten, insbesondere der Lumbriculiden. *Arch. Naturgesch.*, 86.

Michaelsen, W. (1922) Die Verbreitung der Oligochäten in Lichte der Wegener'schen Theorie der Frontinentverschiebung. *Verh. Ver. naturs. Unterh. Hamburg*, 3, 29.

Michaelsen, W. (1926) *Pelodrilus bureschi*, ein Süsswasser Höhlenoligochät aus Bulgarien. *Arb. Bulgar. Naturf. Ges.*, 12, 57–66.

Michon, J. (1949) Influence of desiccation on diapause in Lumbricids. *C.r. hebd. Séanc. Acad. Sci., Paris*, 228(18), 1455–6.

Michon, J. (1951) Supernumerary regeneration in *A. terrestris f. typica*. *C. r. hebd. Séanc. Acad. Sci., Paris*, 232, 1449–51.

Michon, J. (1954) Influence de l'isolement à partir de la maturité sexuelle sur la biologie des Lumbricidae. *C.r. hebd. Séanc. Acad. Sci., Paris*, 238, 2457–8.

Michon, J. (1957) Contribution experimentale à étude de la biologie des Lumbricidae. *Année Biol.*, 33(7–8), 367–76.

Miles, H. B. (1963a) Soil protozoa and earthworm nutrition. *Soil Sci.*, 95, 407–9.

Miles, H. B. (1963b) Heat-death temperature in *Allolobophora terrestris* f. *longa* and *Eisenia foetida*. *Nature, Lond.*, **199**, 826.

Mill, P. J. (1982) Recent developments in earthworm neurobiology. *Comp. Biochem. Physiol.*, **73A**(4), 641–61.

Millott, N. (1944) The visceral nerves of the earthworm. 3. Nerves controlling secretion of protease in the anterior intestine. *Proc. R. Soc.*, **132**, 200–12.

Milne, D. L. and du Toit, W. (1976) The effect of citrus nematicides on the earthworm population in the soil. *Citrus Grow. Sub-Trop. Fruit J.*, **13**, 15.

Mishra, P. C. and Dash, M. C. (1984) Population dynamics and respiratory metabolism of earthworm population in a sub tropical dry woodland of western Orissa, India. *Trop. Ecol.*, **25**, 103–16.

Mitchell, M. J. (1978) Role of invertebrates and microorganisms in sludge decomposition, in *Utilization of Soil Organisms in Sludge Management*, (ed. R. Hartenstein), Natl. Tech. Inf. Services, PB286932, Springfield, Virginia, pp. 35–50.

Moeed, A. (1975) Effects of isobenzan, fensulfothion and diazinon on invertebrates and microorganisms. *NZ J. Exp. Agric.*, **3**, 181–5.

Moeed, A. (1976) Birds and their food resources at Christchurch International Airport, New Zealand. *N. Z. J. Zool.*, **3**, 373–90.

Moment, G. B. (1953a) The relation of body level, temperature and nutrition to regenerative growth. *Physiol. Zool.*, **26**, 108–17.

Moment, G. B. (1953b) A theory of growth limitation. *Am. Nat.*, **88**(834), 139–53.

Moment, G. B. (1979) Growth, posterior regeneration and segment number in *Eisenia foetida*. *Megadrilogica*, **3**, 167–75.

Monnig, H. O. (1927) The anatomy and life history of the fowl tapeworm *Amoebotaenia spheroides*. *Rep. Dir. Vet. Educ. Res.*, **11–12**, 199–206.

Moore, A. R. (1923) Muscle tension and reflexes in the earthworm. *J. Gen. Physiol.*, **5**, 327–33.

Moore, B. (1922) Earthworms and soil reaction. *Ecology*, **3**, 347–8.

Morgan, A. J. (1986) Calcium-lead interactions involving earthworms: a hypothesis. *Chem. Ecol.*, **2**, 251–61.

Morgan, A. J. and Winters, C. (1991) Diapause in the earthworm *Aporrectodea langa*: Morphological and quantitative X-ray microanalysis of cryosectioned chloragogenous tissue. *Scanning Microscopy*, **5**(1), 219–28.

Morgan, A. J., Morris, B., James, N. *et al.* (1986) Heavy metals in terrestrial macroinvertebrates: species differences within and between trophic levels. *J. Chem. Ecol.*, **2.**, 319–34.

Moriarty, F. (1983) *Ecotoxicology*. Academic Press, New York.

Morris, D. E. and Pivnick, K. A. (1991) Earthworm mucus stimulates oviposition in a predatory fly (Diptera: Anthomyiidae). *J. Chem. Ecol.*, **17**(11), 2045–52.

Morris, H. M. (1922) Insect and other invertebrate fauna of arable land at Rothamsted. *Ann. Appl. Biol.*, **9**(3–4), 282–305.

Morrison, F. O. (1950) The toxicity of BHC to certain micro organisms, earthworms and arthropods. *Ont. Ent. Soc. Ann. Rep.*, **80**, 50–7.

Mouat, M. C. H. and Keogh, R. G. (1987) Adsorption of water-soluble phosphate from earthworm casts. *Plant and Soil*, **97**, 233–41.

Mountford, M. D. (1962) An index of similarity and its application to classificatory problems, in *Progress in Soil Zoology*, (ed. P. W. Murphy), Butterworths, London, pp. 43–50.

Mozgovoy, A. A. (1952) The biology of *Porrocaecum crassum*, a nematode of aquatic birds. *Trudy gelmint. labot.*, **6**, 114–25.

Muldal, S. (1949) Cytotaxonomy of British earthworms. *Proc. Linn. Soc. Lond.*, **161**, 116–18.

Muldal, S. (1952a) A new species of earthworm of the genus *Allolobophora*. *Proc. Zool. Soc. Lond.*, **122**, 463–5.

Muldal, S. (1952b) The chromosomes of earthworms. 1. The evolution of polyploidy. *Heredity*, **6**, 55–76.

Muller, G. (1965) *Bodenbiologie*, Verlag VEB Gustav Fischer, Jena.

Müller, P. E. (1878) Nogle Undersogelser af Skovjord. *Tidsskr. Landoko*, **4**, 259–83.

Müller, P. E. (1884) Studier over Skovjord. II. Om Muld og Mor i Egeskove og paa Heder. *Tidsskr. Skovbrug*, **7**, 1–232.

Müller, P. E. (1950) Forest-soil studies, a contribution to silvicultural theory. III. On compacted ground deficient in mull, especially in beach forests. *Dansk Skovforen. Tidsskr.*, **I**, 10–61.

Murchie, W. R. (1955) A contribution on the natural history of *Allolobophora minima*, Muldal. *Ohio J. Sci.*, **55**(4), 241–4.

Murchie, W. R. (1956) Survey of the Michigan earthworm fauna. *Mich. Acad. Sci. Arts. Let.*, **151**, 53–72.

Murchie, W. R. (1958a) Biology of the oligochaete *Eisenia rosea* (Savigny) in an upland forest soil of Southern Michigan. *Am. Midl. Nat.*, **66**(I), 113–31.

Murchie, W. R. (1958b) A new megascolecid earthworm from Michigan with notes on its biology. *Ohio J. Sci.*, **58**(5), 270–2.

Murchie, W. R. (1959) Redescription of *Allolobophora muldali* Omodco. *Ohio J. Sci.*, **59**(6), 229–32.

Murchie, W. R. (1960) Biology of the oligochaete *Bimastos zeteki* Smith and Gittins (Lumbricidae) in Northern Michigan. *Am. Midl. Nat.*, **64**(1), 194–215.

Murchie, W. R. (1961a) A new species of *Diplocardia* from Florida. *Ohio J. Sci.*, **61**(3), 175–7.

Murchie, W. R. (1961b) A new diplocardian earthworm from Illinois. *Ohio J. Sci.*, **61**(6), 367–71.

Murchie, W. R. (1963) Description of a new diplocardian earthworm, *Diplocardia longiseta*. *Ohio J. Sci.*, **63**(1), 15–18.

Murchie, W. R. (1965) *Diplocardia gatesi*, a new earthworm from North Carolina. *Ohio J. Sci.*, **65**(4), 208–11.

Murchie, W. R. (1967) Chromosome numbers of some diplocardian earthworms (Megascolecidae - Oligochaeta). *Am. Midl. Nat.*, **87**, 534–7.

Murray, A. C. Jr. and Hinckley, L. S. (1992) Effect of the earthworm (*Eisenia foetida*) on *Salmonella enteritidis* in horse manure. *Bioresource Tech.*, **41**(2), 97–100.

Murray, P. M., Feest, A. and Madelin, M. F. (1985) The numbers of viable myxomycete cells in the alimentary tracts of earthworms and in earthworm casts. *Bot. J. Linn. Soc.*, **91**, 359–66.

Muys, B., Lust, N. and Granval, Ph. (1992) Effects of grassland afforestation with different tree species on earthworm communities, litter decomposition and nutrient status. *Soil Biol. Biochem.*, **24**, 1459–66.

Nakamura, Y. (1968a) Studies on the ecology of terrestrial Oligochaetae. I. Seasonal variation in the population density of earthworms in alluvial soil grassland in Sapporo, Hokkaido. *Jap. J. Appl. Ent. Zool.*, **3**(2), 89–95.

Nakamura, Y. (1968b) Population density and biomass of the terrestrial earthworm in the grasslands of three different soil types near Sapporo. *Jap. J. Appl. Ent. Zool.*, **11**, 164–8.

Nakatsugawa, T. and Nelson, P. A. (1972) Insecticide detoxication in invertebrates. Enzymological approach to the problem of biological magnification, in *Environmental Toxicology of Pesticides*, (eds F. Matsumara, G. M. Boush and T. Misato), Academic Press, New York, pp. 501–24.

Needham, A. E. (1957) Components of nitrogenous excreta in the earthworms *L. terrestris* and *E. foetida*. *J. Exp. Biol.*, **34**(4), 425–46.

Needham, A. E. (1962) Distribution of arginase activity along the body of earthworms. *Comp. Biochem. Physiol.*, **5**, 69–82.

Nelson, J. M. and Satchell, J. E. (1962) The extraction of Lumbricidae from soil with special reference to the hand-sorting method, in *Progress in Soil Zoology*, (ed. P. Murphy), Butterworths, London, p. 2949.

Neuhauser, E. F. and Hartenstein, R. (1978) Reactivity of macroinvertebrate peroxidases with lignins and lignin model compounds. *Soil Biol. Biochem.*, **10**, 341–2.

Neuhauser, E. F., Kaplan, D. L. and Hartenstein, R. (1979) Life history of the earthworm *Eudrilus eugeniae*. *Rev. Ecol. Biol. Sol*, **16**, 525–34.

Neuhauser, E. F., Kaplan, D. L., Malecki, M. R. and Hartenstein, R. (1980) Materials supporting weight gain by the earthworm, *E. foetida* in waste conversion systems. *Agric. Wastes*, **2**, 43–60.

Neuhauser, E. F., Malecki, M. R. and Loehr, R. C. (1984) Growth and reproduction of the earthworm *E. fetida* after exposure to sublethal concentrations of metals. *Pedobiologia*, **27**, 89–97.

Neuhauser, E. F., Durkin, P. R., Malecki, M. R. and Anatra, M. (1986) Comparative toxicity of ten organic chemicals to four earthworm species. *Comp. Biochem. Physiol.*, **83C**, 197–200.

Neuhauser, E. F., Loehr, R. C. and Malecki, M. R. (1988) The potential of earthworms for managing sewage sludge, in *Earthworms and Waste Management*, (ed. C. A. Edwards and E. F. Neuhauser), SPB Acad. Publ., The Netherlands, pp. 9–20.

Neuhauser, E. F., Zoran, V. C., Malecki, M. R. *et al.*, (1995) Bioconcentration and biokinetics of heavy metals in the earthworm. *Environ. Pollut. Ser. A.*, **89**(3), 293–301.

Newell, G. E. (1950) The role of the coelomic fluid in the movements of earthworms. *J. Exp. Biol.*, **21**(1), 110–21.

Nielsen, C. O. (1953) Studies on Enchytraeidae. 1. A technique for extracting Enchytraeidae from soil samples. *Oikos*, **4**(2), 187–96.

Nielsen, M. G. and Gissel-Nielsen, G. (1975) Selenium in soil–animal relationships. *Pedobiologia*, **15**, 65–7.

Nielson, G. E. and Hole, F. E. (1964) Earthworms and the development of coprogenous A_1 horizons in forest soils of Wisconsin. *Soil Sci. Soc. Am. Proc.*, **28**, 426–30.

Nielson, R. L. (1951) Effect of soil minerals on earthworms. *N. Z. J. Agric.*, **83**, 433–5.

Nielson, R. L. (1952) Earthworms and soil fertility. *N. Z. Grassl. Assoc. Proc.*, pp. 158–67.

Nielson, R. L. (1953) Recent research work. Earthworms. *N. Z. J. Agric.*, **86**, 374.

Nielson, R. L. (1965) Presence of plant growth substances in earthworms demonstrated by paper chromatography and the Went pea test. *Nature Lond.*, **208**, 1113–14.

Nijhawan, S. D. and Kanwar, J. S. (1952) Physiochemical properties of earthworm castings and their effect on the productivity of soil. *Indian J. Agric. Sci.*, **22**, 357–73.

Niklas, J. von (1979) Histochemische Untersuchungen zur Wirkung von Pestiziden als Cholinesterase-Inhibitoren bei *Lumbricus terrestris*. *Z. Angew. Zool.*, **66**, 359–68.

Niklas, J. and Kennel, W. (1978) Lumbricid populations in orchards of W. Germany and the influence of fungicides based on copper compounds and benzimidazole derivatives upon them. *Z. Pflanzenkrank. Pflanzenschutz.*, **85**, 705–13.

Noble, J. C., Gordon, W. T. and Kleinig, C. R. (1970) The influence of earthworms on the development of mats of organic matter under irrigated pasture in Southern Australia. *Proc. 11th Int. Grassl. Conf.*, Brisbane, pp. 465–8.

Nordström, S. and Rundgren, S. (1973) Associations of lumbricids in Southern Sweden. *Pedobiologia*, **13**, 301–26.

Nordström, S. and Rundgren, S. (1974) Environmental factors and lumbricid associations in Southern Sweden. *Pedobiologia*, **14**, 1–27.

Nowak, E. (1975) Population density of earthworms and some elements of their production in several grassland environments. *Ekol. Pol.*, **23**, 459–91.

Nuutinen, V. (1992) Earthworm community response to tillage and residue management on different soil types in southern Finland. *Soil Till. Res.*, **23**(3), 221–39.

Nye, P. H. (1955) Some soil-forming processes in the humid tropics. IV. The action of soil fauna. *J. Soil Sci.*, **6**, 78.

Oades, J. M. (1993) The role of biology in the formation, stabilization and degradation of soil structure. *Geoderma*, **56**, 377–400.

O'Brien, B. J. and Stout, J. D. (1978) Movement and turnover of soil organic matter as indicated by carbon isotope measurements. *Soil Biol. Biochem.*, **10**, 309–17.

Ogg, W. G. and Nicol, H. (1945) Balanced manuring. *Scot. J. Agric.*, **25**(2), 76–83.

Ogleoby, L. C. (1969) Inorganic components and metabolism: ionic and osmotic regulation, in *Chemical Zoology*, (eds M. Florkin and B. T. Scheer), Academic Press, New York, vol. 4, pp. 211–310.

Ogren, R. E. and Sheldon, J. K. (1991) Ecological observations on the land planarian *Bipalium pennsylvanicum* Ogren, with references to phenology, reproduction, growth rate and food niche. *J. Penn. Acad.Sci.*, **65**, 3–9.

Oka, K., Takeda, N. and Hasimoto, T. (1984) Examination of water balance and ionic regulation in the earthworm *Eisenia foetida*. *Comp. Biochem. Physiol. A*, **79A**(3), 405–8.

Oldham, C. (1915) *Testacella scutulum* in Hertfordshire. *Trans. Herts Nat. Hist. Soc.*, **15**, 193–4.

Olive, P. J. W. and Clark, R. B. (1978) Physiology of reproduction, in *Physiology of Annelids*, (ed. P. J. Mill), Academic Press, London, pp. 271–368.

Oliver, J. H. (1962) A mite parasite in the cocoons of earthworms. *J. Parasit.*, **48**, 120–3.

Olson, H. W. (1928) The earthworms of Ohio. *Ohio Biol. Surv. Bull.*, **17**, 47–90.

Omodeo, P. (1952a) Cariologia Dei Lumbricidae. *Inst. Biol. Zool. Gen. Univ. Siena*, **4**, 173–275.

Omodeo, P. (1952b) Lumbricidae. In Materiali Zoologica Roccolti dal Dr. Marcuzzi sulla Alpi Dolomitiche. *Arch. Zool. Ital.*, **37**, 29–59.

Omodeo, P. (1956) Contributo alla revisione dei Lumbricidae. *Arch. Zool. Ital.*, **41**, 129–212.

Omodeo, P. (1958) La réserve naturelle integrale du Mont Nimba. I. Oligochaetes. *Mem. Inst. Fr. Afr. Noire*, **53**, 9–10.

Omodeo, P. (1959) Oligochati dell' Afghanistan. Contribution a l'etude de la faune d'Afghanistan 5. *Boll. Zool.*, **26**, 1–20.

Omodeo, P. (1963) Distribution of the terricolous oligochaetes on the two shores of the Atlantic, in *North Atlantic Biota and Their History*, (eds A. Love and D. Love), Pergamon Press, New York, pp. 127–51.

Otanes, F. G. and Sison, P. L. (1947) Pests of Rice. *Philip. J. Agric.*, **13**, 36–88.

Pal, S., Nanda, T., Pal, U. and Nanda, D. K. (1992) Anterior regeneration in earthworm, *Metaphire peguana*: I. Replenishment of lost segments and concomitant development of cephalic nerve ring. *Proc. Zool. Soc. (Calcutta)*, **45**(1), 33–7.

Pallas, S. L. and Drewes, C. D. (1981) The rapid tail flattening component of MGF-mediated escape behavior in the earthworm, *Lumbricus terrestris. Comp. Biochem. Physiol. (A)*, **70**(1), 57–64.

Park, S. C., Smith, T. J. and Bisesi, M. S. (1992) Activities of phosphomonoesterase and phophodiesterase from *Lumbricus terrestris. Soil Biol. Biochem.*, **24**, 873–6.

Parker, G. H. and Parshley, H. M. (1911) The reactions of earthworms to dry and moist surfaces. *J. Exp. Zool.*, **11**, 361–3.

Parle, J. N. (1959) Activities of micro-organisms in soil and influence of these on soil fauna. Ph.D. Thesis, University of London.

Parle, J. N. (1963a) Micro-organisms in the intestines of earthworms. *J. Gen. Microbiol.*, **31**, 1–13.

Parle, J. N. (1963b) A microbiological study of earthworm casts. *J. Gen. Microbiol.*, **31**, 13–23.

Parmelee, R. W. and Crossley, D. A. Jr (1988) Earthworm production and role in the nitrogen cycle of a no-tillage agroecosystem on the Georgia piedmont. *Pedobiologia*, **32**, 351–61.

Parmelee, R. W., Beare, M. H., Cheng, W. *et al.*(1990) Earthworms and enchytraeids in conventional and no-tillage agroecosystems: a biocide approach to assess their role in organic matter breakdown. *Biol. Fertil. Soils*, **10**, 1–10.

Patel, H. K. (1960) Earthworms in tobacco nurseries and their control. *Indian Tobacco*, **10**(I), 56.

Patel, H. K. and Patel, R. M. (1959) Preliminary observations on the control of earthworms by soapdust (*Sapindus laurifolius* Vahl) extract. *Indian J. Ent.*, **21**, 251–5.

Peachey, J. E. (1963) Studies on the Enchytraeidae (Oligochaeta) of moorland soil. *Pedobiologia*, **2**, 81–95.

Peredel'sky, A. A. (1960a) Effect of earthworms and wireworms on absorption by plants of the radioactive isotopes Ca^{45} and Sr^{90} from soil. *Dokl. Akad. Nauk.*, **134**, 1450–2.

Peredel'sky, A. A. (1960b) Dispersion of radioactive isotopes in the soil by earthworms. *Dokl. Akad. Nauk.*, **135**, 185–8.

Peredel'sky, A. A., Poryadkova, N. A. and Rodionova, L. Z. (1957) The role of earthworms in purification of soil contaminated with radioactive isotopes. *Dokl. Akad. Nauk.*, **115**(4), 809–12.

Perel, T. S. (1977) Differences in lumbricid organization connected with ecological properties, in *Soil Organisms as Components of Ecosystems*, (eds U. Lohm and T. Perssom), *Biol. Bull.*, **25**, 56–63.

Perel, T. S. and Sokolov, D. F. (1964) Quantitative evaluation of the participation of the earthworm *Lumbricus terrestris* Linné (Lumbricidae-Oligochaeta) in the transformation of forest litter. *Zool. Zh.*, **53**, 1618–25.

Perel, T. S., Karpachevskii, L. O. and Yegorova, S. V. (1966) Experiments for studying the effect of earthworms on the litter horizon of forest soils. *Pedobiologia*, **6**, 269–76.

Perfect, J. (1980) The environmental impact of DDT in a tropical agroecosystem. *Ambio*, **9**, 16–21.

Persson, T. and Lohm, U. (1977) Energetical significance of the annelids and arthropods in a Swedish grassland soil. *Ecol. Bull. (Stockholm)*, **23**, 1–211.

Petal, J., Nowak, E., Jakubczyk, H. and Czerwinski, Z. (1977) Effect of ants and earthworms on soil habitat modification, in *Soil Organisms as Components of Ecosystems*, (eds U. Lohm and T. Persson), *Ecol. Bull. (Stockholm)*, **25**, 501–3.

Peterson, A. E. and Dixon, R. M. (1971) Water movement in large soil pores: validity and utility of the channel system concept. *Coll. Agr. Life Sci. Univ. Wisc. Res. Rep.*, **75**.

Petrov, B. C. (1946) The active reaction of soil (pH) as a factor in the distribution of earthworms. *Zool. J.*, **25**(1), 107–10.

Phillips, E. F. (1923) Earthworms, plants and soil reactions. *Ecology*, **4**, 89.

Phillips, V. R. (1988) Engineering problems in the breakdown of animal wastes by earthworms, in *Earthworms in Waste and Environmental Management*, (eds C. A. Edwards and E. F. Neuhauser), SPB Acad. Publ., The Netherlands.

Phillipson, J. and Bolton, P. J. (1976) The respiratory metabolism of selected Lumbricidae. *Oecologia*, **22**, 135–52.

Phillipson, J. and Bolton, P. J. (1977) Growth and cocoon production by *Allolobophora rosea* (Oligochaeta, Lumbricidae). *Pedobiologia*, **17**, 70–82.

Phillipson, J., Abel, R., Steel, J. and Woodell, S. R. J. (1976) Earthworms and the factors that govern their distribution in an English beechwood. *Pedobiologia*, **16**, 258–85.

Piearce, T. G. (1972) The calcium relations of selected Lumbricidae. *J. Anim. Ecol.*, **41**, 167–88.

Piearce, T. G. (1978) Gut contents of some lumbricid earthworms. *Pedobiologia*, **18**, 153–7.

Piearce, T. G. (1983) Functional morphology of lumbricid earthworms with special reference to locomotion. *J. Nat. Hist.*, **17**(1), 95–111.

Piearce, T. G. and Phillips, M. J. (1980) The fate of ciliates in the earthworm gut: an *in vitro* study. *Microb. Ecol.*, **5**, 313–20.

Pietz, R. I., Peterson, J. R., Prater, J. E. and Zenz, D. R. (1984) Metal concentrations in earthworms from sewage sludge-amended soils at a strip mine reclamation site. *J. Environ. Qual.*, **13**(4).

Pigozzi, G. (1991). The diet of the European badger in a Mediterranean coastal area. *Acta Theriol.*, **36**, 293–306.

Pincince, A. B., Donovan, J. F. and Bates, J. E. (1980) Vermicomposting municipal sludge: an economical stabilization alternative. *Sludge*, **23**, 30.

Pincince, A. B., Donovan, J. F. and Bates, J. E. (1981) Vermicomposting of municipal solid wastes and municipal wastewater sludges, in *Proc. Workshop on the*

Role of Earthworms in the Stabilization of Organic Residues, (ed M. Appelhof), Beach Leaf Press, Kalamazoo, Michigan, vol. 1, pp. 207–19.

Pižl, V. (1985) The effect of the herbicide Zeazin 50 on the earthworm infection by monocystid gregarines. *Pedobiologia,* **28**, 399–402.

Pižl, V. and Sterzynska, M. (1991) The influence of urbanization on the earthworm infection by monocystid gregarines. *Frag. Faun. (Warsaw),* **35**, 203–12.

Poinar, G. O. (1978) Associations between nematodes (Nematoda) and oligochaetes (Annelida). *Proc. Helminth. Soc. Washington,* **45**, 202–10.

Pokarzhevskii, A. D. and Titisheva, N. G. (1982) Population dynamics of the earthworm *Eisenia nordenskioldi* in meadow steppe habitats in the USSR. *Pedobiologia,* **23**, 266–7.

Polivka, J. B. (1951) Effect of insecticides upon earthworm populations. *Ohio J. Sci.,* **51**, 195–6.

Polivka, J. B. (1953) More about the effect of insecticides on earthworm populations. Unpublished mimeo, Ohio Acad. Sci.

Pomerat, G. M. and Zarrow, M. T. (1936) The effect of temperature on the respiration of the earthworm. *Proc. Natl Acad. Sci., USA,* **22**, 270–2.

Ponge, J. F. (1991) Succession of fungi and fauna during decomposition of needles in a small area of Scots pine litter. *Plant and Soil,* 138, 99–114.

Ponomareva, S. I. (1950) The role of earthworms in the creation of a stable structure in ley rotations. *Pochvovedenie,* 476–86.

Ponomareva, S. I. (1952) The importance of biological factors in increasing the fertility of sod-podzolic soils. *Z. PflErnähr Düng,* **97**, 205–15.

Ponomareva, S. I. (1953) The influence of the activity of earthworms on the creation of a stable structure in a sod-podolized soil. *Trudy pochv. Inst. Dokuchaeva,* **41**, 304–78.

Ponomareva, S. I. (1962) Soil macro and micro-organisms and their role in increasing fertility. *Vtoraya Zoologischeskaya Konfereniya Litovskoi SSR,* 97–9.

Potter, D. A., Powell, A. J. and Smith, M. S. (1990) Degradation of turfgrass thatch by earthworms (Oligochaeta: Lumbricidae) and other soil invertebrates. *J. Econ. Ent.,* **83**, 205–11.

Powers, W. L. and Bollen, W. B. (1935) The chemical and biological nature of certain forest soils. *Soil Sci.,* **40**, 321–9.

Prabhoo, N. R. (1960) Studies on Indian Enchytraeidae (Oligochaeta: Annelida). Description of three new species. *J. Zool. Soc. India,* **12**(2), 125–32.

Price, J. S. (1987) Development of a vermicomposting system. *Proc. 4th Int. CIEC Sym. Agric. Waste Manage. Env. Protection,* **1**, 294–300.

Prisyaznyuk, A. A. (1950) Use of 666 for control of chafer grubs. *Agrobiologiya,* **5**, 141–2.

Prosser, C. L. (1935) Impulses in the segmental nerves of the earthworm. *J. Exp. Biol.,* **12**, 95–104.

Puh, P. C. (1941) Beneficial influence of earthworms on some chemical properties of the soil. *Contr. Biol. Lab. Sci. Soc. China,* **15**, 147–55.

Pulliainen, E., Lajunen, L. H. J. and Itämies, J. (1986) Lead and cadmium in earthworms (Oligochaeta, Lumbricidae) in northern Finland. *Ann. Zool. Fennici,* **23**, 303–6.

Puttarudriah, M. and Sastry, K. S. S. (1961) A preliminary study of earthworm damage to crop growth. *Mysore Agric. J.,* **36**, 2–11.

Rabatin, S. C. and Stinner, B. R. (1989) The significance of vesicular-arbuscular-mycorrhizal faunal–soil–macroinvertebrate interactions in agroecosystems. *Agric., Ecosyst. and Environ*, **27**, 195–204.

Raffy, A. (1930) La respiration des vers de terre dans l'eau. Action de la teneur en oxygene et de la lumiéu sur l'intensite de la respiration pendant l'immersion. *C.R. Hebd. Séanc. Acad. Sci., Paris*, **105**, 862–4.

Ragg, J. M. and Ball, D. F. (1964) Soils of the ultra-basic rocks of the Island of Rhum. *J. Soil Sci.*, **15**(1), 124–34.

Ralph, C. L. (1957) Persistent rhythms of activity and O_2 consumption in the earthworm. *Physiol. Zoöl.*, **30**, 41–55.

Rambke, J. and Verhaagh, M. (1992) About the earthworm community in a recently developed pasture, in comparison with a rain forest site in Peru. *Amazoniana*.

Ramsay, J. A. (1949) Osmotic relations of worms. *J. Exp. Biol.*, **26**(1), 65–75.

Randall, R., Butler, J. D. and Hughes, T. D. (1972) The effect of pesticide on thatch accumulation and earthworm populations in Kentucky bluegrass turf. *Hortsci.*, **7**, 64–5.

Rao, B. R. C. (1979) Studies on the biological and ecological aspects of certain Indian earthworms. Synopsis, Ph.D. Thesis, Mysore University, Kasturba Medical College, Manipa 576119.

Ravasz, K., Zicsi, A., Contreras, E. *et al.* (1986) Über die Darmaktinomyceten-gemeinschaften einiger Regenwurm-arten. *Opusc. Zool.*, **22**, 85–102.

Ravasz, K., Contreras, E. and Máriaglieti, K. (1987a) The influence of the composition of food materials on the gut flora of *Eisenia lucens* (Waga 1857), in *Soil Fauna and Soil Fertility*, (ed. B. Striganova), Nauka, Moscow, pp. 443–5.

Ravasz, K., Zicsi, A., Contreras, E. and Szabó, I. M. (1987b) Comparative bacteriological analyses of the faecal matter of different earthworm species, in *On Earthworms*, (eds A. M. B. Pagliai and P. Omodeo), Mucchi Editore, Modena, pp. 389–99.

Raw, F. (1959) Estimating earthworm populations by using formalin. *Nature, Lond.*, **184**, 1661.

Raw, F. (1960a) Observations on the effect of hexoestrol on earthworms and other soil invertebrates. *J. Agric. Sci.*, **55**(1), 189–90.

Raw, F. (1960b) Earthworm population studies: a comparison of sampling methods. *Nature, Lond.*, **187**(4733), 257.

Raw, F. (1961) The effect of Bordeaux mixture on the earthworm population of apple orchards. *Rep. Rothamsted Exp. Sta.*, 1960, Harpenden, pp. 37–9.

Raw, F. (1962) Studies of earthworm populations in orchards. I. Leaf burial in apple orchards. *Ann. Appl. Biol.*, **50**, 389–404.

Raw, F. (1965) Current work on side effects of soil applied organophosphorus insecticides. *Ann. Appl. Biol.*, 55, 342–3.

Raw, F. (1966) The soil fauna as a food source for moles. *J. Zool., Lond.*, **149**, 50–4.

Raw, F. and Lofty, J. R. (1959) Earthworm populations in orchards. *Rep. Rothamsted Exp. Stn for 1958*, pp. 134–5.

Raw, F. and Lofty, J. R. (1962) The effect of chemical control on pests and on other arthropods and worms in the soil. *Rep. Rothamsted Exp. Sta.*, 1961, p. 146.

Raw, F. and Lofty, J. R. (1964) The side-effects of toxic chemicals in the soil on arthropods and earthworms. *Rep. Rothamsted Exp. Sta.*, 1963, p. 149.

Reddell, P. and Spain, A. V. (1991a) Earthworms as vectors of viable propagules of mycorrhizal fungi. *Soil Biol. Biochem.*, **23**, 767–74.

Reddell, P. and Spain, A. V. (1991b) Transmission of infective *Frankia* (Actinomycetales) propagules in casts of the endogenic earthworm *Pontoscolex corethrurus* (Oligochaeta: Glossoscolecidae). *Soil Biol. Biochem.*, **23**, 775–8.

Reddy, M. V. (1980) Mass migration and mortality of *Amynthas* (=*Pheretima*) *alexandri* (Beddard) (Megascolecidae: Oligochaeta). *Curr. Sci. (Bangalore)*, **49**, 606.

Reddy, M. V. (1983) Effects of fire on the nutrient content and microflora of casts of *Pheretima alexandri*, in *Earthworm Ecology, from Darwin to Vermiculture*, (ed. J. E. Satchell), Chapman & Hall, London, pp. 209–13.

Reddy, M. V. (1988) The effect of casts of *Pheretima alexandri* on the growth of *Vinca rosea* and *Oryza sativa*, in *Earthworms in Environmental and Waste Management*, (eds C. A. Edwards and E. F. Neuhauser), SPB Bakker, The Netherlands, pp. 241–8.

Reddy, M. V. and Pasha, M. (1993) Influence of rainfall, temperature and some soil physiochemical variables on seasonal population structure and vertical distribution of earthworms in two semi- and tropical grassland soils. *Int. J. Biotech.*, **37**, 19–26.

Reest, P. J. van der and Rogaar, H. (1988) The effect of earthworm activity on the vertical distribution of plant seeds in newly-reclaimed soils in the Netherlands. *Pedobiologia*, 31, 211–18.

Reinecke, A. J. (1974) The upper lethal temperature of *Eisenia rosea* (Oligochaeta). *Natuurwetenskappe*, **62**, 1–14.

Reinecke, A. J. (1975) The influence of acclimation and soil moisture on the temperature preference of *Eisenia rosea* (Lumbricidae), in *Progress in Soil Zoology*, (ed. J. Vanek), Proc. 5th Intl. Colloq. Soil Zool., Junk, The Hague/Academia, Prague, pp. 341–9

Reinecke, A. J. and Kriel, J. R. (1981) Influence of temperature on the reproduction of the earthworms *Eisenia foetida* (Oligochaeta). *S. Afr. J.Zool.*, 16, 96–100.

Reinecke, A. J. and Ljungström, P. O. (1969) An ecological study of the earthworms from the banks of the Mooi River in Potchefstroom, South Africa. *Pedobiologia*, **9**, 106–11

Reinecke, A. J. and Nash, R. G. (1984) Toxicity of 2,3,7,8 TcDD and short term bioaccumulation by earthworms (Oligochaeta). *Soil Biol. Biochem.*, 16, 45–9.

Reinecke, A. J. and Ryke, P. A. J. (1970) Casting activity of a South African endemic earthworm, *Microchaetus modestus*. Wer. Bydraes. *PUCHO Reeks B. Natuurwet.* No. 16.

Reinecke, A. J. and Viljoen, S. A. (1991) Vertical deposition of cocoons by the compost worm *Eisenia fetida*. *Pedobiologia*, **35**.

Reinecke, A. J. and Viljoen, S. A. (1993) Effects of worm density on growth and cocoon production of the African nightcrawler (*Eudrilus eugeniae*) Oligochaeta. *Eur. J. Soil Biol.*, **29**(1), 29–34.

Ressler, R. H., Cialdini, R. B., Ghoca, M. L. and Kleist, S. M. (1968) Alarm pheromone in the earthworm *Lumbricus terrestris*. *Science, NY*, **161**, 59–79.

Reynolds, J. W. (1972) The relationship of earthworm (Oligochaeta: Acanthodrilidae and Lumbricidae) distribution and biomass in six heterogenous woodlot sites in Tippecanoe Country, Indiana. *J. Tenn. Acad. Sci.*, **47**, 63–7.

Reynolds, J. W. (1973a) The earthworms of Delaware. *Megadrilogica*, **1**(5), 1–4.

Reynolds, J. W. (1973b) The earthworms of Rhode Island (Oligochaeta: Lumbricidae). *Megadrilogica*, **1**(6), 1–4.

Reynolds, J. W. (1973c) The earthworms of Connecticut. *Megadrilogica*, **1**(7), 1–6.

Reynolds, J. W. (1974) The earthworms of Maryland, Oligochaeta: Acanthodrilidae, Lumbricidae, Megascolecidae and Sparganophilidae. *Megadrilogica*, 1(11), 1–12.

Reynolds, J. W. (1976) Un apercu des vers de terre dans les forêts nord-americaines, leurs activités et leurs repartition. *Megadrilogica*, 2, 1–11.

Reynolds, J. W. (1977) Earthworm populations as related to woodcock habitat usage in Central Maine. *Proc. Woodcock Symp.*, 6, 135–46.

Reynolds, J. W. and Cook, D. G. (1976) *Nomenclatura Oligochaetologica*. A catalogue of the names, descriptions and type specimens of the Oligochaeta. Fredericta, University of New Brunswick.

Reynolds, J. W. and Cook, D. G. (1981a) *Nomenclatura Oligochaetologica. Supplementum Primum*, Centennial Print Ltd., Canada.

Reynolds, J. W. and Cook, D. G. (1981b) *Nomenclatura Oligochaetologica. Supplementum Secundum*, Centennial Print Ltd., Canada.

Reynolds, J. W. and Cook, D. G. (1981c) *Nomenclatura Oligochaetologica. Supplementum Tertium*, New Brunswick Museum Monograph Series No. 9, Canada.

Reynolds, J. W. and Cook, D. C. (1989) *Nomenclatura Oligochaetologica Supplementum Secundum*, New Brunswick Museum Monograph No. 8, Canada.

Reynolds, J. W. and Cook, D. C. (1993) *Nomenclatura Oligochaetologica Supplementum Tertium*, New Brunswick Museum Monograph No. 9, Canada.

Reynolds, J. W. and Reynolds, W. M. (1972) Earthworms in medicine. *Am. J. Nursing*, 72, 1273.

Reynolds, J. W., Clebsch, E. E. C. and Reynolds, W. W. (1974) The earthworms of Tennessee (Oligochaeta). I. Lumbricidae. Contributions on North American Earthworms (Oligochaeta), no. 13, *Bull. Tall Timbers Res. Stn*, 17.

Reynoldson, T. B. (1955) Observations on the earthworms of North Wales. *N. Wales Nat.*, 3, 291–304.

Reynoldson, T. B. (1966) The ecology of earthworms with special reference to North Wales habitats. *Rep. Welsh Soils Discuss Grp.*, 25–32.

Reynoldson, T. B., O'Connor, F. B. and Kelly, W. A. (1955) Observations on the earthworms of Bardsey. *Bardsey Obs. Rep.*, 9.

Rhee, J. A. van (1963) Earthworm activities and the breakdown of organic matter in agricultural soils, in *Soil Organisms*, (eds J. Doeksen and J. van der Drift), North Holland, Amsterdam, pp. 55–9.

Rhee, J. A. van (1965) Earthworm activity and plant growth in artificial cultures. *Pl. Soil*, 22, 45–8.

Rhee, J. A. van (1967) Development of earthworm populations in orchard soils, in *Progress in Soil Biology*, (eds O. Graff and J. Satchell), North Holland, Amsterdam, pp. 360–71.

Rhee, J. A. van (1969a) Inoculation of earthworms in a newly-drained polder. *Pedobiologia*, 9, 128–32.

Rhee, J. A. van (1969b) Development of earthworm populations in polder soils. *Pedobiologia*, 9, 133–40.

Rhee, J. A. van (1969c) Effects of biocides and their residues on earthworms. *Meded. Rijksfac. Landbouwwet. Gent.*, 34, 682–9.

Rhee, J. A. van (1971) Some aspects of the productivity of orchards in relation to earthworm activities. *Ann. Zool. Ecol. Anim. Special Publ.*, 4, 99–108.

Rhee, J. A. van (1975) Copper contamination effects on earthworms by disposal of pig waste in pastures, in *Progress in Soil Zoology*, (ed. J. Vanek), Academia, Prague, pp. 451–7.

Rhee, J. A. van (1977) A study of the effect of earthworms on orchard productivity. *Pedobiologia*, **17**, 107–14.

Rhee, J. A. van and Nathans, S. (1961) Observations on earthworm populations in orchard soils. *Neth. J. Agric. Sci.*, **9**(2), 94–100.

Rhett, R. G., Simmers, J. W. and Lee, C. R. (1988) *Eisenia foetida* used as a biomonitoring tool to predict the potential bioaccumulation of contaminants from contaminated dredged material, in *Earthworms in Environmental and Waste Management*, (eds C. A. Edwards and E. F. Neuhauser), SPB Bakker, The Netherlands, pp. 321—8.

Rhoades, W. C. (1963) A synecological study of the effects of the imported fire ant (*Solenopsis saevissima* Vichteri) eradication program. II. Light trap, soil sample, litter sample and sweep net methods of collecting. *Fla. Ent.*, **46**, 301–10.

Ribaudcourt, E. and Combault, A. (1907) The role of earthworms in agriculture. *Bull. Soc. For. Belg.*, 212–23.

Richards, J. G. (1955) Earthworms (recent research work). *N. Z. J. Agric.*, **91**, 559.

Richardson, H. C. (1938) The nitrogen cycle in grassland soils: with special reference to the Rothamsted Park grass experiment. *J. Agric. Sci., Camb.*, **28**, 73–121.

Richter, G. (1953) The action of insecticides on soil macrofauna. *NachBl. dt. PflSchutzdienst, Berl.*, **7**, 61–72.

Rivero-Hernandez, R. (1991) Influence of pH on the production of *Eisenia foetida*. *Avanc. Aliment. Anim.*, **31**(5), 215–17.

Roark, J. H. and Dale, J. L. (1979) The effect of turf fungicides on earthworms. *Ark. Acad. Sci. Proc.*, **33**, 71–4.

Roberts, B. L. and Dorough, H. W. (1984) Relative toxicities of chemicals to the earthworm *Eisenia foetida*. *Environ. Toxicol. Biochem.*, **3**, 67–78.

Robertson, J. D. (1936) The function of the calciferous glands of earthworms. *J. Exp. Biol.*, **13**, 279–97.

Robinson, C. H., Ineson, P., Piearce, T. G. and Rowland, A. P. (1992) Nitrogen mobilization by earthworms in limed peat soils under *Picea sitchensis*. *J. Appl. Ecol.*, **29**, 226–37.

Robinson, J. S. (1953) Stimulus substitution and response learning in the earthworm. *J. Comp. Physiol. Psychol.*, **46**, 262–6.

Rodale, R. (1948) Do chemical fertilizers kill earthworms? *Org. Gard.*, **12**(2), 12–17.

Rodale, R. (1961) *The challenge of earthworm research*, S. & H. Foundation, Penn.

Rogaar, H. and Boswinkel, J. A. (1978) Some soil morphological effects of earthworm activity, field data and x-ray radiography. *Neth. J. Agric. Sci.*, **26**, 145–60.

Rognes, K. (1991) Revision of the cluster-flies of the *Pollenia viatica* species-group (Diptera: Calliphoridae). *Syst. Ent.*, **16**, 439–98.

Roots, B. I. (1955) The water relations of earthworms. I. The activity of the nephridiostome cilia of *L. terrestris* L. and *A. chlorotica* (Sav.) in relation to the concentration of the bathing medium. *J. Exp. Biol.*, **32**, 765–74.

Roots, B. I. (1956) The water relations of earthworms. II. Resistance to desiccation and immersion and behaviour when submerged and when allowed choice of environment. *J. Exp. Biol.*, **33**, 29–44.

Roots, B. I. (1957) Nature of chloragogen granules. *Nature, Lond.*, **179**, 679–80.

Roots, B. I. (1960) Some observations on the chloragogenous tissue of earthworms. *Comp. Biochem. Physiol.*, **I**, 218–26.

Rose, C. J. and Wood, A. W. (1980) Some environmental factors affecting earthworm populations and sweet potato production in the Tari Basin, Papua New Guinea Highlands. *Papua New Guinea Agric. J.,* **31,** 1–13.

Rosenkoetter, J. S. and Boice, R. (1973) *Earthworm pheromones and T-maze learning,* Psychonomic Society, St. Louis, p. 17.

Rosswall, T. and Paustian, K. (1984) Cycling of nitrogen in modern agricultural systems. *Pl. Soil,* **76,** 3–21.

Roth, C. H. and M. Joschko (1991) A note on the reduction of runoff from crusted soil by earthworm burrows and artificial channels. *Pflanzen. Bodenk.,* **154,** 101–6.

Rouelle, J. (1983) Introduction of an amoeba and *Rhizobium japonicum* into the gut of *Eisenia fetida* (Sav.) and *Lumbrucus terrestris* L., in *Earthworm Ecology, From Darwin to Vermiculture,* (ed. J. E. Satchell), Chapman & Hall, New York, 375–81.

Roy, S. K. (1957) Studies on the activities of earthworms. *Proc. Zool. Soc. Bengal,* **10,** 81–98.

Rundgren, S. (1975) Vertical distribution of lumbricids in southern Sweden. *Oikos,* **26,** 299–306.

Ruppel, R. F. and Laughlin, C. W. (1977) Toxicity of some soil pesticides to earthworms. *J. Kansas Ent. Soc.,* **50,** 113–18.

Ruppel, R. F., Laughlin, C. W. and Fogg, R. (1973) Toxicities of some insecticides to earthworms. *Proc. North Cent. Branch Entomol. Soc. Am.,* **28,** 189.

Ruschmann, G. (1953) Antibioses and symbioses of soil organisms and their significance in soil fertility. Earthworm symbioses and antibioses. *Z. Acker. PflBau,* **96,** 201–18.

Rushton, S. P. and Luff, M. L. (1984) A new electrical method for sampling earthworm populations. *Pedobiologia,* **26**(1), 15–19.

Russell, E. J. (1910) The effect of earthworms on soil productiveness. *J. Agric. Sci., Camb.,* **2,** 245–57.

Russell, E. J. (1950) *Soil Conditions and Plant Growth,* 8th edn, Longman, London.

Ruz-Jerez, B. E., Ball, P. R. and Tillman, R. W. (1992) Laboratory assessment of nutrient release from a pasture soil receiving grass or clover residues, in the presence or absence of *Lumbricus rubellus* or *Eisenia fetida. Soil Biol. Biochem.,* **24,** 1529–34.

Ryder, M. H. and Rovira, A. D. (1993) Biological control of take-all in glasshouse-grown wheat using strains of *Pseudomonas corrugata* isolated from wheat field soil. *Soil Biol. Biochem.,* **25,** 311–20.

Ryšavý, B. (1964) Some notes of the life history of the cestode *Dilepis undula* Shrank. *Helminthologia,* **5,** 173–6.

Ryšavý, B. (1969) Lumbricidae – an important parasitological factor in helminthoses of domestic and wild animals. *Pedobiologia,* **9**(1/2), 171–4.

Ryzhikov, K. M. (1949) *Syngamidae of domestic and wild animals,* Moskva, pp. 1–165.

Sabine, J. R. (1978) The nutritive value of earthworm meal, in *Utilization of Soil Organisms in Sludge Management,* (ed. R. Hartenstein), Natl. Tech. Inf. Services, PB286932, Springfield, VA, pp. 285–96.

Sabine, J. R. (1981) Vermiculture as an option for resource recovery in the intensive animal industries, *Proc. Workshop on the Role of Earthworms in the Stabilization of Organic Residues,* (ed. M. Appelhof), Beach Leaf Press, Kalamazoo, Michigan, vol. 1, pp. 241–54.

Sabine, J. R. (1983) Earthworms as a source of food and drugs, in *Earthworm Ecology: From Darwin to Vermiculture,* (ed. J. E. Satchell), Chapman & Hall, London, pp. 285–96.

Salisbury, E. J. (1925) The influence of earthworms on soil reaction and the stratification of undisturbed soils. *J. Linn. Soc. (Bot)*, **46**, 415–25.

Saroja, K. (1959) Studies on oxygen consumption in tropical poikilotherms. 2. Oxygen consumption in relation to body size and temperature in the earthworm *Megascolex mauritii* when kept submerged in water. *Proc. Indian Acad. Sci. B.*, **49**, 183–93.

Satchell, J. E. (1955a) Some aspects of earthworm ecology, in *Soil Zoology*, (ed. D. K. Mc. E. Kevan), Butterworths, London, pp. 180–201.

Satchell, J. E. (1955b) *Allolobophora limicola*. An earthworm new to Britain. *Ann. Mag. Nat. Hist.*, **8**(12), 224.

Satchell, J. E. (1955c) The effects of BHC, DDT and parathion on the soil fauna. *Soils and Fert.*, **18**(4), 279–85.

Satchell, J. E. (1955d) An electrical method of sampling earthworm populations, in *Soil Zoology*, (ed. D. K. Mc E. Kevan), Butterworths, London, pp. 356–64.

Satchell, J. E. (1958) Earthworm biology and soil fertility. *Soils Fert.* **21**, 209–19.

Satchell, J. E. (1960) Earthworms and soil fertility. *New Sci.*, **7**, 79–81.

Satchell, J. E. (1963) Nitrogen turnover by a woodland population of *Lumbricus terrestris*, in *Soil Organisms*, (eds J. Doeksen and J. van der Drift), North Holland, Amsterdam, pp. 60–6.

Satchell, J. E. (1967) Lumbricidae, in *Soil Biology*, (eds A. Burgess and F. Raw), Academic Press, London, pp. 259–322.

Satchell, J. E. (1969) Studies on methodical and taxonomical questions. *Pedobiologia*, **9**, 20–5.

Satchell, J. E. (1980) R worms and K worms: A basis for classifying lumbricid earthworm strategies, in *Soil Biology as Related to Land Use Practices*, (ed. D. L. Dindal), Proc. 7th Intl Coll. Soil Zool. EPA, Washington, DC, pp. 848–54.

Satchell, J. E. (ed.) (1983) *Earthworm Ecology: From Darwin to Vermiculture*, Chapman & Hall, London.

Satchell, J. E. and Lowe, D. G. (1967) Selection of leaf litter by *Lumbricus terrestris*, in *Progress in Soil Biology*, (eds O. Graff and J. E. Satchell), North Holland, Amsterdam, pp. 102–19.

Satchell, J. E. and Martin, K. (1984) Phosphatase activity in earthworm faeces. *Soil Biol. Biochem.*, **16**, 191–4.

Satchell, J. E. and Martin, K. (1985) *A Bibliography of Earthworm Research*, Institute of Terrestrial Ecology, Grange-over-Sands, UK.

Satchell, J. E., Martin, K. and Krishnamoorthy, R. V. (1984) Stimulation of microbiol phosphatases produced by earthworm activity. *Soil Biol. Biochem.*, **16**, 195.

Saunders, D. G. and Forgie, C. D. (1977) Some effects of phorate on earthworm populations. *Proc. 30th NZ Weed Pest Contr. Conf.*, pp. 222–6.

Saussey, M. (1957) A case of commensalism in the lumbricids. *Bull. Soc. Ent. Fr.*, **62**(1/2), 15–19.

Saussey, M. (1959) Observations sur les relations entre la composition physico-chimique du sol et son peuplement en Lumbricides. *Arch. Zool Exp. Gen.*, **93**, 123–4.

Saussey, M. (1966) Zoologie experimentale-relations entre la régéneration caudale et la diapause chez *Allolobophora icterica* (Savigny) (Oligochaete lombricien). *C.R. Hebd. Séanc. Acad. Sci., Paris*, **263**, 1092–4.

Savigny, E. (1826) Enterion chlorotieum. *Mem. Acad. Sci. Inst. France*, **5**, 183.

Scharpenseel, H. W. and Gewehr, H. (1960) Studien zur Wasserbewegung im Boden mit Tritium-Wasser. *Z. PflErnähr. Düng.*, **88**, 35–49.

Scheu, S. (1987a) Microbial activity and nutrient dynamics in earthworm casts (Lumbricidae). *Biol. Fert. Soils*, **5**, 230–4.

Scheu, S. (1987b) The role of substrate-feeding earthworms (Lumbricidae) for bioturbation in a beechwood soil. *Oecologia*, **72**, 192–6.

Scheu, S. (1991) Mucus excretion and carbon turnover of endogenic earthworms. *Biol. Fert. Soils*, **12**, 217–20.

Scheu, S. (1992) Automated measurement of the respiratory response of soil microcompartments: active microbial biomass in earthworm feces. *Soil Biol. Biochem.*, **24**, 1113–18.

Scheu, S. (1993a) Cellulose and lignin decomposition in soils from different ecosystems on limestone as affected by earthworm processing. *Pedobiologia*, **37**, 167–77.

Scheu, S. (1993b) Litter microflora–soil macrofauna interactions in lignin decomposition: a laboratory experiment with ^{14}C-labelled lignin. *Soil Biol. Biochem.* **25**, 1703–11.

Scheu, S. (1994) There is an earthworm-mobilizable nitrogen pool in soil. *Pedobiologia*, **38**, 243–9.

Scheu, S. and Wolters, V. (1991a) Influence of fragmentation and bioturbation on the decomposition of carbon-14-labelled beech leaf litter. *Soil Biol. Biochem.*, **23**, 1029–34.

Scheu, S. and Wolters, V. (1991b) Buffering of the effect of acid rain on decomposition of carbon-14-labelled beech leaf litter by saprophagous invertebrates. *Biol. Fertil. Soils*, **11**, 285–9.

Schmid, L. A. (1947) Induced neurosecretion in *Lumbricus terrestris*. *J. Exp. Zool.*, **104**, 365–77.

Schmidt, H. (1955) Behaviour of two species of earthworm in the same maze. *Science, NY*, **121**, 341–2.

Schmidt, P. (1918) Anabiosis of the earthworm. *J. Exp. Zool.*, **27**, 55–72.

Schneider, K. C. (1908) *Histologisches Prakticum det Tiere*, Jena.

Schread, J. C. (1952) Habits and control of the oriental earthworm. *Bull. Conn. Agric. Exp. Stn*, **556**, 5–15.

Schultz, E. and Graff, O. (1977) Zur Berwertung von Regenwurmmehl aus *Eisenia foetida* als Eiweissfuttermittel. *Landb. Forsch. Vdkr.*, **27**, 216–18.

Schultz, W. and Felber, E. (1956) Welche Mikroorganismen spielen im Regenwurmdarm bei der Bildung von Bodenkrumeln eine Rolle? *Z. Acher. Pflbau.*, **101**, 471–6.

Schwartz, B. and Alicata, J. E. (1931) Concerning the life history of lungworms of swine. *J. Parasit.*, **18**, 21–7.

Schwert, D. P. (1980) Active and passive dispersal of lumbricid earthworms, in *Soil Biology as Related to Land Use Practices*, (ed. D. L. Dindal), Proc. 7th Intl. Coll. Soil Zool., EPA, Washington, DC, pp. 182–9.

Schwert, D. P. (1992) Oligochaeta: Lumbricidae, in *Soil Biology Guide*, (ed. D. L. Dindal), Wiley and Sons, New York, pp. 341–56.

Schwert, D. P. and Dance, K. W. (1979) Earthworm cocoons as a drift component in a southern Ontario stream. *Can. Field Nat.*, **93**, 180–3.

Scott, H. E. (1960) Control of mites in earthworm beds. *North Carolina State Agr. Ext. Serv. Ext. Folder*, **181**.

Scott, M. A. (1988) The use of worm-digested animal waste as a supplement to peat in loamless composts for hardy nursery stock, in *Earthworms in*

Environmental and Waste Management, (eds C. A. Edwards and E. F. Neuhauser), SPB Acad. Publ., The Netherlands, pp. 231–9.

Scrickhande, J. C. and Pathak, A. N. (1951) A comparative study of the physico-chemical characters of the castings of different insects. *Indian J. Agric. Sci.*, **21**, 401–7.

Scullion, J. (1984) The assessment of experimental techniques developed to assist the rehabilitation of restored open-cast coal-mining land. Ph.D. thesis, University College of Wales, Aberystwyth.

Scullion, J. and Mohammed, A. R. A. (1991) Effects of subsoiling and associated incorporation of fertilizer on soil rehabilitation after opencast mining for coal. *J. Agric. Sci.*, **116**(2), 256–74.

Shapiro, D. I., Berry, E. C. and Lewis, L. C. (1993) Interactions between nematodes and earthworms: enhanced dispersal of *Steinernema carpocapsae*. *J. Nematol.*, **25**(2), 189–92.

Sharpley, A. N. and Syers, J. K. (1976) Potential role of earthworm casts for the phosphorus enrichment of run-off waters. *Soil Biol. Biochem.*, **8**, 341–6.

Sharpley, A. N. and Syers, J. K. (1977) Seasonal variation in casting activity and in the amounts and release to solution of phosphorus forms in earthworm casts. *Soil Biol. Biochem.*, **9**, 227–31.

Sharpley, A. N., Syers, J. K. and Springett, J. A. (1979) Effect of surface-casting earthworms on the transport of phosphorus and nitrogen in surface runoff from pasture. *Soil Biol. Biochem.*, **11**, 459–62.

Shaw, C. and Pawluk, S. (1986a) Faecal microbiology of *Octolasion tyrtaeum*, *Apporectodea turgida* and *Lumbricus terrestris* and its relation to carbon budgets of three artificial soils. *Pedobiologia*, **29**, 377–89.

Shaw, C. and Pawluk, S. (1986b) The development of soil structure by *Octolasion tyrtaeum*, *Aporrectodea turgida* and *Lumbricus terrestris* in parent materials belonging to different textural classes. *Pedobiologia*, **29**, 327–39.

Sheppard, P. S. (1988) Species differences in cocoon and hatchling production in *E. fetida* and *E. andrei*, in *Earthworms in Environmental and Waste Management*, (eds C. A. Edwards and E. F. Neuhauser), SPB Acad. Publ., The Netherlands, pp. 83–4.

Shindo, B. (1929) On the seasonal and depth distribution of some worms in soil. *J. Coll. Agric., Tokyo*, **10**, 159–71.

Shipitalo, M. J. and Protz, R. (1988) Factors influencing the dispersibility of clay in worm casts. *Soil Sci. Soc. Am. J.*, **52**, 764–9.

Shipitalo, M. J. and Protz, R. (1989) Chemistry and micromorphology of aggregation in earthworm casts. *Geoderma*, **45**, 357–74.

Shipitalo, M. J., Edwards, W. M., Dick, W. A. and Owens, L. B. (1990). Initial storm effects on macropore transport of surface-applied chemicals in no-till corn. *Soil Sci. Soc. Am. J.*, **54**, 1530–6.

Shipitalo, M. J., Edwards, W. M. and Redmond, C. E. (1994) Comparison of water movement and quality in earthworm burrows and pan lysimeters. *J. Environ. Qual.*, **23**, 345–51.

Shiraishi, K. (1954) On the chemotaxis of the earthworm to carbon dioxide. *Sci. Rep. Tōhohu Univ.*, **20**(4), 356–61.

Shrikhande, J. C. and Pathak, A. N. (1951) A comparative study of the physico-chemical characters of the castings of different insects. *Indian J. Agric. Sci.*, **21**, 401–7.

Shumway, D. L. and Koide, R. T. (1994) Seed preferences of *Lumbricus terrestris*. *Appl. Soil Ecol.*, 11–14.

Šimek, M. and V. Pizl (1989) The effects of earthworms (Lumbricidae) on nitrogenase activity in soil. *Biol. Fertil. Soils*, **7**, 370–3.

Simmers, J. W. *et al.* (1983) Application of a terrestrial bioassay for determining toxic metal uptake from dredged material, in *Cong. Heavy Metals in Env.*, Heidelberg, Germany.

Sims, R. W. (1963a) Oligochaeta (Earthworms). *Proc. S. Lond. Ent. Nat. Hist. Soc.*, (2) 53.

Sims, R. W. (1963b) A small collection of earthworms from Nepal. *J. Bombay Nat. Hist. Soc.*, **60**(I), 84–91.

Sims, R. W. (1964a) Oligochaeta from Ascension Island and Sierra Leone including records of *Pheretima* and a new species of *Dichogaster*. *Ann. Mag. Nat. Hist.*, 7(13), 107–13.

Sims, R. W. (1964b) Internal fertilization and the functional relationship of the female and the spermathecal systems in new earthworms from Ghana (Eudrilidae: Oligochaeta). *Proc. Zool. Soc. Lond.*, **143**(4), 587–608.

Sims, R. W. (1966) The classification of the megascolecid earthworms: an investigation of Oligochaete systematics by computer techniques. *Proc. Linn. Soc. Lond.*, **177**, 125–41.

Sims, R. W. (1967) Earthworms (Acanthrodilidae and Eudrilidae: Oligochaeta) from Gambia. *Bull. Br. Mus. Nat. Hist. Zool.*, **16**, 1–43.

Sims, R. W. (1969) Outline of an application of computer techniques to the problem of the classification of the megascolecoid earthworms. *Pedobiologia*, **9**,(5), 35–41.

Sims, R. W. (1983) The scientific names of earthworms, in *Earthworm Ecology from Darwin to Vermiculture*, (ed. J. E. Satchell), Chapman & Hall, London, pp. 467–74.

Sims, R. W. and Easton, E. G. (1972) A numerical revision of the earthworm genus *Pheretima* (Megacolecidae: Oligochaeta) with the recognition of new genera and an appendix on the earthworms collected by the Royal Society North Borneo Expedition. *Biol. J. Linn. Soc.*, **4**, 169–268.

Sims, R. W. and Gerard, B. M. (1985) Earthworms: synopses of British fauna. *Nat. Hist. Mus.*, **31**, 1–171.

Skarbilovic, T. S. (1950) The study of the biology of *Capillaria mucronata* and on the epizootology of capillarioses of the urinary bladder of sable and mink. *Trudy vsesoyuznogo inst. gelmintologii im akad. K.I. Skrianbina*, **4**, 27–33.

Skoczen, S. (1970) Food storage of some insectivorous mammals (Insectivora). *Przegl. Zool.*, **14**, 243–8.

Slater, C. S. (1954) *Earthworms in Relation to Agriculture*, U.S.D.A.A.R.C. Circ.

Slater, C. S. and Hopp, H. (1947) Leaf protection in winter to worms. *Proc. Soil Sci. Soc. Am.*, **12**, 508–11.

Smallwood, W. M. (1923) The nerve net in the earthworm: preliminary report. *Proc. Nat. Acad. Sci. Washington*, p. 9.

Smallwood, W. M. (1926) The peripheral nervous system of the common earthworm, *Lumbricus terrestris*. *J. Comp. Neurol.*, **42**, 35–55.

Smirnoff, W. A. and Heimpel, A. M. (1961) Notes on the pathogenicity of *Bacillus thuringiensis* var. *thuringiensis* Berliner for the earthworm, *Lumbricus terrestris* L. *J. Insec. Pathol.*, **3**, 403–8.

Smith, F. (1915) Two new varieties of earthworms with a key to described species in Illinois. *Bull. Ill. State Lab. Nat. Hist.*, **10**(8), 551–9.

Smith, F. (1928) An account of changes in the earthworm fauna of Illinois. *Bull. Ill. State Nat Hist. Survey*, **17**(10), 347–62.

Smith, M. (1951) *The British Amphibians and Reptiles*, Collins, London.

Smith, R. D. and Glasgow, L. L. (1965) Effects of heptachlor on wildlife in Louisiana. *Proc. 17th Ann. Conf. S/E Ass. Game and Fish Comm.*, **17**, 140–54.

Smith, W. W. (1893) Further notes on New Zealand earthworms with observations on the known aquatic species. *Trans. N. Z. Inst.*, **25**, 111–46.

Soni, R. and Abbasi, S. A. (1981) Mortality and reproduction in earthworm *Pheretima posthuma* exposed to chromium (IV). *Int. J. Environ. Stud.*, **17**, 147–9.

Southwood, T. R. E. (1966) *Ecological Methods*, Methuen, London.

Spain, A. V., Lavelle, P. and Mariotti, A. (1992) Stimulation of plant growth by tropical earthworms. *Soil Biol. Biochem.*, **24**, 1629–33.

Spiers, G. A., Gagnon, D., Nason, G. E. *et al.* (1986) Effects and importance of indigenous earthworms on decomposition and nutrient cycling in coastal forest ecosystems. *Can. J. For. Res.*, **16**, 983–9.

Springett, J. A. and Syers, J. K. (1979) The effect of earthworm casts on ryegrass seedlings, in *Proceedings of the 2nd Australasian conference on grassland invertebrate ecology*, (eds T. K. Crosby and R. P. Pottinger), Government Printer, Wellington, pp. 44–7.

Springett, J. A., Gray, R. A. J. and Reid, J. B. (1992) Effect of introducing earthworms into horticultural land previously denuded of earthworms. *Soil Biol. Biochem.*, **16**, 1615–22.

Staaf, H. (1987) Foliage litter turnover and earthworm populations in three beech forests of contrasting soil and vegetation types. *Oecologia*, **72**, 58–64.

Stafford, E. A. and Tacon, A. G. J. (1988) The use of earthworms as food for rainbow trout *Salmo gairdneri*, in *Earthworms in Waste and Environmental Management*, (eds C. A. Edwards and E. F. Neuhauser), SPB Acad. Publ., The Netherlands, pp. 193–208.

Standen, V., Stread, C. B. and Dunning, A. (1982) Lumbricid populations in open cast reclamation sites and colliery spoil heaps in Co. Durham U.K. *Pedobiologia*, **24**, 57–64.

Stein, A., Bekker, R. M., Blom, J. H. C. and Rogaar, H. (1992) Spatial variability of earthworm populations in a permanent polder grassland. *Biol. Fert. Soils,* **14**(4), 260–6.

Stenersen, J. (1979) Action of pesticides on earthworms. Part I. The toxicity of cholinesterase-inhibiting insecticides to earthworms as evaluated by laboratory tests. *Pestic. Sci.*, **10**, 66–74.

Stenersen, J., Gilman, A. and Vardanis, A. (1973) Carbofuran: its toxicity to and metabolism by earthworms (*Lumbricus terrestris*). *J. Agric. Food Chem.*, **21**, 166–71.

Stephens, P. M. and Davoren, C. W. (1995) Influence of the earthworms *Aporrectodea trapezoides* and *A. rosea* on the disease severity of *Rhizoctonia solani* on subterrarrean cloves and ryegrass. *Soil Biol. Biochem.*, in press.

Stephens, P. M., Davoren, C. W., Doube, B. M. *et al.* (1993). Reduced severity of *Rhizoctonia solani* disease on wheat seedling associated with the presence of the earthworm *Aporrectodea trapezoides* (Lumbricidae). *Soil Biol. Biochem.*, **11**, 1477–84.

Stephens, P. M., Davoren, C. W., Ryder, M. H. and Doube, B. M. (1994a) Ability of the earthworms *Aporrectodea rosea* and *Aporrectodea trapezoides* (Lumbricidae) to influence the colonization of alfalfa (*Medicago sativa* L.) roots by *Rhizobium meliloti* strain L5-30R and the survival of *Rhizobium meliloti* L5-30R in soil. *Biol. Fertil. Soils*, **18**, 150–4.

Stephens, P. M., Davoren, C. W., Ryder, M. H. and Doube, B. M. (1994b) Influence of the earthworm *Aporrectodea trapezoides* (Lumbricidae) on the colonisation of wheat (*Triticum aestivum* cv. Spear) roots by *Pseudomonas corrugata* strain 2140R and survival of 2140R in soil. *Soil Biol. Biochem.*

Stephens, P. M., Davoren, C. W., Doube, B. M. and Ryder, M. H. (1995) Ability of the earthworms *Aporrectodea rosea* and *Aporrectodea trapezoides* to increase plant growth and the foliar concentration of elements in wheat (*Triticum aestivum* cv. Spear) in a sandy loam soil. *Biol. Fert. Soils*, in press.

Stephenson, J. (1929) Oligochaeta: in reports of an expedition to Brazil and Paraguay, 1926–7. *J. Linn. Soc. (Zool.)*, **37**, 291–325.

Stephenson, J. (1930) *The Oligochaeta*, Oxford University Press.

Stephenson, J. (1945) Concentration regulation and volume control in *Lumbricus terrestris* L. *Nature, Lond.*, **155**, 635.

Stewart, V. I. and Scullion, J. (1988) Earthworms, soil structure and the rehabilitation of former open-cast coal-mining land, in *Earthworms in Waste and Environmental Management*, (eds C. A. Edwards and E. F. Neuhauser), SPB Acad. Press, The Hague, The Netherlands, pp. 263–72.

Stickel, W. H., Mayne, D. W. and Stickel, L. F. (1965) Effects of heptachlor-contaminated earthworms on woodcocks. *J. Wildl. Manage*, **29**, 132–46.

Stockdill, S. M. J. (1959) Earthworms improve pasture growth. *N. Z. J. Agric.*, **98**, 227–33.

Stockdill, S. M. J. (1966) The effect of earthworms on pastures. *Proc. N. Z. Ecol. Soc.*, **13**, 68–74.

Stockdill, S. M. J. (1982) Effect of introduced earthworms on the productivity of New Zealand pastures. *Pedobiologia*, **24**, 29–35.

Stockdill, S. M. J. and Cossens, G. G. (1966) The role of earthworms in pasture production and moisture conservation. *Proc. N. Z. Ecol. Soc.*, **13**, 68–83.

Stöckli, A. (1928) Studien über den Einfluss der Regenwürmer auf die Beschaffenheit des Bodens. *Landw. Jb. Schweiz.*, **42**(1).

Stöckli, A. (1949) Einfluss der Mikroflora und Fauna auf die Beschaffenheit des Bodens. *Z. PflErnähr. Düng.*, **45**(90), 41–53.

Stöckli, A. (1958) Die Regenwurmarten in Landwirtschaftlich genutzten Böden des schweizerischen Mittellandes. *Separatabdruck Landwirtschaft Jbuch Schweiz.*, **72**(7), 699–725.

Stokes, B. M. (1958) The worm-eating slugs *Testacella scutulum* Sowerby and *T. haliotidea* Drapernaud in captivity. *Proc. malac. Soc. Lond.*, **33**(1), 11–20.

Stolte, H. A. (1962) Oligochaeta. *Bronn's Klassen und Ordnungen des Tierreichs*, **4**(3), 891–1141.

Stone, P. C. and Ogles, G. D. (1953) *Uropoda agitans*, a mite pest in commercial fishworm beds. *J. Econ. Ent.*, **46**, 711.

Stop-Bowitz, C. (1969) Did lumbricids survive the quarternary glaciations in Norway? *Pedobiologia*, **9**, 93–8.

Stough, H. B. (1926) Giant nerve fibres of the earthworm. *J. Comp. Neurol.*, **40**.

Stout, J. D. and Goh, K. M. (1980) The use of radiocarbon to measure the effects of earthworms on soil development. *Radiocarbon*, **22**, 892–6.

Stranden, V., Stead, G. B. and Dunning, A. (1982) Lumbricid populations in open cast reclamation sites and spoil heaps in Co. Durham, U.K. *Pedobiologia*, **24**, 57–64.

Striganova, B. R., Marfenina, O. E. and Ponomarenko, V. A. (1989) Some aspects of the effect of earthworms on soil fungi. *Biol. Bull. Acad. Sci. USSR*, **15**, 460–3.

Stringer, A. and Lyons, C. H. (1974) The effect of benomyl and thiophanate-methyl on earthworm populations in apple orchards. *Pestic. Sci.*, **5**, 189–96.

Stringer, A. and Lyons, C. H. (1977) The effect on earthworm populations of methods of spraying benomyl in an apple orchard. *Pestic. Sci.*, **8**, 647–50.

Stringer, A. and Pickard, J. A. (1963) The DDT content of soil and earthworms in an apple orchard at Long Ashton. *Long Ashton Res. Stn Rep.*, pp. 127–31.

Stringer, A. and Pickard, J. A. (1964) The DDT content of soil and earthworms in an apple orchard at Long Ashton. *Rep. Agr. Hort. Research Station Univ. Bristol*, 127.

Stringer, A. and Wright, M. A. (1973) The effect of benomyl and some related compounds on *Lumbricus terrestris* and other earthworms. *Pestic. Sci.*, **4**, 165–70.

Stringer, A. and Wright, M. A. (1976) The toxicity of benomyl and some related 2-substituted benzimidazoles to the earthworm *Lumbricus terrestris*. *Pestic. Sci.*, **7**, 459–64.

Stringer, A. and Wright, M. A. (1980) The toxicity of methiocarb and its breakdown products to earthworms. *Rep. Long Ashton Res. Sta.*, 1979, pp. 120–212.

Subler, S., Baranski, T., Edwards, C. A. and Edwards, W. M. (1995) The influence of earthworms on soil microbial and physical processes regulating nitrogen availability and leaching in agricultural ecosystems. *Soil Biol. Biochem.*, in press.

Sugi, Y. and Tanaka, M. (1978a) Population study of an earthworm, *Pheretima sieboldi*, in *Biological Production in a Warm-temperate Evergreen Oak Forest of Japan*, (eds T. Kira, Y. Ono and T. Hosokawa), J.I.P.B. Synthesis, no. 18, Univ. of Tokyo Press, pp. 163–71.

Sugi, Y. and Tanaka, M. (1978b) Number and biomass of earthworm populations, in *Biological Production in a Warm-temperate Evergreen Oak Forest of Japan*, (eds T. Kira, Y. Ono and T. Hosokawa), J.I.P.B. Synthesis no. 18, Univ. of Tokyo Press, pp. 171–8.

Sun, K. H. and Pratt, K. C. (1931) Do earthworms grow by adding segments? *Am. Nat.*, **65**, 31–48.

Svendsen, J. A. (1955) Earthworm population studies: a comparison of sampling methods. *Nature, Lond.*, **175**, 864.

Svendsen, J. A. (1957a) The distribution of Lumbricidae in an area of Pennine Moorland (Moor House, Nature Reserve). *J. Anim. Ecol.*, **26**(2), 409.

Svendsen, J. A. (1957b) The behaviour of lumbricids under moorland conditions. *J. Anim. Ecol.*, **26**(2), 423–39.

Svensson, B. H., Boström, U. and Klemedston, L. (1986) Potential for higher rates of denitrification in earthworm casts than in the surrounding soil. *Biol. Fertil. Soils*, **2**, 147–9.

Swaby, R. J. (1949) The influence of earthworms on soil aggregation. *J. Soil Sci.*, I(2), 195–7.

Swartz, R. D. (1929) Modification of the behaviour of earthworms. *J. Comp. Psychol.*, **9**, 17–33.

Syers, J. K. and Springett, J. A. (1983) Earthworm ecology in grassland soils, in *Earthworm Ecology: From Darwin to Vermiculture*, (ed. J. E. Satchell), Chapman & Hall, London, pp. 67–83.

Syers, J. K., Sharpley, A. N. and Keeney, D. R. (1979) Cycling of nitrogen by sur-face-casting earthworms in a pasture ecosystem. *Soil Biol. Biochem.*, **11**, 181–5.

Szabó, I. M., Prauser, H., Bodnar, G. *et al.* (1990) The indigenous intestinal bacteria of soil arthropods and worms, in *Microbiology in Poecilotherms*, (ed. R. Lessel), Elsevier, Amsterdam, pp. 109–17.

Taboga, L. (1980) The nutritional value of earthworms for chickens. *Br. Poult. Sci.*, **21**, 405–10.

Tacon, A. G. J., Stafford, E. A. and Edwards, C. A. (1983) A preliminary investiga-tion of the nutritive value of three terrestrial worms for rainbow trout. *Aquaculture*, **35**, 187–99.

Takahashi, K. and Sakai, Y. (1982) The effect of the surfactants to use with herbi-cides in the earthworms in citrus orchards. *Weed Res. Jap.*, **27**, 10–15.

Takano, S. and Nakamura, I. (1968) A new host earthworm, *Allolobophora japonica* Michaelsen, (Oligochaeta: Lumbricidae), of the calypterate muscoid fly, *Onesia subalpina* Kurahashi (Diptera: Calliphoridae). *Appl. Ent. Zool.*, 3(I), 51–2.

Tembe, V. B. and Dubash, P. J. (1961) The earthworms: a review. *J. Bombay Nat. Hist. Soc.*, **58**(1), 171–201.

Tenney, F. G. and Waksman, S. A. (1929) Composition of natural organic materials and their decomposition in the soil. IV. The nature and rapidity of decomposi-tion of the various organic complexes in different plant materials, under aero-bic conditions. *Soil Sci.*, **28**, 55–84.

Teotia, S. P., Duley, F. L. and McCalla T VI. (1950) Effect of stubble mulching on number and activity of earthworms. *Neb. Agric. Exp. Stn Res. Bull.*, **165**, 20.

Terhivuo, J., Pankakoski, E., Hyvarinen, H. and Koivisto, I. (1994) Pb uptake by ecologically dissimilar earthworm (Lumbricidae) species near a lead smelter in south Finland. *Environ. Poll.*, **85**, 87–96.

Theilemann, U. (1986) Elektrischer Regenwurmstand mit der Oktett-Methode. *Pedobiologia*, **29**, 295–302.

Thompson, A. R. (1971) Effects of nine insecticides on the numbers and biomass of earthworms in pasture. *Bull. Env. Contam. Toxicol.*, 5(6), 577–86.

Thompson, A. R. (1973) Pesticide residues in soil invertebrates, in *Environmental Pollution by Pesticides*, (ed. C. Edwards), Plenum, London, pp. 87–133.

Thompson, A. R. and Edwards C. A. (1974) Effects of pesticides on non-target invertebrates in freshwater and soil, in *Pesticides in Soil and Water*, (ed. W. D. Guezi), Publ. Soil Science Society of America, Wisconsin, USA, pp. 341–86.

Thompson, A. R. and Sans, W. W. (1974) Effects of soil insecticides in southwest-ern Ontario on non-target invertebrates: earthworms in pasture. *Environ. Entomol.*, **3**, 305–8.

Thompson, K. (1987) Seed and seed banks. *New Physiol.*, **106**(supplement), 23–4.

Thompson, L., Thomas, C. D., Radley, J. M. A. *et al.* (1993). The effects of earth-worms and snails in a simple plant community. *Oecologia*, **95**, 171–8.

Thomson, A. J. and Davies, D. M. (1974) Mapping methods for studying soil fac-tors and earthworm distribution. *Oikos*, **25**, 199–203.

Thornton, A. R. (1970) Transport of soil-dwelling aquatic phycomycetes by earth-worms. *Trans. Br. Mycol. Soc.*, **55**, 391–7.

Tian, G., Brussaard, L. and Kang, B. T. (1993) Biological effects of plant residues with contrasting chemical compositions under humid tropical conditions: effects on soil fauna. *Soil Biol. Biochem.*, **25**, 731–7.

Tillinghast, E. K. (1967) Excretory pathways of ammonia and urea in the earth-worm *Lumbricus terrestris* L. *J. Exp. Zool.*, **166**, 295–300.

Tillinghast, E. K., McInnes, D. C. and Djuffill, R. A. Jr (1969) The effect of temperature and water availability on the output of ammonia and urea by the earthworm *Lumbricus terrestris*. *Comp. Biochem. Physiol.*, **29**, 1087–92.

Tischler, W. (1955) Effect of agricultural practice on the soil fauna, in *Soil Zoology*, (ed. D. K. Mc. E. Kevan), Butterworths, London, pp. 125–37.

Tiwari, S. C. and Mishra, R. R. (1993) Fungal abundance and diversity in earthworm casts and in uningested soil. *Biol. Fertil. Soils*, **16**, 131–4.

Tiwari, S. C., Tiwari, B. K. and Mishra, R. R. (1989) Microbial populations, enzyme activities and nitrogen-phosphorus-potassium enrichment in earthworm casts and in the surrounding soil of a pineapple plantation. *Biol. Fertil. Soils*, **8**, 178–82.

Tiwari, S. C., Tiwari, B. K. and Mishra, R. R. (1990) Microfungal species associated with the gut content and casts of *Drawida assamensis* Gates. *Proc. Indian Acad. Sci. (Plant Sci)*, **100**, 379–82.

Tiwari, S. C., Tiwari, B. K. and Mishra, R. R. (1992) Relationship between seasonal populations of earthworms and abiotic factors in pineapple plantations. *Proc. Nat. Acad. Sci. Ind. Sec. B: (Biol. Sci.)*, **62**(2), 223–6.

Tomati, U. and Grappelli, A. (eds) (1984) *Proc. Int. Sym. on Agricultural and Environmental Prospects in Earthworm Farming*, Publ. Minist. Ric. Sci. Tech., Rome, p. 183.

Tomati, U., Grappelli, A. and Galli, E. (1983) Fertility factors in earthworm humus. *Proc. Int. Sym. on Agricultural and Environmental Prospects in Earthworm Farming*, Publ. Minist. Ric. Sci. Tech., Rome, pp. 49–56.

Tomati, U., Grappelli, A. and Galli, E. (1987) The presence of growth regulators in earthworm-worked wastes, in *On Earthworms*, (ed. A. M. Bonvicini-Pagliai and P. Omodeo), Selected Symposia and Monographs 2, Mucchi Editore, Modena, Italy, pp. 423–36.

Tomati, U., Grappelli, A. and Galli, E. (1988) The hormone-like effect of earthworm casts on plant growth. *Biol. Fertil. Soils*, **5**, 288–94.

Tomlin, A. D. (1983) The earthworm bait market in North America, in *Earthworm Ecology: From Darwin to Vermiculture*, (ed. J. E. Satchell), Chapman & Hall, London, pp. 331–8.

Tomlin, A. D. and Gore, F. L. (1974) Effects of six insecticides and fungicide on the numbers and biomass of earthworms in pasture. *Bull. Environ. Contam. Toxicol.*, **12**, 487–92.

Tomlin, A. D., Tolman, J. H. and Thorn, G. D. (1981) Suppression of earthworm *Lumbricus terrestris* populations around an airport by soil application of the fungicide benomyl. *Prot. Ecol.*, **2**, 319–23.

Tomlin, A. D., Shipitalo, M. J., Edwards, W. M. and Protz, R. (1995) Earthworms and their influence on soil structure and infiltration, in *Earthworm Ecology and Biogeography in North America*, (ed. P. F. Hendrix), Lewis Publishers, Chelsea, pp. 159–84.

Tracey, M. V. (1951) Cellulase and chitinase in worms. *Nature, Lond.*, **167**, 776–7.

Tréhen, P. and Bouché, M. B. (1983) Place des lumbriciens dans les processes de reestauvaja les sols de lande, in *Proc. 8th Int. Coll. Soil Zool.*, (eds Ph. Lebrun, H. M. Audné, A. de Meds, C. Gregoire-Wibo, and G. Wauthy), Louvain la Neuve, Belgium, pp. 471–86.

Trifonov, D. (1957) Über die Bekämpfing der Maulwurfurgrille und des Regenwurms mit dem Präparat. *Alon Kombi Bulgar. Tiutium*, **2**, 114–15.

Trigo, D. and Lavelle, P. (1993) Changes in respiration rate and some physiochemical properties of soil during gut transit through *Allolobophora molleri*
(Lumbricidae, Oligochaeta). *Biol. Fertil. Soils*, **15**, 185–8.

Trojan, M. D. and Linden, D. R. (1992) Microrelief and rainfall effects on water
and solute movement in earthworm burrows. *Soil Sci.Soc. Am. J.*, **56**, 727–33.

Tromba, F. G. (1955) Role of the earthworm *Eisenia foetida*, in the transmission of
Stephanurus dentatus. *J. Parasit.*, **41**, 157–61.

Tsukamoto, J. (1985) Soil macro-animals on a slope in a deciduous broad-leaved
forest. II. Earthworms of Lumbricidae and Megascolecidae. *Jap. J. Ecol.*, **35**(1),
37–48.

Tsukamoto, J. and Watanabe, H. (1977) Influence of temperature on hatching and
growth of *Eisenia foetida* (Oligochaeta, Lumbricidae). *Pedobiologia*, 17, 338–42.

Tucker, G. M. (1992) Effects of agricultural practices on field use by invertebrate-
feeding birds in winter. *J. Appl. Ecol.*, **29**, 779–90.

Tyler, G., Balsberg Påhlsson, A. M., Bengtsson, G. *et al.* (1989) Heavy-metal ecology of terrestrial plants, microorganisms and invertebrates: a review. *Water,
Air, Soil Poll.*, **47**, 189–215.

Uhlen, G. (1953) Preliminary experiments with earthworms. *Landbr. Hogsk. Inst.
Jordkiltur Meld.*, **37**, 161–83.

Urbasek, F. (1990) Cellulase activity in the gut of some earthworms. *Rev. Biol. Sol.*,
27(1), 21–8.

Urquhart, A. T. (1887) On the work of earthworms in New Zealand. *Trans. N.Z.
Inst.*, **19**, 119–23.

Vail, V. A. (1972) Contributions to North American earthworms. I. Natural history
and reproduction in *Diplocardia mississippiensis*. *Bull. Tall Timbers Res. Stn*, **11**,
1–39.

Vail, V. A. (1974) Contributions on North American earthworms (Annelida). II.
Observations on the hatchings of *Eisenia foetida* and *Bimastos tumidus*
(Oligochaeta: Lumbricidae). *Bull. Tall Timbers Res. Stn*, **16**, 1–8.

Vakili, N. G. (1993) *Exophiala jeanselmei*, a pathogen of earthworm species. *J. Med.
Vet. Mycol.*, **31**, 343–6.

Velasquez, L., Ibanez, I., Herrera, C. and Oyarzun, M. (1991) A note on the nutritional evaluation of worm meal (*Eisenia foetida*) in diets for rainbow trout. *Anim.
Prod.*, **53**(1), 119–22.

Venables, B. J., Fitzpatrick, L. C. and Goven, A. J. (1992) Earthworms as indicators
of ecotoxicity, in *Exotoxicology of Earthworms*, (eds P. W. Greig-Smith, H. Becker,
P. J. Edwards and F. Heimbach), Intercept, Andover, UK, pp. 197–208.

Venter, J. M. and Reinecke, A. J. (1988) Sublethal ecotoxicological studies with the
earthworm *Eisenia fetida* (Lumbricidae), in *Earthworms in Waste and
Environmental Management*, (eds, C. Edwards and E. F. Neuhauser), SPB Acad.
Publ., The Hague, The Netherlands, pp. 337–54.

Viljoen, S. A. and Reinecke, A. J. (1992) The temperature requirements of the
epigeic earthworm species *Endrilus eugeniae* (Oligochaeta)—a laboratory study.
Soil Biol. Biochem. **24**, 1345–50.

Viljoen, S. A., Reinecke, A. J. and Hartman, L. (1992) The influence of temperature
on the life-cycle of *Dendrobaena venata* (Oligochaeta). *Soil Biol. Biochem.*, **24**,
1341–4.

Villot, F. C. A. (1883) Memoire sur les cystiques des ténias. *Ann. Sci. Nat. Zool.*, **15**,
1–61.

Vimmerstedt, J. P. (1983) Earthworm ecology in reclaimed opencast coal mining sites in Ohio, in *Earthworm Ecology: From Darwin to Vermiculture*, (ed. J. E. Satchell), Chapman & Hall, London, pp. 229–40.

Vimmerstedt, J. P. and Finney, J. H. (1973) Impact of earthworm introduction on litter burial and nutrient distribution in Ohio strip mine spoil banks. *Soil Sci. Soc. Am. Proc.*, **37**, 388–91.

Volz, P. (1962) Contributions to a pedo-zoological study of sites based on observations in the south-eastern Palatinate. *Pedobiologia*, **1**, 242–90.

Voronova, L. D. (1968) The effect of some pesticides on the soil invertebrate fauna in the south Taiga zone in Perm region, U.S.S.R. *Pedobiologia*, **8**, 507–25.

Waite, R. H. (1920) Earthworms. The important factor in the transmission of gapes in chickens. *Maryland Agric. Exp. Bull.*, **234**, 103–18.

Waksman, S. A. and Martin, J. P. (1939) The conservation of the soil. *Science, NY*, **90**, 304–5.

Wallwork, J. A. (1983) *Earthworm Biology*, Studies in Biology No. 161, Institute of Biology, Camelot Press, Southampton, UK.

Walther, P. B. and Snider, R. M. (1984) Techiques for sampling earthworms and cocoons from leaf litter, humus and soil. *Pedobiologia*, **27**(4), 293–7.

Walton, W. R. (1928) *Earthworms as pests and otherwise*. USDA Farmers' Bulletin 1569, Washington, DC, p. 14.

Walton, W. R. (1933) The reaction of earthworms to alternating currents of electricity in the soil. *Proc. Ent. Soc. Wash.*, **35**, 24–7.

Wasawo, D. P. S. and Visser, S. A. (1959) Swampworms and tussock mounds in the swamps of Teso, Uganda. *East Afr. Agric. J.*, **25**, 86–90.

Watanabe, H. (1975) On the amount of cast production by the megascolecid earthworm *Pheretima hupeiensis*. *Pedobiologia*, **15**, 20–8.

Watanabe, H. and Tsukamoto, J. (1976) Seasonal change in size, class and stage structure of lumbricid *Eisenia foetida* population in a field compost and its practical application as the decomposer of organic waste matter. *Rev. Ecol. Biol. Sol*, **13**, 141–6.

Waters, R. A. S. (1951) Earthworms and the fertility of pasture. *Proc. N. Z. Grassl. Ass.*, pp. 168–75.

Waters, R. A. S. (1955) Numbers and weights of earthworms under a highly productive pasture. *N. Z. J. Sci. Technol.*, **36**(5), 516–25.

Watkin, B. R. (1954) The animal factor and levels of nitrogen. *J. Br. Grassld Soc.*, **9**, 35–46.

Way, M. J. and Scopes, N. E. A. (1965) Side effects of some soil applied systemic insecticides. *Ann. App. Biol.*, **55**, 340–1.

Way, M. J. and Scopes, N. E. A. (1968) Studies on the persistence and effects on soil fauna of some soil applied systemic insecticides. *Ann. Appl. Biol.*, **62**, 199–214.

Webb, J. K. and Shine, R. (1993) Dietary habits of Australian blindsnakes (Typhlopidae). *Copeia*, **1993**(3), 762–70.

Weber, G. (1953) The macrofauna of light and heavy arable soils and the effect on them of plant protection substances. *Z. Pflernähr Düng.*, **61**, 107–18.

Wei-Chun, Ma, Brussaard, L. and De Ridder, J. A. (1990) Long-term effects of nitrogenous fertilizers on grassland earthworms and their relation to soil acidification. *Agric. Ecosyst. Environ.*, **30**, 71–80.

Weisbach, W. W. (1962) Regenwürmer und Essbare Erde. *Biol. Jaarb. Dodonea.*, **30**, 225–38.

Went, J. C. (1963) Influence of earthworms on the number of bacteria in the soil, in *Soil Organisms*, (eds J. Doeksen and J. van der Drift), North Holland, Amsterdam, pp. 260–5.

Wentsel, R. S. and Gueltar, M. A. (1987) Toxicity of brass powder in soil to the earthworm *Lumbricus terrestris*. *Environmental Toxicology and Chemistry*, 741–3.

West, L. T., Hendrix, P. F. and Bruce, R. R. (1987) Micromorphic observation of soil alteration by earthworms. *Agric. Ecosys. Environ.*, **34**, 363–70.

Westeringh, W. van de (1972) Deterioration of soil structure in worm-free orchard soils. *Pedobiologia*, **12**, 615.

Westernacher-Dotzler, E. and Dumbeck, G. (1992) Earthworm occurrence in agricultural reclamation areas of the Lower Rhine basin. *J. Agron. Crop Sci.*, **169**(5), 298–309.

Wheatley, G. A. and Hardman, J. A. (1968) Organochlorine insecticide residues in earthworms from arable soils. *J. Sci. Fd Agric.*, **19**, 219–25.

Wherry, E. T. (1924) Soil acidity preferences of earthworms. *Ecology*, **5**, 89–90.

Whiston, R. A. and Seal, K. J. (1988) The occurrence of cellulases in the earthworm *Eisenia foetida*. *Biol. Wastes*, **25**(3), 239–42.

White, G. C. (1980) Effects of dinoseb sprays on earthworms. *Rep. East Malling Res. Sta.*, 1979, p. 46.

Whitney, W. K. (1967) Laboratory tests with dursban and other insecticides in soil. *J. Econ. Ent.*, **60**, 68–74.

Wiecek, C. S. and Messenger, A. S. (1972) Calcite contributions by earthworms to forest soils in northern Illinois. *Soil Sci. Soc. Am.Proc.*, **36**, 478–80.

Wilcke, D. E. von (1952) On the domestication of the 'soilution' earthworm. *Anz. Schädlingsk.*, **25**, 107–9 (G).

Wilcke, D. E. von (1953) Zur Kenntnis der Lumbricidenfauna Deutschlands. *Zool. Anz.*, **151**, 104–6.

Wilcke, D. E. von (1955) Critical observations and proposals on the quantitative analysis of earthworm populations in soil zoology studies. *Z. PflErnähr Düng*, **68**, 44–9 (G).

Wilkinson, G. E. (1975) Effect of grass fallow rotations on the infiltration of water into a savanna zone soil of Northern Nigeria. *Trop. Agric.*, **52**, 97–104.

Witkamp, M. (1966) Decomposition of leaf litter in relation to environment, microflora and microbial respiration. *Ecology*, **47**, 194–201.

Wittich, W. (1953) Untersuchungen über den Verlauf der Streuzersetzung auf einem Boden mit Regenwurmtätigkeit. *Schrift Reige forstl. Fak. Univ. Gottingen*, **9**, 7–33.

Wojewodin, A. W. (1958) Ungefährlichkeit der Herbizide für die Biozönose. *Int. Konf. Herb.*, 97–102.

Wolf, A. V. (1937) Notes on the effect of heat on *L. terrestris*. *Ecology*, **19**, 34–68.

Wolf, A. V. (1938) Studies on the behaviour of *L. terrestris* to dehydration; and evidence for a dehydration tropism. *Ecology*, **19**, 233–42.

Wolf, A. V. (1940) Paths of water exchange in the earthworm. *Physiol Zool.*, **13**, 294–308.

Wolf, A. V. (1941) Survival time of the earthworm as affected by raised temperatures. *J. Cell. Compt. Physiol.*, **13**, 275–8.

Wollny, E. (1890) Untersuchungen über die Beeinflussung der Fruchtbarkeit der Ackerkrume durch die Tätigkeit der Regenwürmer. *Forsch. Agrik. Physik.* **13**,381–95.

Wolters, V. and Joergensen, R. G. (1992) Microbial carbon turnover in beech forest soils worked by *Aporrectodea caliginosa* (Savigny) (Oligochaeta: Lumbricidae). *Soil Biol. Biochem.*, **24**, 171–7.

Wolters, V. and Stickan, W. (1991) Resource allocation of beech seedlings (*Fagus sylvatica* L.): relationship to earthworm activity and soil conditions. *Oecologia*, **88**, 125–31.

Wong, S. H. and Griffiths, D. A. (1991) Vermicomposting in the management of pig-waste in Hong Kong. *World J. Microbiol. Biotech.*, **7**(6), 593–5.

Wood, T. G. (1974) The distribution of earthworms (Megascolecidae) in relation to soils, vegetation and altitude on the slopes of Mt. Kosciusko, Australia. *J. Anim. Ecol.*, **43**, 87–106.

Woodhead, A. A. (1950) Life history cycle of the giant kidney worm, *Dioctophyma renale* (Nematoda) of man and many other animals. *Trans. Am. Microsc. Soc.*, **69**, 21–46.

Wright, M. A. (1972) Factors governing ingestion by the earthworm *Lumbricus terrestris* (L.), with special reference to apple leaves. *Ann. Appl. Biol.*, **70**, 175–88.

Wright, M. A. (1977) Effects of benomyl and some other systemic fungicides on earthworms. *Ann. App. Biol.*, **87**, 520–4.

Wright, M. A. and Stringer, A. (1973) The toxicity of thiabendazole, benomyl, methyl benzimidazol-2-yl carbamate and thiophanate-methyl to the earthworm, *Lumbricus terrestris. Pestic. Sci.*, **4**, 431–2.

Wright, M. A. and Stringer, A. (1980) Lead, zinc, and cadmium content of earthworms from pasture in the vicinity of an industrial smelting complex. *Environ. Pollut. Ser. A.*, **23**, 313–21.

Yadav, D. V., Pillai M. K. K. and Agarwal, H. C. (1976) Uptake and metabolism of DDT and lindane by the earthworm *Pheretima posthuma. Bull. Environ. Contam. Toxicol.*, **16**, 541–5.

Ydrogo, H. F. B. (1994) Effecto de las lombrices de tierra (*Pontoscolex corethrurus*) en las micorrizas vesiculo arbusculares (M.V.A.) en la etapa de crecimiento de araza (*Eugenia stipitata*), achiote (*Bixa orellana*), pijuayo (*Bactris gasipaes*), en suelos ultisoles de Yurimaguas. Thesis, University Nac. San Martin, Peru.

Yeates, G. W. (1981) Soil nematode populations depressed in the presence of earthworms. *Pedobiologia*, **22**, 191–5.

Yerkes, R. M. (1912) The intelligence of earthworms. *J. Anim. Behav.*, **2**, 332–52.

Yoshida, M. and Hoshii, H. (1978) Nutritional value of earthworms for poultry feed. *Jpn Poult. Sci.*, **15**(6), 308–11.

Zachmann, J. E., Linden, D. R. and Clapp, C. E. (1987) Macroporous infiltration and redistribution as affected by earthworms, tillage, and residue. *Soil Sci. Soc. Am. J.*, **51**, 1580–6.

Zajonc, I. (1970) Earthworm synusiae of meadow stratocenoses and choriocenoses. *Acta Zooltech. Univ. Agr. Nitra.*, **21**, 203–11.

Zajonc, I. (1975) Variations in meadow associations of earthworms caused by the influence of nitrogen fertilizers and liquid manure irrigation, in *Progress in Soil Zoology*, (ed. J. Vanek), Proceedings 5th Int. Colloquium in Soil Zoology, Prague, pp. 497–503.

Zhang, H. and Schrader, S. (1993) Earthworm effects on selected chemical properties of soil aggregates. *Biol. Fertil. Soils*, **15**, 229–34.

Zharikov, G., Fartukov, S. V., Tumanskii, I. M. and Ishchenko, N. V. (1993) Use of the solid wastes of the microbial industry by worm composting. *Biotekhnologiya* (1), 21–3.

Zhinkin, L. (1936) The influence of the nervous system on the regeneration of *Rhynchelmis limosella*. *J. Exp. Zool.*, **73**, 43–65.

Zicsi, A. (1954) The role of earthworms in the soil, as investigated by soil analyses, experiments and survey at the University in Gödöllö. *Agrartud. Egypt. agron. Kar. Kiadv.*, **I** (14), 1–20.

Zicsi, A. (1958a) Einfluss der Trockenheit und der Bodenarbeitung auf das Leben der Regenwürmer in Ackerboden. *Acta Agron*, **7**, 67–74.

Zicsi, A. (1958b) Freilandsuntersuchungen zur Kenntnis der Empfindlichkeit einiger Lumbriciden-Arten gegen Trockenperioden. *Acta Zool. Acad. Sci. Hung.*, **3**, 369–83.

Zicsi, A. (1962a) Determination of number and size of sampling unit for estimating lumbricid populations of arable soils, in *Progress in Soil Zoology*, (ed. P. W. Murphy), Butterworths, London, pp. 68–71.

Zicsi, A. (1962b) Über die Dominanzhaltnisse einheimischer Lumbriciden auf Ackerboden. *Opusc. Zool Budapest.*, **4**(2–4), 157–61.

Zicsi, A. (1969) Über die Auswirkung der Nachfrucht und Bodenbearbeitung auf die Aktivität der Regenwürmer. *Pedobiologia*, **9**(1–2), 141–6.

Zicsi, A. (1983) Earthworm ecology in deciduous forests in central and southeast Europe, in *Earthworm Ecology: from Darwin to Vermiculture*, (ed. J. E. Satchell), Chapman & Hall, London, pp. 171–8.

Ziegler, F. and Zech, W. (1991) Chemical changes of beech litter and straw during decomposition under laboratory conditions. *Z. Pflanzen. Bodenk.*, **154**, 377–85.

Ziegler, F. and Zech, W. (1992) Formation of water-stable aggregates through the action of earthworms: implications from laboratory experiments. *Pedobiologia*, **36**, 91–6.

Zoran, M. J., Heppner, T. J. and Drewes, C. D. (1986) Teratogenic effects of the fungicide benomyl on posterior segmental regeneration in the earthworm *Eisenia fetida*. *Pest. Sci.*, **17**, 641–52.

Zou, X. (1993) Species effects on earthworm density in tropical tree plantations in Hawaii. *Biol. Fertil. Soils*, **15**, 35–8.

Zrazhevskii, A. I. (1957) *Dozhdevye chervi kak fakto plodorodiya lesnykj pochv*, Kiev.

Author index

Systematic index

Subject index

Italic page numbers indicate the more important references in the text whereas **bold** numbers refer to illustrations.